BIOINTENSIVE INTEGRATED PEST MANAGEMENT IN HORTICULTURAL ECOSYSTEMS

P. Parvatha Reddy
Former Director
Indian Institute of Horticultural
Research, Bangalore - 560 089
E-mail: reddy_parvatha@yahoo.com

SCIENTIFIC
PUBLISHERS (INDIA)
www.scientificpub.com

Published by:

Scientific Publishers (India)
5-A, New Pali Road, P.O. Box 91,
Jodhpur – 342 001 (India)

E-mail: info@scientificpub.com
Website: www.scientificpubonline.com

Print: 2021

ISBN: 978-81-7233-858-9

Printed in India

PREFACE

Through 'Green Revolution' in late 1960's, India achieved self sufficiency in food production, which was hailed as a breakthrough on the farm front by international agricultural experts. But still the country has not achieved self sufficiency in production of horticultural crops. Most of the growth in food production during the green revolution period is attributed to the use of improved crop varieties and higher levels of inputs of fertilizers and pesticides. The modern agricultural techniques such as use of synthetic fertilizers and pesticides are continuing to destroy stable traditional ecosystems and the use of high yielding varieties of crops has resulted in the elimination of thousands of traditional varieties with the concurrent loss of genetic resources. The introduction of high yielding varieties changed the agricultural environment leading to numerous pest problems of economic importance. In the process of intensive farming, the environment has been treated in an unfriendly manner.

Prof. Swaminathan (2000) emphasized the need for 'Ever Green Revolution' keeping in view the increase in population. The increase in population and diminishing per capita availability of land demands rise in productivity per unit area. In India, annual crop losses due to pests, diseases and weeds have been estimated to be about Rs. 600,000 million in 2005. Increasing yields from existing land requires effective crop protection to prevent losses before and after harvest. The challenge before the crop protection scientist is to do this without harming the environment and resource base. This can be achieved by adopting eco-friendly Biointensive Integrated Pest Management (BIPM) strategy.

BIPM is defined as "A systems approach to pest management based on an understanding of pest ecology. It begins with steps to accurately diagnose the nature and source of pest problems, and then relies on a range of preventive tactics and biological controls to keep pest populations within acceptable limits. Reduced-risk pesticides are used if other tactics have not been adequately effective, as a last resort, and with care to minimize risks" (Benbrook, 1996).

BIPM incorporates ecological and economic factors into agricultural system design and decision making, and addresses public concerns about environmental quality and food safety. The benefits of implementing BIPM can include reduced chemical input costs, reduced on-farm and off-farm environmental impacts, and more effective and sustainable pest management. An ecology-based IPM has the potential of decreasing inputs of fuel, machinery, and synthetic chemicals-all of which are energy intensive and increasingly costly in terms of financial and environmental impact. Such reductions will benefit the grower and society.

The information on biointensive integrated pest management (insect, mite and nematode pests, and diseases caused by bacteria, fungi, virus/mycoplasma) in horticultural ecosystems (fruits, vegetables, ornamentals, medicinal, aromatic, tuber, plantation and spice crops) is very much scattered. There is no book at present which comprehensively and exclusively deals with the above aspects. The present book deals with the most recent biointensive integrated approaches utilizing components such as bioagents [predators, parasitoids and pathogens (bacteria, fungi, viruses)], botanicals (biofumigation, oil cakes, FYM, compost, crop residues, green manuring and other organic amendments), endomycorrhizae, physical methods (hot water treatment of planting material, soil solarization), cultural methods (crop rotation, summer ploughing, fallowing, intercropping, pruning, mulching, spacing, planting date, trap cropping, *etc.*), biorational chemicals (pheromones) and resistant cultivars. The book is illustrated with excellent quality photographs enhancing the quality of publication. The book is written in lucid style, easy to understand language along with adoptable recommendations for pest management.

This book can serve as a useful reference to policy makers, research and extension workers, practicing farmers and students. The material can also be used for teaching post-graduate courses. Suggestions to improve the contents of the book are most welcome (E-mail: reddy_parvatha@yahoo.com). The Publisher, Scientific Publishers (India), Jodhpur, deserves commendation for their professional contribution.

P. Parvatha Reddy
34, UAS Layout, Ist Main, 7th Cross
Sanjay Nagar, Bangalore-560 094.

Bangalore
December 15, 2012

CONTENTS

Section I

INTRODUCTION

Chapter - 1

INTRODUCTION

Through 'Green Revolution' in late 1960's, India achieved self sufficiency in food production, which was hailed as a breakthrough on the farm front by international agricultural experts. But still the country has not achieved self sufficiency in production of horticultural crops. Most of the growth in food production during the green revolution period is attributed to the use of improved crop varieties and higher levels of inputs of fertilizers and pesticides. The modern agricultural techniques such as use of synthetic fertilizers and pesticides are continuing to destroy stable traditional ecosystems and the use of high yielding varieties of crops has resulted in the elimination of thousands of traditional varieties with the concurrent loss of genetic resources. The introduction of high yielding varieties changed the agricultural environment leading to numerous pest problems of economic importance. In the process of intensive farming, the environment has been treated in an unfriendly manner.

Prof. Swaminathan (2000) emphasized the need for 'Ever green revolution' keeping in view the increase in population. The increase in population and diminishing per capita availability of land demands rise in productivity per unit area. In India, annual crop losses due to pests, diseases and weeds have been estimated to be about Rs. 600,000 million in 2005. Increasing yields from existing land requires effective crop protection to prevent losses before and after harvest. The challenge before the crop protection scientist is to do this without harming the environment and resource base. This can be achieved in horticultural ecosystems by adopting eco-friendly biointensive integrated pest management (BIPM) strategy.

1.1. INTEGRATED PEST MANAGEMENT (IPM)

Integrated pest management is an important principle on which sustainable crop protection can be based. IPM allows farmers to manage pests in a cost effective, environmentally sound and socially acceptable way. According to FAO, IPM is defined as "A pest management system that in the context of the associated environment and the population dynamics of the pest species utilizes all suitable techniques and methods, in a compatible manner as possible and maintains the pest populations at levels below those causing economic injury".

1.2. BIOINTENSIVE INTEGRATED PEST MANAGEMENT (BIPM)

Biointensive IPM incorporates ecological and economic factors into agricultural system design and decision making, and addresses public concerns about environmental quality and food safety. The benefits of implementing biointensive IPM can include reduced chemical input costs, reduced on-farm and off-farm environmental impacts, and more effective and sustainable pest management. An ecology-based IPM has the potential of decreasing inputs of fuel, machinery, and synthetic chemicals-all of which are energy intensive and increasingly costly in terms of financial and environmental impact. Such reductions will benefit the grower and society.

Over-reliance on the use of synthetic pesticides in crop protection programmes around the world has resulted in disturbances to the environment, pest resurgence, pest resistance to pesticides, and lethal and sub-lethal effects on non-target organisms, including humans. These side effects have raised public concern about the routine use and safety of pesticides. At the same time, population increases are placing ever-greater demands upon the "ecological services" i.e. provision of clean air, water and wildlife habitat for a landscape dominated by farms. Although some pending legislation has recognized the costs to farmers of providing these ecological services, it is clear that farmers will be required to manage their land with greater attention to direct and indirect off-farm impacts of various farming practices on water, soil, and wildlife resources. With this likely future in mind, reducing dependence on chemical pesticides in favour of ecosystem manipulations is a good strategy for farmers.

Biointensive IPM is defined as "A systems approach to pest management based on an understanding of pest ecology. It begins with steps to accurately diagnose the nature and source of pest problems, and then relies on a range of preventive tactics and biological controls to keep pest populations within acceptable limits. Reduced-risk pesticides are used if other tactics have not been adequately effective, as a last resort, and with care to minimize risks" (Benbrook, 1996).

The primary goal of biointensive IPM is to provide guidelines and options for the effective management of pests and beneficial organisms in an ecological context. The flexibility and environmental compatibility of a biointensive IPM strategy make it useful in all types of cropping systems. Biointensive IPM would likely decrease chemical use and costs even further.

1.2.1. Components of Biointensive IPM

An important difference between conventional and biointensive IPM is that the emphasis of the latter is on proactive measures to redesign the agricultural ecosystem to the disadvantage of a pest and to the advantage of its parasite and predator complex. At the same time, biointensive IPM shares many of the same components as conventional IPM, including monitoring, use of economic thresholds, record keeping, and planning.

1.2.1.1. Planning

Good planning must precede implementation of any IPM programme, but is particularly important in a biointensive programme. Planning should be done before planting because many pest strategies require steps or inputs, such as beneficial organism habitat management, that must be considered well in advance. Attempting to jump-start an IPM programme in the beginning or middle of a cropping season generally does not work.

When planning a biointensive IPM programme, some considerations include:

- Options for design changes in the agricultural system (beneficial organism habitat, crop rotations).
- Choice of pest-resistant cultivars.
- Technical information needs.
- Monitoring options, record keeping, equipment, etc.

When making a decision about crop rotation, consider the following questions: Is there an economically sustainable crop that can be rotated into the cropping system? Is it compatible? Important considerations when developing a crop rotation are:

- How might the cropping system be altered to make life more difficult for the pest and easier for its natural controls? What two (or three or several) crops can provide an economic return when considered together as a biological and economic system that includes considerations of sustainable soil management?
- What are the impacts of this season's cropping practices on subsequent crops?
- What specialized equipment is necessary for the crops?
- What markets are available for the rotation crops?

Management factors should also be considered. For example, one crop may provide a lower direct return per hectare than the alternate crop, but may also lower management costs for the alternate crop, with a net increase in profit.

1.2.1.2. Pest Identification

A crucial step in any IPM programme is to identify the pest. The effectiveness of both proactive and reactive pest management measures depend on correct identification. Misidentification of the pest may be worse than useless; it may actually be harmful and cost time and money. Help with positive identification of pests may be obtained from university personnel, private consultants, the Cooperative Extension Service, and books and web sites.

After a pest is identified, appropriate and effective management depends on knowing answers to a number of questions. These may include:

- What plants are hosts and non-hosts of this pest?
- When does the pest emerge or first appear?

- Where does it lay its eggs?
- For plant pathogens, where is the source(s) of inoculum?
- Where, how, and in what form does the pest overwinter?

Monitoring (field scouting) and economic injury and action levels are used to help answer these and additional questions.

1.2.1.3. Monitoring

Monitoring involves systematically checking crop fields for pests and beneficials, at regular intervals and at critical times, to gather information about the crop, pests, and natural enemies. Sweep nets, sticky traps, and pheromone traps can be used to collect insects for both identification and population density information. Leaf counts are one method for recording plant growth stages. Records of rainfall and temperature are sometimes used to predict the likelihood of disease infections.

The more often a crop is monitored, the more information the grower has about what is happening in the fields. Monitoring activity should be balanced against its costs. Frequency may vary with temperature, crop, growth phase of the crop, and pest populations. If a pest population is approaching economically damaging levels, the grower will want to monitor more frequently.

1.2.1.4. Economic Injury and Action Levels

The economic injury level (EIL) is the pest population that inflicts crop damage greater than the cost of control measures. Because growers will generally want to act before a population reaches EIL, IPM programmes use the concept of an economic threshold level (ETL or ET), also known as an action threshold. The ETL is closely related to the EIL, and is the point at which suppression tactics should be applied in order to prevent pest populations from increasing to injurious levels.

ETLs are intimately related to the value of the crop and the part of the crop being attacked. For example, a pest that attacks the fruit or vegetable will have a much lower ETL (i.e., the pest must be controlled at lower populations) than a pest that attacks a non-salable part of the plant. The exception to this rule is an insect or nematode pest that is also a disease vector. Depending on the severity of the disease, the grower may face a situation where the ETL for a particular pest is zero, i.e., the crop cannot tolerate the presence of a single pest of that particular species because the disease it transmits is so destructive.

1.2.2. BIPM Options

BIPM options may be considered as proactive or reactive.

1.2.2.1. Proactive Options

Proactive options, such as crop rotations and creation of habitat for beneficial organisms, permanently lower the carrying capacity of the farm for the pest. The carrying capacity is determined by the factors like food, shelter, natural enemy complex and weather, which affect the reproduction and survival of a pest species. Cultural control practices are generally considered to be proactive strategies. Proactive practices include crop rotation, resistant crop cultivars including transgenic plants, disease-free seed and plants, crop sanitation, spacing of plants, altering planting dates, mulches, etc.

The proactive strategies (cultural controls) include:

- Healthy, biologically active soils (increasing below-ground diversity).
- Habitat for beneficial organisms (increasing above-ground diversity).
- Appropriate plant cultivars.

(i) Intercropping: Intercropping is the practice of growing two or more crops in the same, alternate, or paired rows in the same area. This technique is particularly appropriate in vegetable production. The advantage of intercropping is that the increased diversity helps "disguise" crops from insect pests, and if done well, may allow for more efficient utilization of limited soil and water resources.

(ii) Strip Cropping: Strip cropping is the practice of growing two or more crops in different strips across a field wide enough for independent cultivation. It is commonly practiced to help reduce soil erosion in hilly areas. Like intercropping, strip cropping increases the diversity of a cropping area, which in turn may help "disguise" the crops from pests. Another advantage to this system is that one of the crops may act as a reservoir and/or food source for beneficial organisms.

The options described above can be integrated with no-till cultivation schemes and all its variations (strip till, ridge till, etc.) as well as with hedgerows and intercrops designed for beneficial organism habitat. With all the cropping and tillage options available, it is possible, with creative and informed management, to evolve a biologically diverse, pest-suppressive farming system appropriate to the unique environment of each farm.

(iii) Disease-free Seed and Plants: These are available from most commercial sources, and are certified as such. Use of disease-free seed and nursery stock is important in preventing the introduction of disease.

(iv) Resistant Varieties: These are continually being bred by researchers. Growers can also do their own plant breeding simply by collecting non-hybrid seed from healthy plants in the field. The plants from these seeds will have a good chance of being better suited to the local environment and of being more resistant to insects and diseases. Since natural systems are dynamic rather than static, breeding for resistance must be an ongoing process, especially in the case of plant disease, as the pathogens themselves continue to evolve and become resistant to control measures.

Perhaps the greatest single technological achievement is the advance in breeding crops for resistance to pests. Cultivation of resistant varieties is the cheapest and best method of controlling pests. One of the important components of IPM is the use of resistant cultivars to key pests. Under All India Co-ordinated Research Projects of Indian Council of Agricultural Research, large number of highly/ moderately resistant varieties are released to the farmers (Table 1.1).

Table 1.1. Horticultural crop varieties resistant to pests/diseases.

Horticultural crop	Pest/Disease	Resistant varieties
Banana	*Radopholus similis*	Kadali, Pedalimoongil, Ayiramkapoovan, Peykunnan, Kunnan, Pisang Seribu, Tongat, Vennettu Kunnan, Anaikomban.
	Panama wilt (*Fusarium oxysporum* f. sp. *cubense*)	Robusta, Dwarf Cavendish.
Citrus	*Tylenchulus semipenetrans*	Trifoliate Orange, Swingle Citrumelo.
	Gummosis, leaf fall, fruit rot (*Phytophthora* spp.)	Cleopatra mandarin, Rangpur lime, Trifoliate orange rootstocks.
Grapevine	Root-knot nematode, *Meloidogyne incognita*	Black Champa, Dogridge, 1613, Salt Creek, Cardinal, Banquabad.
Papaya	Ring spot virus	Rainbow, Sun Up.
Passion fruit	Root-knot nematode, *M. incognita*	Yellow, Kaveri.
Potato	Late blight	Kufri Sutlej, Kufri Badshah, Kufri Jawahar (in plains), Kufri Jyothi, Kufri Giriraj, Kufri Kanchan, Kufri Meghad (in hills).
Tomato	Bacterial wilt	Arka Abha, Arka Alok, Arka Shreshta, Arka Abhijit, Megha, Shakthi, Sun 7610, Sun 7611.
	Powdery mildew (PM)	Arka Asish.
	Fusarium and *Verticillium* wilt	Vaishali, Rupali, Rashmi
	Leaf curl virus	Avinash-2, Hisar Anmol.
	Root-knot nematode	Hisar Lalit, Pusa Hybrid-2, Arka Vardan.
Brinjal	Bacterial wilt	Arka Nidhi, Arka Keshav, Arka Neelkant, Arka Anand, Swarna Shree, Swarna Shyamali, Surya, Ujjwala.
	Phomopsis blight	Pusa Bhairav.
	Little leaf	Pusa Purple Long, Pusa Purple Cluster (Field resistant).
Chilli	TMV, CMV, leaf curl	Pusa Sada Bahar, Punjab Lal, Pusa Jwala.
	Thrips	NP 46° (T)
	Powdery mildew	Arka Suphala (T).
	Dieback and powdery mildew	Musalwadi (T)
	Mosaic, leaf curl	Pant C-1.
	Leaf curl and fruit rot	Jawahar 218 (T)
	Viruses	Arka Harita, Arka Meghana.
French bean	Angular leaf spot, mosaic	Pant Anupama.
	Rust, bacterial blight	Arka Anoop.

Horticultural crop	Pest/Disease	Resistant varieties
	Rust	Arka Bold, Swarna Priya, Swarna Latha, Arka Anoop.
	Rust, *Alternaria* leaf spot	Arka Bold.
Pea	Powdery mildew	Pusa Pragati, Jawahar Matar 5, Jawahar Peas 83.
	PM, rust	Arka Ajit, Arka Karthik, Arka Sampoorna.
	Fusarium wilt	JP Batri Brown 3, JP Batri Brown 4.
Cowpea	Bacterial blight	Pusa Komal.
Pigeon pea	*Fusarium* wilt	Maruti.
Field bean	Viral diseases, jassid, aphid, pod borer	Pusa Sem-2, Pusa Sem-3.
Cluster bean	PM, *Alternaria* leaf spot	Gomah Manjari.
Okra	YVMV	Pusa Sawani, Arka Abhay, Arka Anamika, Hisar Unnat, DVR-1, DVR-2, IIVR-10, Varsha Upkar, P-7, Pusa A-4, Parbhani Kranti (T), Punjab Kesari, Punjab Padmini, Sun-40, Makhmali.
	YVMV & fruit borer	Pusa A-4.
Cucumber	PM	Swarna Poorna.
	PM, downy mildew (DM), angular leaf spot, anthracnose	Poinsette.
Cabbage	Black rot	Pusa Mukta.
	Black leg	Pusa Drum Head.
Cauliflower	Black rot	Pusa Snowball K-1.
	Black rot & curd blight	Pusa Shubhra.
	Curd blight	Pusa Synthetic.
	DM	Pusa Hybrid-2.
Onion	Purple blotch, basal rot, thrips	Arka Pitamber, Arka Kirtiman, Arka Lalima.
	Purple blotch, *Alternaria porri*	Arka Kalyan.
Garlic	Purple blotch, *Stemphylium* disease	Agri-found White.
Muskmelon	PM	Arka Rajhans, Pusa Madhuras (MR).
	PM, DM	Punjab Rasila.
	Fusarium wilt	Pusa Madhuras, Durgapura Madhu, Arka Jeet, Punjab Sunehari (MR), Harela.
Watermelon	PM, DM, anthracnose	Arka Manik
Pumpkin	Fruit fly	Arka Suryamukhi.
Ridge gourd	PM, DM	Swarna Uphaar.

Horticultural crop	Pest/Disease	Resistant varieties
Bottle gourd	Blossom end rot	Arka Bahar (T).
	CMV	Punjab Komal
Carrot	PM, root-knot nematode	Arka Suraj.
Amaranth	White rust	Arka Arunima, Arka Suguna (MR).
Palak	*Cercospora* leaf spot	Arka Anupama.
China aster	Root-knot nematode, *M. incognita*	Shashank, Poornima (MR).
Tuberose	Root-knot nematode, *M. incognita*	Sringar, Suvasini (T).
Mentha	Root-knot nematode, *M. incognita*	Kukrail, Arka Neera.
Black pepper	Root-knot nematode, *M. incognita*	IISR Pournami (T).
	Foot rot, *Phytophthora capsici*	IISR Shakthi
Cardamom	Mosaic	IISR Vijetha
	Rhizome rot	IISR Avinash
Ginger	Root-knot nematodes	IISR Mahima
	Soft rot	Maran
Cumin	Fusarium wilt	GC-4

MR – Moderately Resistant; T – Tolerant

(v) Biotech Crops: Gene transfer technology is being used by several companies to develop cultivars resistant to insects, diseases and nematodes. An example is the incorporation of genetic material from Bacillus thuringiensis (*Bt*), a naturally occurring bacterium, into brinjal and potatoes, to make the plant tissues toxic to shoot and fruit borer and potato beetle larvae, respectively.

Whether or not this technology should be adopted is the subject of much debate. Opponents are concerned that by introducing *Bt* genes into plants, selection pressure for resistance to the *Bt* toxin will intensify and a valuable biological control tool will be lost. There are also concerns about possible impacts of genetically-modified plant products (i.e., root exudates) on non-target organisms as well as fears of altered genes being transferred to weed relatives of crop plants. Whether there is a market for gene-altered crops is also a consideration for farmers and processors. Proponents of this technology argue that use of such crops decreases the need to use toxic chemical pesticides.

Transgenic crop varieties in horticultural crops (tomato, potato, brinjal, beans, cabbage, cauliflower, musk melon, banana, coffee) have been developed by cloning *Bt* endotoxin genes which are cultivated in large areas. In 2011, India is the 4th largest GM crops growing country (10.6 million ha) in the world only next to USA (69 million ha), Brazil (30.3 million ha) and Argentina (23.7 million ha) (Clive

James, 2011). Combining a host gene for resistance with pathogen derived genes or with genes coding for antimicrobial compounds provide for a broad and effective resistance in many host-pathogen combinations (Table 1.2).

Table 1.2. Development of transgenics in vegetable crops in India

Vegetable crop	Target pathogen	Transgene/s	Institute
Potato	Tuber moth	*Bt* Cry 1Ab	CPRI, Shimla
	Potato virus Y	Coat protein	CPRI, Shimla
Tomato	Leaf curl virus	Leaf curl virus sequence	IIHR, Bangalore IAHS, Bangalore
		Replicase gene	IARI, New Delhi
	Fungal diseases	Chitinase and glucanase	IIHR, Bangalore
		Alfalfa glucanase	IAHS, Bangalore
		OXDC	JNU, New Delhi
	Lepidopteran pests	*Bt* Cry 1Ab	IARI, New Delhi Proagro PG-S (India) Ltd.
Brinjal	Fungal diseases	Chitinase, glucanase and thaumatin encoding genes	
	Lepidopteran pests	*Bt* Cry 1Ab	IARI, New Delhi Proagro PG-S (India) Ltd.
Cabbage	Lepidopteran pests	*Bt* Cry 1Ab	IARI, New Delhi Proagro PG-S (India) Ltd.
		Cry 1H/ Cry 9C	Proagro PG-S (India) Ltd.
Cauliflower	Lepidopteran pests	*Bt* Cry 1Ab	IARI, New Delhi Proagro PG-S (India) Ltd.
		Cry 1H/ Cry 9C	Proagro PG-S (India) Ltd.

(vi) Sanitation: It involves removing and destroying the overwintering or breeding sites of the pest as well as preventing a new pest from establishing on the farm (e.g. not allowing off-farm soil from farm equipment to spread nematodes or plant pathogens to your land). This strategy has been particularly useful in horticultural and tree-fruit crop situations involving twig and branch pests. If, however, sanitation involves removal of crop residues from the soil surface, the soil is left exposed to erosion by wind and water. As with so many decisions in farming, both the short- and long-term benefits of each action should be considered when tradeoffs like this are involved.

(vii) Spacing of Plants: It heavily influences the development of plant diseases. The distance between plants and rows, the shape of beds, and the height of plants influence air flow across the crop, which in turn determines how long the leaves remain damp from rain and morning dew. Generally speaking, better air flow will decrease the incidence of plant disease. However, increased air flow through wider spacing will also allow more sunlight to the ground. This is another instance in which detailed knowledge of the crop ecology is necessary to determine the best pest management strategies. How will the crop react to increased spacing between rows and between plants? Will yields drop because of reduced crop density? Can this be offset by reduced pest management costs or fewer losses from disease?

(viii) Altered Planting Dates: This can at times be used to avoid specific insects or diseases. For example, squash bug infestations on cucurbits can be decreased by the delayed planting strategy, i.e., waiting to establish the cucurbit crop until overwintering adult squash bugs have died. To assist with disease management decisions, the Cooperative Extension Service (CES) will often issue warnings of "infection periods" for certain diseases, based upon the weather.

In some cases, the CES also keeps track of "degree days" needed for certain important insect pests to develop. Insects, being cold-blooded, will not develop below or above certain threshold temperatures. Calculating accumulated degree days, i.e., the number of days above the threshold development temperature for an insect pest, makes the prediction of certain events, such as egg hatch, possible. University of California has an excellent web site that uses weather station data from around the state to help California growers predict pest emergence.

Some growers gauge the emergence of insect pests by the flowering of certain non-crop plant species native to the farm. This method uses the "natural degree days" accumulated by plants. For example, a grower might time cabbage planting for three weeks after the Amelanchier species (also known as saskatoon, shad bush, or service berry) on their farm are in bloom. This will enable the grower to avoid peak egg-laying time of the cabbage maggot fly, as the egg hatch occurs about the time Amelanchier species are flowering (Couch, 1994). Using this information, cabbage maggot management efforts could be concentrated during a known time frame when the early instars (the most easily managed stage) are active.

(ix) Optimum Growing Conditions: Plants that grow quickly and are healthy can compete with and resist pests better than slow-growing, weak plants. Too often, plants grown outside their natural ecosystem range must rely on pesticides to overcome conditions and pests to which they are not adapted.

(x) Mulches: Living or non-living mulches are useful for suppression of insect pests, and some plant diseases. Hay and straw, for example, provide habitat for spiders. Research in Tennessee showed a 70% reduction in damage to vegetables by insect pests when hay or straw was used as mulch. The difference was due to spiders, which find mulch more habitable than bare ground (Reichert and Leslie, 1989). Other researchers have found that living mulches of various clovers reduce insect pest damage to vegetables and orchard crops. Again, this reduction is due to natural predators and parasites provided habitat by the clovers.

Mulching helps in minimizing the spread of soil-borne plant pathogens by preventing their spread through soil splash. Winged aphids are repelled by silver- or aluminum-coloured mulches. Recent springtime field tests at the Agricultural Research Service in Florence, South Carolina, have indicated that red plastic mulch suppresses root-knot nematode damage in tomatoes by diverting resources away from the roots (and nematodes) and into foliage and fruit (Adams, 1997).

1.2.2.2. Reactive Options

The reactive options mean that the grower responds to a situation, such as an economically damaging population of pests, with some type of short-term suppressive action. Reactive methods generally include inundative releases of biological control agents, mechanical and physical controls, botanical pesticides and chemical controls.

(i) Biological Controls

Biological control is the use of living organisms - parasites, predators, or pathogens - to maintain pest populations below economically damaging levels, and may be either natural or applied. A first step in setting up a biointensive IPM programme is to assess the populations of beneficials and their interactions within the local ecosystem. This will help to determine the potential role of natural enemies in the managed horticultural ecosystem. It should be noted that some groups of beneficials (e.g., spiders, ground beetles, bats) may be absent or scarce on some farms because of lack of habitat. These organisms might make significant contributions to pest management if provided with adequate habitat.

(a) Natural Biological Control: It results when naturally occurring enemies maintain pests at a lower level than would occur without them, and is generally characteristic of biodiversity systems. Mammals, birds, bats, insects, fungi, bacteria, and viruses all have a role to play as predators, parasites and pathogens in a horticultural system. By their very nature, pesticides decrease the biodiversity of a system, creating the potential for instability and future problems. Pesticides, whether synthetically or botanically derived, are powerful tools and should be used with caution.

Creation of habitat to enhance the chances for survival and reproduction of beneficial organisms is a concept included in the definition of natural biocontrol. **Farmscaping** is a term coined to describe such efforts on farms. Habitat enhancement for beneficial insects, for example, focuses on the establishment of flowering annual or perennial plants that provide pollen and nectar needed during certain parts of the insect life cycle. Other habitat features provided by farmscaping include water, alternative prey, perching sites, overwintering sites, and wind protection. Beneficial insects and other beneficial organisms should be viewed as mini-livestock, with specific habitat and food needs to be included in farm planning.

The success of such efforts depends on knowledge of the pests and beneficial organisms within the cropping system. Where do the pests and beneficials overwinter? What plants are hosts and non-hosts? When this kind of knowledge informs planning, the ecological balance can be manipulated in favour of beneficials and against the pests.

It should be kept in mind that ecosystem manipulation is a two-edged sword. Some plant pests (such as the tarnished plant bug and lygus bug) are attracted to the same plants that attract beneficials. The development of beneficial habitats with a mix of plants that flower throughout the year can help prevent such pests from migrating en masse from farmscaped plants to crop plants.

(b) Applied Biological Control: It is also known as augmentative biocontrol, involves supplementation of beneficial organism populations, for example through periodic releases of parasites, predators, or pathogens. This can be effective in many situations-well-timed inundative releases of Trichogramma egg wasps for codling moth control, for instance.

Most of the beneficial organisms used in applied biological control today are insect parasites and predators. They control a wide range of pests from caterpillars to mites. Some species of biocontrol organisms, such as Eretmocerus californicus, a parasitic wasp, are specific to one host-in this case the sweet potato whitefly. Others, such as green lacewings, are generalists and will attack many species of aphids and whiteflies.

Information about rates and timing of release is available from suppliers of beneficial organisms. It is important to remember that released insects are mobile; they are likely to leave a site if the habitat is not conducive to their survival. Food, nectar, and pollen sources can be "farmscaped" to provide suitable habitat.

The quality of commercially available applied biocontrols is another important consideration. For example, if the organisms are not properly labeled on the outside packaging, they may be mishandled during transport, resulting in the death of the organisms. A recent study by Rutgers University noted that only two of six suppliers of beneficial nematodes sent the expected numbers of organisms, and only one supplier out of the six provided information on how to assess product viability.

While augmentative biocontrols can be applied with relative ease on small farms and in gardens, applying some types of biocontrols evenly over large farms has been problematic. New mechanized methods that may improve the economics and practicality of large-scale augmentative biocontrol include ground application with "biosprayers" and aerial delivery using small-scale (radio-controlled) or conventional aircraft.

Inundative releases of beneficials into greenhouses can be particularly effective. In the controlled environment of a greenhouse, pest infestations can be devastating; there are no natural controls in place to suppress pest populations once an infestation begins. For this reason, monitoring is very important. If an infestation occurs, it can spread quickly if not detected early and managed. Once introduced, biological control agents cannot escape from a greenhouse and are forced to concentrate predation/parasitism on the pest(s) at hand.

An increasing number of commercially available biocontrol products are made up of microorganisms, including fungi, bacteria, nematodes, and viruses.

Of late, biological suppression of pests has become an intensive area of research because of environmental concerns. About 60% of the natural control of insect pests is by the natural enemies of pests such as parasitoids, predators and pathogens. The Australian lady bird beetle, *Cryptolaemus motrouzieri* has been found very effective against mealy bugs infesting grapes, guava, citrus, mango, pomegranate, ber and custard apple. The encyrtid parasite, *Leptomastix dactylopii* is effective against mealy bug, *Planococcus citri* on guava, citrus, pomegranate, ber and custard apple (Mani, 2001). *Bacillus thuringiensis* (*Bt*) is effective against tomato fruit borer, okra fruit borer and diamondback moth on cabbage and cauliflower.

Several methods of enrichment and conservation of natural enemies include providing nesting boxes for wasps and predatory birds; retaining pollen and nectar bearing flowering plants like Euphorbia, wild clover on bunds to provide supplementary food for natural enemies; placing bundles of paddy straw in fields for attracting predatory spiders. In addition, erecting perching sites, water pans, retaining bushes (Acalypha, Hibiscus, Crotons) helps in retention of predatory birds.

The last decade has witnessed a tremendous break through in biological control of diseases and nematodes like *Rhizoctonia, Pythium, Fusarium, Macrophomina, Ralstonia* and *Meloidogyne* in banana, tomato, egg plant, pea, grapes, cucumber, black pepper, cardamom, ginger and turmeric, especially by using species of *Trichoderma, Pochonia, Pseudomonas* and *Bacillus* (Tables 1.3 to 1.12).

Table 1.3. Biological control of fruit crop pests.

Fruit crop	Pest	Biocontrol agent/Dosage
Apple	Woolly aphid, *Eriosoma lanigerum*	*Aphelinus mali*-1000 adults or mummies/ infested tree.
	San Jose scale, *Quadraspidiotus perniciosus*	*Encarsia periniciosi*-2000 adults/infested tree.
	Codling moth, *Cydia pomonella*	*Chilocorus infernalis*-20 adults or 50 grubs/tree; *Trichogramma embryophagum*-2000 adults/tree; *Steinernema carpocapse*
Citrus	Cottony cushion scale, *Icerya purchasi*	*Rodolia cardinalis* -10 beetles/infested plant.
	Mealy bug, *Planococcus citri*	*Cryptolaemus montrouzieri*-10 beetles / infested plant; *Leptomastix dactylopii* 3000 adults/ha.
	Red scale, *Aonidiella aurantii*	*Chilocorus nigrita*-15 adults/infested tree
	Scale insect, *Coccus viridis*	*Verticillium lecanii*-16 x 10^4 spores/ml + 0.05% Teepol.
	Leaf miner, *Phyllocnistis citrella*	*S. carpocapse*
Grapevine	Mealy bug, *Maconellicoccus hirsutus*	*C. montrouzieri*-2500-3000 beetles/ha or 10 beetles/vine.
Guava	Green shield scale, *Chloropulvinaria psidii*	*C. montrouzieri*-10-20 beetles /infested plant.
	Aphid, *Aphis gossypii*	*V. lecanii*-10^9 spores/ml + 0.1% Teepol.

Table 1.4. Biological control of fruit crop diseases.

Fruit Crop	Disease/Pathogen/s	Potential biocontrol agent/s
Banana	Panama wilt, *Fusarium oxysporum* f. sp. *cubense*	*Trichoderma viride, Aspergillus niger, Pseudomonas fluorescens, T. viride* + *P. fluorescens* – sucker treat.
Citrus	Root rot, *Phytophthora* spp.	*T. viride /T. harzianum* at 100 kg/ha; *Pencillium funiculosum, Pythium nunn*-soil treat.
	Canker, *Xanthomonas campestris* pv. *citri*	*Aspergillus niger* AN 27
Strawberry	Grey mold, *Botrytis cineria*	*T. harzianum*
Mulberry	Leaf spot, *Cercospora moricola*	*T. viride, T. harzianum, Pseudomonas fluorescens*
	Cutting rot, *F. solani*	*T. virens, T. harzianum, T. pseudokoningii*
Grapevine	Powdery mildew, *Uncinula necator*	*Ampelomyces quisqualis*-dispersal from wick cultures at 15 cm of shoot growth & bloom
	Downy mildew, *Plasmopara viticola*	*Fusarium proliferatum* weekly spray starting from 15 cm of shoot growth – 10^6 spores/ml
Guava	Anthracnose, *Pestalotia psidii, Colletotrichum gloeosporoides*	*T. harzianum*
	Wilt, *Gliocladium roseum* & *F. solani*	*Pencillium citrinum, Aspergillus niger* AN 27, *T. harzianum*
Mango	Anthracnose, *Colletotrichum gloeosporoides*	*T. harzianum, Streptosporangium pseudovulgare*
	Powdery mildew, *Oidium mangiferae*	*S. pseudovulgare*
	Bacterial canker, *Xanthomonas campestris* pv. *mangiferaeindicae*	*Bacillus coagulans*
Apple	Scab, *Venturia inaequalis*	*Chaetomium globosum, A. pullulans, Microsphaeropsis* sp., *Cladosporium* spp., *Trichothecium roseum*-Foliar spray.
	Collar rot, *Phytophthora cactorum*	*Enterobacter aerogenes, Bacillus subtilis*-Soil treatment; *T. virens*-soil treat.
	White root rot, *Dematophora necatrix*	*T. viride, T. harzianum, T. virens*-soil treat.
Pear	Blue mold, *Pencillium expansum*; Grey mold, *Botrytis cineria*	*C. infirmo-miniatus* YY6, *C. laurentii* RR87-108, *R. glutinis* HRB6-fruit spray 3 week or 1 day prior to harvest - 10^8 cfu/ml;

Fruit Crop	Disease/Pathogen/s	Potential biocontrol agent/s
		Pantoea agglomerans CPA-2-post-harvest fruit dipping in 8 x10^8 cfu/ml
	Fire blight, *Erwinia amylovora*	*Pseudomonas fluorescens*-foliar spray
Peach	Brown rot, *Monilia fructicola*	*Bacillus subtilis* (B-3)-post-harvest fruit line spray at 5 x 10^8 cfu/g); *Pseudomonas syringae*-post-harvest fruit dipping in 10^7 cfu/ml.
	Twig blight, *Monilia laxa*	*Pencillium frequentans*-spray shoots in early growing season - $10^{8\text{-}9}$ spores /ml
	Crown gall, *Agrobacterium tumefaciens*	*Agrobacterium radiobacter* K84, K1026-root dip treat.
Strawberry	Grey mold, *Botrytis cineria*	*Trichoderma* products (BINAB TF & BINAB T), *Bacillus pumilus, Pseudomonas fluorescens, Gliocladium roseum* - spray flowers & fruits – white flower bud to pink fruit-10^6 spores/ml; *G. roseum*-bee vectoring of flowers - 10^9 cfu/g of powder.
Passion fruit	Collar rot, *Rhizoctonia solani*	*T. harzianum, Trichoderma* sp.
Amla	Bark splitting, *Rhizoctonia solani*	*Aspergillus niger* AN 27

Table 1.5. Biological control of vegetable crop pests.

Vegetable crop	Pest	Biocontrol agent/Dosage
Beans	Mite, *Tetranychus* spp.	*Phytoseiulus persimilis*-10 adults/plant or release 1-6 leaves with predatory mites.
Pigeon pea	Pod borer, *Helicoverpa armigera*	*Ha* NPV-250 LE/ha
Potato	Cut worm, *Agrotis ipsilon, A. segetum*	*Steinernema carpocapse, S. bicornutum, Heterorhabditis indica*
Tomato	Fruit borer, *Helicoverpa armigera*	*Trichogramma brasiliensis/ T. chilonis/ T. pretiosum*-50,000/ha; *Ha* NPV-250 LE/ha
Brinjal	Fruit & shoot borer, *Leucinodes orbonalis*	*S. carpocapse, H. indica*
Chilli	Fruit borer, *H. armigera*	*Ha* NPV-250 LE/ha
Cabbage	Diamondback moth, *Plutella xylostella*	*S. carpocapse, S. glaseri, S. carpocapse, S. feltiae, S. bicornutum, H. bacteriophora*
Mushroom	*Lycoriella auripila, L. mali, L. solani, Megaselia halterata*	*S. feltiae*

Table 1.6. Biological control of vegetable crop diseases.

Crop	Disease/Pathogen/s	Biocontrol agent/ Mode of application
French bean	Dry root rot, *Macrophomina phaseolina*	*Pseudomonas cepacia* UPR5C-seed treat.
	Wilt, *Fusarium oxysporum* f. sp. *phaseoli*	*Streptomyces* spp.- seed treat.
Pea	Root rot, *Aphanmomyces euteiches*	*Pseudomonas cepacia, P. fluorescens* PRA25, AMMD-seed treat.
	Damping-off, *Pythium ultimum*	*Pseudomonas cepacia* AMMD, *P. putida* NIR-seed treat.
	Wilt, *F. oxysporum* f.sp. *udum*	*T. viride, T. harzianum, T. koningii* - seed treat., *Aspergillus niger* AN 27
Cluster bean	Bacterial blight, *Xanthomonas axonopodis* pv. *cyamopsidis*	*Aspergillus niger* AN 27
Cabbage	Damping-off, *R. solani*	*T. viride, T. harzianum, T. koningii* -seed treat.
Cauliflower	Blight, *Alternaria brassicola*	*Streptomyces gresioviridis*-seed treat.
Okra	*Rhizoctonia solani*	*Bradyrhizobium japonicum, Rhizobium* spp.- seed treat.
Tomato	Damping-off, *Pythium aphanidermatum*	*T. viride, T. harzianum, Pseudomonas aeruginosa* 7NSK2
	Wilt, *Fusarium oxysporum* f. sp. *lycopersici*	*T. viride, T. harzianum, Aspergillus niger,* non-pathogenic *F. oxysporum, F. oxysporum* f. sp. *dianthi, P. fluorescens* strains Pf1, *P. putida, Pencillium oxalicum, Pythium oligandrum, Bacillus subtilis* strain FZB-G, *Streptomyces* spp.- seed treat, seed & soil treat.
Potato	Black scurf, *R. solani*	*T. harzianum, T. viride*- tuber treat, *Aspergillus niger* AN 27, *Verticillium biguttatum*-soil treat, *Laetisaria arvalis*-tuber treatment, Binucleate *Rhizoctonia.*
	Wilt, *Ralstonia solanacearum*	*Bacillus cereus, B. subtilis*
Bell pepper	*Phytophthora capsici*	*T. viride, T. harzianum*-fruit treat.
	Damping-off, *Pythium aphanidermatum*	*Streptomyces griseoviridis*-seed & soil treat.
Brinjal	Damping-off, Wilt, *Phytophthora* sp., *Pythium aphanidermatum, F. solani*	*T. viride, T. harzianum, T. koningii*-seed & soil treat.
	Collar rot, *Sclerotinia sclerotiorum*	*T. viride, T. virens, Bacillus subtilis*-soil treat.
Carrot	Soft rot, *Sclerotinia sclerotiorum*	*Coniothyrium minitans*-soil treat.

Crop	Disease/Pathogen/s	Biocontrol agent/ Mode of application
	Root rot, *R. solani*	*T. virens* GL-21
Radish	Wilt, *F. oxysporum* f.sp. *raphani*	*P. fluorescens* strains WCS374, WCS417r – soil treat.
	Root rot, *R. solani*	*Laetisaria rosiepellis, Pythium acanthicum*-soil treat.
Beet root	Damping-off, *Pythium debarryanum, P. ultimum*	*Pencillium* spp + *P. fluorescens*-seed treat., *Pythium oligandrum*
Cucumber	Wilt, *Fusarium oxysporum* f. sp. *cucumerinum, R. solani*	*Colletotrichum orbiculare, F. oxysporum* f. sp. *niveum, Pseudomonas putida* 89B-27, *Serratia marcescens,* Tobacco necrosis virus.
	Powdery mildew	*Ampelomyces quisqualis*-foliar spray
	Cucumber mosaic virus	*P. fluorescens* strain 89B-27
Water melon	Wilt, *Fusarium oxysporum* f. sp. *solani, F. o.* f. sp. *niveum*	*T. viride, Aspergillus niger*-seed & soil treat., *Pencillium janczewskii.*
Musk melon	Wilt, *F. oxysporum, F. solani, R. solani*	*T. harzianum, Aspergillus niger* – seed treat.
Onion	Soft rot, *Sclerotium cepivorum*	*Chaetomium globosum, Trichoderma* sp. C62 -soil treat.

Table 1.7. Biological control of ornamental crop diseases.

Crop	Disease / Pathogen/s	Biocontrol agent/ Mode of application
Rose	Grey mold, *Botrytis cineria*	*T. viride, T. harzianum*-cutting treat.
Gladiolus	Yellows & corm rot, *F. oxysporum* f. sp. *gladioli*	*T. virens, T. harzianum*-corm & soil treat.
Chrysanthemum	Wilt, *Fusarium oxysporum*	*T. harzianum* at 160 kg/ha - soil appln.
	Rhizoctonia solani	*Aspergillus niger* AN 27
Carnation	Wilt, *F. oxysporum* f. sp. *dianthi*	*Pseudomonas fluorescens* strain WCS 417r - soil appln; *P. putida* WCS 358r-root dip treat.; *Alcaligenes* sp., *Bacillus* sp.; *Arthrobacter* sp., *Hafnia* sp.; *Serratia liquifaciens.*
Gerbera	*Phytophthora cryptogea*	*Trichoderma* spp.-soil treat.
Narcissus	Wilt, *F. oxysporum* f. sp. *narcissi*	*Streptomyces griseoviridis, Minimedusa polyspora*-bulb treat.
Zinnia	*Rhizoctonia solani*	*T. virens* GL-21, *T. virens* GL-20
Marigold	*Pythium ultimum*	*Glomus intraradices, G. mosseae* -soil treat.

Table 1.8. Biological control of medicinal and aromatic crop diseases.

Medicinal/ aromatic crop	**Disease / Pathogen/s**	**Biocontrol agent**
Opium poppy	Sclerotinia rot & blight, *Sclerotinia sclerotiorum*	*T. harzianum, T. viride, T. koningii, T. virens*- soil treat.
	Downy mildew, *Peronospora arborescens*	*Trichoderma* spp.-seed treat.
Periwinkle	*Phytophthora parasitica*	*Phytophthora parasitica* var. *nicotianae*- soil treat.
Jasmine	Root rot, *Macrophomina phaseolina*	*T. viride, T. harzianum*- cutting treat.
Chinese rose	Wilt, *Fusarium oxysporum*	*Aspergillus niger*- soil treat.
Menthol mint	Stolon decay, *Sclerotinia sclerotiorum*	*T. harzianum, T. virens*- sucker treat.
	Verticillium dahliae	*Verticillium nigrescens*

Table 1.9. Biological control of tuber crop diseases.

Tuber crop	**Disease / Pathogen/s**	**Biocontrol agent**
Yam	*Botrytis theobromae*	*Trichoderma viride*
Cassava	*Phytophthora dechsleri*	*Trichoderma viride*
Elephant foot yam	*Sclerotium rolfsii*	*Trichoderma harzianum, T. pseudokoningii*

Table 1.10. Biological control of plantation crop pests.

Plantation crop	**Pest**	**Biocontrol agent/Dosage**
Coconut	Black headed caterpillar, *Opisinia arenosella*	*Goniozus nephantidis*-3000 adults/ha.
	Rhinoceros beetle, *Oryctes rhinoceros*	Baculovirus-10 infected beetles/tree.
Areca nut	*Ischnaspis longirostris*	*Chilocorus nigrita*-20 to 50 adults/plant.
Coffee	Mealy bugs, *Planococcus* & *Pseudococcus* spp.	*Cryptolaemus montrouzieri*-2-10 beetles/ infested plant.

Table 1.11. Biological control of plantation crop diseases

Crop	**Disease/Causal agent**	**Biocontrol agents reported**
Coconut	Stem bleeding, *Thielaviopsis paradoxa (Ceratostomella paradoxa)*	*Trichoderma virens, T. Harzianum*, phosphobacteria
	Basal stem rot, *Ganoderma lucidum*	*T. virens, T. harzianum*
Areca nut	Bud rot, *Phytophthora* spp.	*Trichoderma* spp.
	Fruit rot, *Phytophthora arecae, Colletotrichum capsici*	*Trichoderma* spp., *P. fluorescens*

Crop	Disease/Causal agent	Biocontrol agents reported
	Foot rot/anabe, *Ganoderma lucidum*	*T. harzianum*
Tea	Red root rot, *Poria hypolateritia*	*T. harzianum*
	Brown root, *Fomes noxius*	*T. virens, T. harzianum*
	Black root rot, *Rosellina arcuata*	*T. virens, T. harzianum*
Coffee	Black root, *Pellicularia koleroga*	*T. virens, T. harzianum*
	Brown root, *Fomes noxius*	*T. virens, T. harzianum*
	Santhaveri wilt, *F. oxysporum* f. sp. *coffeae*	*T. virens, T. harzianum*
Rubber	Brown rot, *Phellinus noxius*	*T. viride, T. harzianum, T. hamatum*
Betel vine	Foot & root rot, *Phytophthora parasitica* pv. *piperina*	*T. viride, T. harzianum*-soil treat.
	Collar rot, *Sclerotium rolfsii*	*T. harzianum, T. viride, T. koningii, T. virens*-soil treat.

Table 1.12. Biological control of spice crop diseases.

Spice crop	Disease/Causal organism	Effective biocontrol agents/Mode of application
Black pepper	Foot rot, *Phytophthora capsici*	*Trichoderma harzianum, T. virens, Glomus fasciculatum*-soil treat.; *Pseudomonas fluorescens, Bacillus* sp.-foliar spray
	Anthracnose, *Colletotrichum gloeosporoides*	*P. fluorescens*-foliar spray
	Slow decline, *Radopholus similis, Meloidogyne incognita, Phytophthora capsici*	*T. harzianum, T. virens, Paecilomyces lilacinus, Pochonia chlamydosporia*- soil treat.
Cardamom	Damping off, *Pythium vexans*	*T. harzianum, T. viride*-soil treat. in solarized nursery beds.
	Clump rot/rhizome rot, *Pythium vexans, Rhizoctonia solani, Meloidogyne incognita*	*T. harzianum*-soil treat.
	Capsule rot, *Phytophthora meadii, P. nicotianae* var. *nicotianae*	*T. harzianum, T. viride, T. virens, T. hamatum*-soil treat.
Ginger	Rhizome rot, *Pythium aphanidermatum, P. myriotylum*	*T. harzianum, T. virens*-soil solarization + soil treat.; *T. viride, P. fluorescens*-seed treat.; *Aspergillus niger* AN 27-soil treat.
	Yellows, *Fusarium oxysporum* f. sp. *zingiberi, M. incognita*	*T. harzianum, T. virens , T. hamatum*-soil solarization + soil treat., Rhizome treat.
	Bacterial wilt, *Ralstonia solanacearum*	Avirulent *Ralstonia solanacearum, P. fluorescens*, endophytic bacteria-soil treat.

Spice crop	Disease/Causal organism	Effective biocontrol agents/Mode of application
Turmeric	Rhizome rot, *Fusarium* sp., *Pythium graminicolum, P. aphanidermatum, R. similis, M. incognita*	*T. harzianum, T. viride, T. virens* -soil treat.
Fenugreek	Root rot, *R. solani*	*T. viride, P. fluorescens*-seed treat.
Coriander	Root rot/wilt, *Fusarium oxysporum* f. sp. *corianderii*	*T. viride, T. harzianum, Streptomyces* sp.-seed treat.
Cumin	Wilt, *F. oxysporum* f. sp. *cumini*	*Trichoderma* spp. , *T. virens*-soil treat.
Vanilla	Root rot, *Phytophthora meadii, F. oxysporum* f. sp. *vanillae*	*T. harzianum, P. fluorescens*-soil treat.
Mustard	Damping-off, *P. aphanidermatum*	*T. viride, T. harzianum*-seed & soil treat.

(c) Avermectins: The avermectins are a new class of macro-cyclic lactones derived from mycelia of the soil actinomycete, *Streptomyces avermitilis* (soil inhabiting which is ubiquitous in nature). These compounds were reported to be possessing insecticidal, acaricidal and nematicidal properties (Putter *et al.* 1981). They are commonly distributed in most of the cultivated soils and are in widespread use, especially as agents affecting plant parasitic nematodes, mites and insect pests. The water solubility of avermectin B1 is approximately 6-8 ppb and its leaching potential through many types of soil is extremely low. These physical properties also confer many advantages upon the use of avermectins as pesticides. Their rapid degradation in soil and poor leaching potential suggest that field applications would not result in persistent residues or contamination of ground water.

Avermectins offer an outstanding alternative to any of the available synthetic pesticides. Their novel mode of action, high potency and specific physico-chemical properties makes the avermectins excellent candidates for further insecticidal, acaricidal and nematicidal studies.

Scientists at the Indian Institute of Horticultural Research, Bangalore, for the first time in India, have isolated 6 strains of *S. avermitilis* and showed their effectiveness for the management of root-knot nematodes infecting tomato, egg plant, chilli, carnation and gerbera; and redspider mite on carnation and gerbera (Parvatha Reddy and Nagesh, 2002) (Table 1.13). Avermectins are also effective against other insect pests (potato leaf miner, *Liriomyza huidobrensis*; chilli thrips, *Scirtothrips dorsalis*; cabbage diamondback moth, *Plutella xylostella*; bean leaf miner, *Liriomyza huidobrensis*; rose thrips, *Rhipiphorothrips cruentatus, Scirtothrips dorsalis*; poinsettia whitefly, *Trialeurodes vaporariorum*); mite pests (chilli yellow mite, *Polyphagotarsonemus latus*; bean spider mite, *Tetranychus urticae*; rose red spider mite, *Tetranychus urticae*); and nematode *pests* (banana nematodes, Meleidogyne javanica, Radopholus similis; citrus nematode, *Tylenchulus semipenetrans*; tomato reniform nematode, Rotylenchulus reniformis; cucumber root-knot nematode, *Meloidogyne incognita*; garlic stem and bulb nematode, Ditylenchus dipsaci).

Table 1.13. Management of horticultural crop nematodes using avermectins

Horticultural crop	Nematode	Avermectin/dose
Tomato, egg plant, chilli	*M. incognita*	Aqueous solution of avermectins (250 ml of 0.001%/m^2 nursery bed); charcoal formulations of *S. avermitilis* at 100 g/m^2
Carnation and gerbera (commercial polyhouses)	*M. incognita*	Post-plant treatment at 250 ml/m^2 at two intervals (6 and 12 months after planting)
	Tetranychus urticae	Avermectins at 0.001% achieved 92% mortality

(ii) Mechanical and Physical Controls

Methods included in this category utilize some physical component of the environment, such as temperature, humidity, or light, to the detriment of the pest. Common examples are tillage, flaming, flooding, soil solarization and plastic mulches to kill pests.

Heat or steam sterilization of soil is commonly used in greenhouse operations for control of soil-borne pests. Floating row covers over vegetable crops exclude flea beetles, cucumber beetles, and adults of the onion, carrot, cabbage, and seed corn root maggots. Insect screens are used in greenhouses to prevent aphids, thrips, mites, and other pests from entering ventilation ducts. Large, multi-row vacuum machines have been used for pest management in strawberries and vegetable crops. Cold storage reduces post-harvest disease problems on produce.

Although generally used in small or localized situations, some methods of mechanical/physical control are finding wider acceptance because they are generally more friendly to the environment.

(iii) Chemical Controls (Reduced-Risk Pesticides)

Included in this category are both synthetic pesticides and botanical pesticides.

(a) Synthetic Pesticides: They comprise a wide range of man-made chemicals used to control insects, mites, nematodes, plant diseases, and vertebrate and invertebrate pests. These powerful chemicals are fast acting and relatively inexpensive to purchase.

Pesticides are the option of last resort in BIPM programmes because of their potential negative impacts on the environment, which result from the manufacturing process as well as from their application on the farm. Pesticides should be used only when other measures, such as biological or cultural controls, have failed to keep pest populations from approaching economically damaging levels.

If chemical pesticides must be used, it is to the grower's advantage to choose the least-toxic pesticide that will control the pest but not harm non-target organisms such as birds, fish, and mammals. Pesticides that are short-lived or act on one or a few specific organisms are in this class. Examples include insecticidal soaps, horticultural oils, copper compounds (e.g., Bordeaux mixture), sulfur, boric acid, and sugar esters.

(b) Biorational Pesticides: Biorational pesticides are generally considered to be derived from naturally occurring compounds or are formulations of microorganisms. Biorationals have a narrow target range and are environmentally benign. Formulations of *Bacillus thuringiensis*, commonly known as *Bt*, are perhaps the best-known biorational pesticide. Other examples include silica aero gels, insect growth regulators, and particle film barriers.

A relatively new technology, particle film barriers are currently available under the trade name Surround WP Crop Protectant. The active ingredient is kaolin clay, an edible mineral long used as an anti-caking agent in processed foods, and in such products as toothpaste and kaopectate. There appears to be no mammalian toxicity or any danger to the environment posed by the use of kaolin in pest control. The kaolin in Surround is processed to a specific particle size range, and combined with a sticker-spreader. Non-processed kaolin clay may be phytotoxic. Surround is sprayed on as a liquid, which evaporates, leaving a protective powdery film on the surface of leaves, stems, and fruits. Conventional spray equipment can be used and full coverage is important. The film works to deter insects in several ways. Tiny particles of the clay attach to the insects when they contact the plant, agitating and repelling them. Even if particles do not attach to their bodies, the insects may find the coated plant or fruit unsuitable for feeding and egg-laying. In addition, the highly reflective white coating makes the plant less recognizable as a host.

(c) Sugar Esters: Sugar esters have performed as well as or better than conventional insecticides against mites and aphids in apple orchards; psylla in pear orchards; whiteflies, thrips, and mites on vegetables. However, sugar esters are not effective against insect eggs. Insecticidal properties of sugar esters were first investigated a decade ago when a scientist noticed that tobacco leaf hairs exuded sugar esters for defense against some soft-bodied insect pests. Similar to insecticidal soap in their action, these chemicals act as contact insecticides and degrade into environmentally benign sugars and fatty acids after application.

(d) Inorganic Chemicals: Spray application of K_2HPO_4 or KH_2PO_4 at 3.5 g/litre of water has been reported to control powdery mildew in rose and carnation. Similarly, the above treatment was also found effective for the management of powdery mildew on mango, grapes and cucurbits.

(e) Strobilurin Fungicides: Strobilurin fungicides are also called Qo inhibitors as they act on cytochrome Qo of the fungi. The Basidiomycetous fungus, *Strobilurus tenacellus* produces antibiotics to ward off competition from other fungi. Based on this principle, several fungicides have been developed namely Azoxystrobin, Kresoxy methyl, Metominostrobin, Trifloxystrobin, Picoxystrobin, Pyraclostrobin, Famoxadone and Fenomidone during 1996-2001. Within 4 years, sale of these fungicides reached $620 million, accounting for 10% of total fungicide market in the world. This success is unparalleled in the history of fungicides.

Strobilurin fungicides are naturally occurring compounds and hence eco-friendly, highly systemic, have unique mode of unisite action, hence development of resistance is common. They have broad spectrum of activity on all groups of fungi and registered in 72 countries on 84 crops representing over 400 crop/disease systems.

(f) Botanical Pesticides: They can be as simple as pureed plant leaves, extracts of plant parts, or chemicals purified from plants. Pyrethrum, neem formulations, and rotenone are examples of botanicals. Some botanicals are broad-spectrum pesticides. Others, like ryania, are very specific. Botanicals are generally less harmful to the environment than synthetic pesticides because they degrade quickly, but they can be just as deadly to beneficials as synthetic pesticides. However, they are less hazardous to transport and in some cases can be formulated on-farm. The manufacture of botanicals generally results in fewer toxic by-products.

Neem products such as cake, oil, neem seed kernel extract (NSKE), neem seed powder extract (NSPE), pulverized NSPE and soaps are being used extensively to manage horticultural crop pests (bean fly, *Ophiomyia phaseoli*; serpentine leaf miner, *Liriomyza trifolii* on several crops; cucurbit fruit fly, *Bactrocera cucurbitae*; tomato fruit borer, *Helicoverpa armigera*; brinjal fruit and shoot borer, *Leucinodes orbonalis*; water melon and chilli thrips, *Thrips* spp.; chilli yellow mite, *Polyphagotarsonemus latus* and okra leaf hopper, *Amrasca biguttulla biguttulla*) (Krishna Moorthy and Krishna Kumar, 2002).

The soap sprays were highly effective on leaf hoppers, aphids, red spider mites and white flies in many vegetables, but moderately effective on thrips in water melon and chillies (Table 1.14).

Table 1.14. Insect pests on which soaps were found effective

Crop	Insect pests
Cabbage and cauliflower	DBM, leaf webber, aphids, young *Spodoptera* larva
Tomato	Whitefly, red spider mites, fruit borer (egg laying stage), leaf miner
Okra	Leaf hopper, whitefly, aphids
Cucurbits	Fruitfly, leaf miner
Mango	Leaf hopper
Ornamental crops	Mites, whitefly

(g) Compost Teas: They are most commonly used for foliar disease control and applied as foliar nutrient sprays. The idea underlying the use of compost teas is that a solution of beneficial microbes and some nutrients is created, then applied to plants to increase the diversity of organisms on leaf surfaces. This diversity competes with pathogenic organisms, making it more difficult for them to become established and infect the plant.

An important consideration when using compost teas is that high-quality, well-aged compost be used, to avoid contamination of plant parts by animal pathogens found in manures that may be a component of the compost. There are different techniques for creating compost tea. The compost can be immersed in the water, or the water can be circulated through the compost. An effort should be made to maintain an aerobic environment in the compost/water mixture.

Section - II

BIOINTENSIVE INTEGRATED PEST MANAGEMENT IN FRUIT CROPS

Chapter - 2

TROPICAL FRUIT CROPS

2.1. BANANA, *Musa* spp.

2.1.1. Diseases

2.1.1.1. Panama Wilt, *Fusarium oxysporum* f. sp. *cubense*

Panama wilt is one of the most devastating diseases of banana in the world. The disease is prevalent in Australia, Coasta Rica, Hawaii, India, Jamaica, Panama, South America, Surinam and West Africa. In India, it became widespread in Tamil Nadu, Kerala, Karnataka, Bihar and Assam. The disease is also prevalent in West Bengal, Maharashtra and Andhra Pradesh, especially where cvs. Rasabale, Amritpani, Malbhog and Martban are grown belonging to Rasthali group.

(i) Symptoms: The entry of the fungus is facilitated by root damage caused by the nematodes (*Radopholus similis, Meloidogyne incognita*, etc.). The fungus blocks the vascular system and causes wilting (Fig. 2.1). The infected plants show characteristic yellowing of leaf blades developing as a band along the margin and spreading towards midrib. The leaf wilts and the petiole buckles. The leaf hangs between the pseudo stem while the middle of lamina is still green. All leaves eventually collapse, whereas the petioles join the pseudo stem and die. Often the emerging heart leaf gets affected. Four to six weeks after the appearance of first symptoms, only the pseudo stem with dead leaves hanging round it remains. Young and old plants show dwarfing or stunting. When an affected rhizome is cut transversely, the disease is seen localized in the vascular strands (Fig. 2.1.). Individual strand appears yellow. Red or brown dots or streaks are also seen. The cut stem smells like rotten fish. The suckers growing out of diseased corms wilt and eventually the whole mat dies.

(ii) Epidemiology: Light textured loam and sandy loam soils which are acidic favours the disease development. Such soils are referred as conducive soils for fusarial infection. Soil infestation by nematodes (*Radopholus similis, Meloidogyne incognita*, etc.) predisposes the plant to infection. The pathogen is soil-borne and survives for long periods in soil as chlamydospores (even up to 20 years in the absence of its host). Ramakrishnan and Damodaran (1956) found that liming of soil reduced the survival period to two months. The primary spread of the disease is through infected rhizomes and secondary spread is through irrigation water. Continuous cultivation of banana in the same field results in abundant build up of inoculum in the soil.

Fig. 2.1. Panama wilt symptoms on banana

(iii) Integrated Management

(a) Cultural, Bioagents and Botanicals: Fallowing for 21 days before planting, soil application of bioagents such as *Trichoderma viride* and *Pseudomonas fluorescens* and incorporation of cassava residue at 10 MT/ha plus rice bran was found to check the occurrence of the disease effectively.

(b) Bioagents and Botanicals: Combination of *T. viride* isolate 6 and neem cake/pongamia cake had inhibitory effect against panama wilt pathogen and increased plant growth parameters (leaf number, leaf area, pseudo stem girth and root weight). Red soil with pH between 7-8 maintained at 60% moisture holding capacity was found to favour the survival of the antagonist (Satish, 1996).

2.1.2. Nematodes

2.1.2.1. Burrowing Nematode, *Radopholus similis*

This nematode in India was first reported on banana from the Palghat district of Kerala by Nair *et al.* (1966). *R. similis* causes 'rhizome rot' or 'toppling' or 'black head' disease of banana and is becoming a serious problem.

(i) Economic Importance and Losses: The burrowing nematode is responsible for 30.76 to 41 pr cent yield loss in banana (Rajagopalan and Naganathan, 1977b; Nair, 1979; Parvatha Reddy *et al.*, 1996d; Vadivelu *et al.*, 1987). Root population of *R. similis* is indirectly correlated with the yield (Charles *et al.*, 1985).

(ii) Symptoms: It causes retarded growth and extensive root and rhizome necrosis. Wounding of banana roots by the burrowing nematode usually induce reddish-brown cortical lesions which are diagnostic of the disease (Fig. 2.2). These lesions are clearly seen when an affected root is split longitudinally and examined immediately. Root and rhizome necrosis is manifested by varying degrees of retarded growth, leaf yellowing and falling of mature plants.

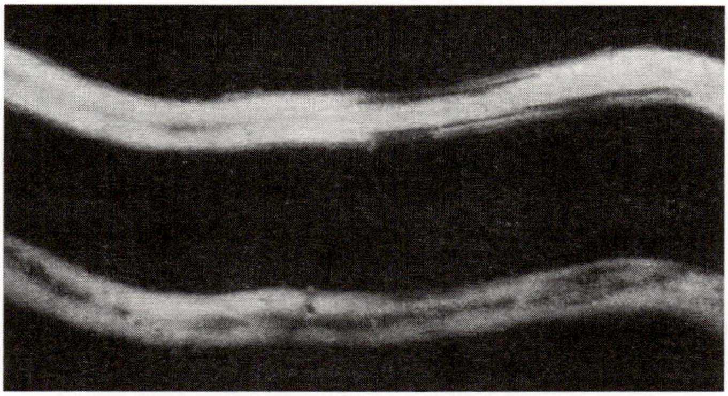

Fig. 2.2. Banana roots infected with *Radopholus similis*. Upper – Longitudinally cut root; Lower – Complete root (Courtesy: Union Carbide Agril. Products Co. Inc., 1986).

With the increase in nematode population, feeding roots are invaded and destroyed as fast as they are formed. The resulting setback in the uptake of plant nutrients leads to debility of the plant and production of smaller fruits. The lesioning of the primary roots together with the girdling and death of these anchor roots makes the plant prone to 'tip over' by wind action (Fig. 2.3).

Fig. 2.3. Premature fall of banana plants due to infection of *Radopholus similis*. Front – Toppled plants from non-treated plots; Back – Treated plants (Courtesy: Union Carbide Agril. Products Co. Inc., 1986).

(iii) Integrated Management

(a) Botanicals and AMF: Integration of neem cake at 200 g/plant with *Glomus mosseae* at 100 g/plant (containing 25-30 chlamydospores/g of inoculum) was most effective in reducing the *R. similis* population both in soil and roots, while karanj cake with *G. mosseae* gave maximum increase in fruit yield of banana. Mycorrhizal root colonization and number of chlamydospores of *G. mosseae* were maximum in neem cake amended soil (Table 2.1) (Parvatha Reddy *et al.*, 2002).

Table 2.1. Effect of *Glomus mosseae* and oil cakes on population of *Radopholus similis* and yield of banana

Treatment	Dose (g)/ plant	Popn. of *R. similis*		Yield(kg)/ plant
		Roots (10 g)	Soil (250 ml)	
G. mosseae	200	112	122	8.64
Castor cake	400	146	132	8.18
Karanj cake	400	118	128	10.34
Neem cake	400	118	112	8.91
G. mosseae + Castor cake	100 + 200	90	108	12.68
G. mosseae + Karanj cake	100 + 200	76	80	16.61
G. mosseae + Neem cake	100 + 200	48	62	14.80
Control	---	218	184	5.45
CD (P = 0.05)		**11.97**	**8.31**	**0.84**

(b) Cultural and Chemical: Double paring of banana suckers along with dipping in 0.5% monocrotophos for 45 min gave maximum yield (63.283 MT/ha) and recorded higher benefit : cost ratio (2.92) (Patil *et al.*, 1999) (Table 2.2).

Table 2.2. Effect of paring of suckers and dipping in insecticide solution on nematode population and yield of banana

Treatment	Nema popn in 200 ml soil	Nema popn in 5g roots	Yield (MT/ha)	Benefit: cost ratio
Unpared sucker	365	27	57	---
Paring + drying for 72 hr	209	23	59	1.17
Double paring + drying for 72 hr	187	21	59	1.72
Paring + dipping in 0.5% monocrotophos for 45 min	109	18	61	2.01
Paring + dipping in 0.75% monocrotophos for 45 min	89	16	60	0.30
Double paring + dipping in 0.5% monocrotophos for 45 min	85	15	63	2.92
Double paring + dipping in 0.75% monocrotophos for 45 min	76	12	63	2.22
Paring + pralinage with carbofuran at 40g/sucker	76	12	63	0.15

Integration of paring of banana suckers, dipping in 0.5% monocrotophos solution for 45 min along with intercropping with marigold or sunnhemp gave higher fruit yield (62.838 and 61.816 MT/ha, respectively) and benefit cost ratio (1.70 and 1.28, respectively) (Patil *et al.*, 1999) (Table 2.3).

Table 2.3. Effect of paring of suckers, dipping in insecticide solution and intercropping on nematode population and yield of banana

Treatment	Nema popn in 200 ml soil	Nema popn in 5g roots	Yield (MT/ha)	Benefit: cost ratio
Untreated control	454	29	58.439	---
Paring + dipping in 0.5% monocrotophos	283	18	60.927	1.20
Paring + dipping in 0.5% monocrotophos + incorporation of sunnhemp at 10g/m^2	252	14	61.816	1.28
Paring + dipping in 0.5% monocrotophos + incorporation of cowpea at 10g/m^2	274	17	60.483	1.04
Paring + dipping in 0.5% monocrotophos + incorporation of marigold at 2g/m^2	239	15	62.838	1.70
Carbofuran at 1.25g a.i./plant	240	14	61.860	1.58

(c) Bioagents and Botanicals: Integration of neem cake at 400 g per plant with *P. penetrans* at 100 g soil (300 spores/g)/*T. harzianum*/*T. viride* at 250 g/plant while planting was found effective in reducing nematode population both in soil and roots of banana by more than 50 per cent and increased plant growth parameters. The treatment should be repeated 4 months after planting. The burrowing nematode on banana were effectively managed by integration of neem cake at 500 g and FYM enriched with *T. harzianum* at 2 kg/plant. The above treatment increased the fruit yield to 45 kg/plant and bunches came to harvest 65 to 75 days earlier (Fig. 2.4).

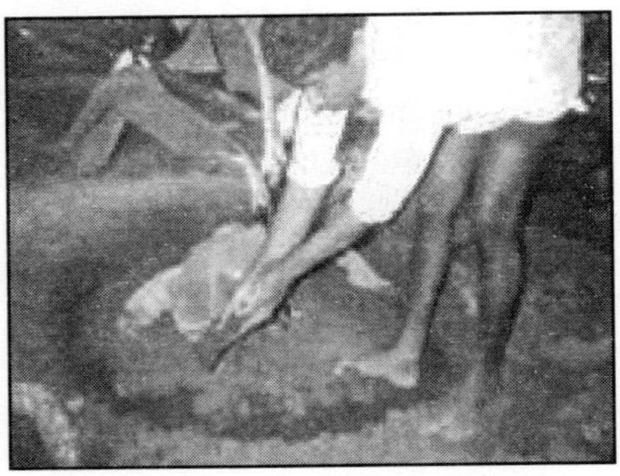

Fig. 2.4. Application of FYM enriched with bioagent to banana plant.

Soil application of FYM enriched with *Paecilomyces lilacinus* (10^6 cfu/g) and *Pseudomonas fluorescens* (10^9 cfu/g) at 2 kg/plant at the time of planting and subsequent application for four times at an interval of six months reduced the root population of *R. similis* by 64.5% and increased the fruit yield by 21%. Benefit: cost ratio (calculated for marginal cost of biopesicides and returns accrued by application of biopesicides) was 3.6.

Soil application of 2 kg FYM with *P. fluorescens* (with 1 x 10^9 spores/g) and *P. chlamydosporia* (with 1 x 10^6 spores/g) per plant at the time of planting and at an interval of 4 months significantly reduced the burrowing nematode by 64% compared to control.

(d) Botanicals, Chemicals, Bioagents and AMF: Integration of neem cake at 400 g/plant, carbofuran at 20 g/plant, *Glomus fasciculatum* at 50 g/plant and *P. penetrans* at 100 g soil/plant was most effective in reducing the population of *R. similis* in banana (Channabasappa *et al.*, 1995).

Integration of neemark, carbofuran, *P. penetrans* and *G. fasciculatum* was found effective in enhancing the plant growth and yield of banana besides raising the benefit cost ratio (2.65) and reducing the *R. similis* population both in soil and roots (Vidya and Reddy, 1998).

2.1.2.2. Spiral Nematode, *Helicotylenchus multicinctus*

(i) Economic Importance: The spiral nematode causes serious decline of banana yield to the tune of 34 to 56% with delayed flowering. It is responsible for 33.83 per cent loss in yield, 55.88 per cent loss in number of fruits per bunch and delayed fruiting by 134 days (Vadivelu *et al.*, 1987).

(ii) Symptoms: The spiral nematode incites discrete, relatively shallow, necrotic lesions on banana roots. *H. multicinctus* causes extensive root necrosis, die-back and dysfunction leading eventually to debility of the entire plant.

(iii) Integrated Management

(a) Bioagents/AMF and Botanicals: Significant reduction in population of *H. multicinctus* was observed in banana plants treated with *P. fluorescens* at 2g/*Glomus mosseae* at 25g along with press mud at 3 kg/plant. These treatments also enhanced the plant height, pseudo stem girth, number of leaves, leaf area and bunch weight (Jonathan and Cannayane, 2002).

2.1.2.3. Root-knot Nematode, *Meloidogyne incognita*

The root-knot nematodes, *M. incognita* and *M. javanica* attack bananas and may interact with other nematodes or soil pests. *M. incognita* caused 30.90% loss in fruit yield of banana (Jonathan and Rajendran, 2000).

(i) Symptoms: The infected plants were stunted, having small chlorotic leaves and galled roots (Fig. 2.5). Galling of the plant roots is most commonly found in areas previously planted with sugarcane. The galls vary in size and occur at the tips as well as in other areas along the root. Roots with galled tips cease to grow and sometimes develop secondary roots above the gall. Swollen female nematodes were found inside the galled and sometimes non-galled roots.

(ii) Survival and Spread: Root-knot nematodes survive on dicotyledonous plants, which are usually present in most soils in which bananas are growing. Survival and spread also occurs with the planting material on infected roots and corms.

Fig. 2.5. Root-knot nematode on banana roots. Left – Infected; Right – Healthy

(iii) Integrated Management

(a) Bioagents and Botanicals: Soil application of 2 kg FYM with *P. fluorescens* (with 1 x 10^9 spores/g) and *P. chlamydosporia* (with 1 x 10^6 spores/g) per plant at the time of planting and at an interval of 4 months significantly reduced the root-knot nematode by 76% compared to control.

Application of *T. harzianum* along with FYM has reduced the root-knot nematode population in soil (from 175/100 ml soil in control plot to 42/100 ml soil in treated plot) and roots (from 17/g root in control plot to 4/g root treated plot). Further, application of the bioagent also reduced *Fusarium* wilt incidence and increased the yield by 4.53 MT/ha over control.

Jonathan and Rajendran (2000) reported that application of *P. lilacinus* multiplied on neem cake at 15-20 g/plant significantly reduced root gall index, egg masses, eggs, females and soil population of *M. incognita* in banana.

2.1.2.4. Burrowing, Spiral and Root-knot Nematode Complex

Radopholus similis, Helicotylenchus multicinctus and *Meloidogyne incognita* cause nematode complex in banana.

(i) Integrated Management

(a) Physical, Cultural, Chemical and Botanical: Integration of paring and hot water treatment (at 55°C for 20 min) of banana suckers along with application of carbofuran 3G (at 16.6 g/plant) and neem cake (at 1 kg/plant) at the time of planting reduced the soil and root population of nematodes besides increasing growth, development and yield of banana (Table 2.4).

Table 2.4. Management of nematode complex in banana

Treatment*	Fruit yield/ plant (kg)	Cost: benefit ratio	Nema popn/200 ml soil**			Nema popn/5g roots		
			Rs	*Hm*	*Mi*	*Rs*	*Hm*	*Mi*
Untreated check	4.2	1: 1.49	213	150	133	62	51	3
Paring + hot water treat. + carbofuran	5.8	1: 2.01	97	78	40	30	19	21
Paring + hot water treat. + carbofuran + neem cake	7.6	1: 2.12	74	53	48	20	11	10
CD (P = 0.05)	**0.6**	**---**	**10.6**	**10.6**	**13.9**	**3.0**	**3.5**	**3.3**

* Hot water treatment at 55°C for 20 min, Carbofuran 3G at 16.6 g/plant, Neem cake at 1 kg/plant.

** Rs – *Radopholus similis*, Hm – *Helicotylenchus multicinctus*, Mi – *Meloidogyne incognita*.

2.1.2.5. Burrowing Nematode, *Radopholus similis* and Panama Wilt, *Fusarium oxysporum* f. sp. *cubense* Disease Complex

Incidence and losses due to Panama wilt caused by *F. oxysporum* f. sp. *cubense* is enhanced in association with the burrowing nematode, *R. similis* under high nematode population. This clearly indicates the existence of synergistic interaction between the burrowing nematode and Fusarium wilt pathogen in banana.

(i) Symptoms: Newhall (1958) showed that the incidence of Panama wilt in banana caused by *F. oxysporum* f. sp. *cubense* (Fig. 2.6) was doubled in the presence of *R. similis* during the experimental period of three months. When Gros Michel bananas infected with *R. similis* were inoculated with *F. oxysporum* f. sp. *cubense*, the period between inoculation and the onset of wilt was also considerably shortened (Loos, 1959).

Fig. 2.6. Burrowing nematode and Panama wilt disease complex in banana.

Lesions formed after inoculation with both *R. similis* and *F. oxysporum* f. sp. *cubense* were more extensively necrotic and increased in size more rapidly than when *R. similis* alone was used (Blake, 1966). *F. oxysporum* f. sp. *cubense* readily establishes itself in the feeder roots of banana when they are invaded by the nematode *R. similis*, but the fungus has seldom been recovered from nematode-free roots (Blake, 1966).

Rishbeth (1960) suggested that nematodes breakdown resistance to Panama wilt in Lacatan bananas.

(ii) Integrated Management

(a) Bioagents, Botanicals and Chemicals: Soil application of neem cake + *T. viride* + carbendazim was found effective in reducing the burrowing nematode (*R. similis*) and wilt (*F. oxysporum* f sp. *cubense*) disease complex and in increasing the banana fruit yield (15.147 MT/ha as compared to 9.887 MT/ha in control). This treatment also gave minimum lesion index (1.1) and root-knot index (1.0) as compared to control (4.0) with benefit: cost ratio of 2.72 (Ravi *et al.*, 2001).

2.2. CITRUS, *Citrus* spp.

2.2.1. Insect Pests

2.2.1.1. Leaf Miner, *Phyllocnists citrella*

(i) Damage: It is a serious pest of nursery and young plants but even attacks the grown up citrus trees all over India. The larvae feed on the epidermis of tender leaves making serpentine mines, which are silvery in colour (Fig. 2.7). The affected leaves become distorted and crumpled. The larvae may also mine the epidermis of tender shoots. Severe infestation may cause defoliation. Since new flush is attacked, the growth is severely hampered. In case of twig attack in young plants, 'die-back' also occurs. Ventral surface is preferred by the pest, but due to high population pressure, dorsal infestation is also seen. Citrus leaf miner helps in spreading mealy bug infestation and also acts as foci of citrus canker.

Fig. 2.7. Leaf miner infestation on leaves of citrus, adult

Of the total damage caused by the pest complex in citrus, 30% is claimed by the leaf miner alone. Moderate infestation of 1-2 larvae of leaf miners per leaf on seven year old trees was sufficient to reduce leaves and lower yields by 30-40% in the following year. A reduction in yield up to 50% and fruit weight from 120 to 70 g was observed.

(ii) Integrated Management

(a) Two Bioagents: Combined release of *Mallada boninensis* (30 larvae/tree) and *Tamarixia radiata* (40 adults/tree) resulted in 23-26% reduction in leaf miner population.

2.2.1.2. Black Fly, *Aleurocanthus woglumi*

This is a regional endemic pest in parts of Maharashtra and Karnataka. The epidemic proportion of the pest during late 70's and 80's brought havoc on the citrus industry in Central India. The monitory loss was estimated to the tune of Rs. 25 to 50 million annually.

(i) Damage: Both nymphs and adults suck cell sap and secrete voluminous honeydew on which sooty mold grows wildly that leads to fungal manifestation (*Capnodium* sp.) covering entire plant due to which photosynthesis is affected. The adults lay eggs in spiral fashion on new leaves. The nymphs, which are black in colour, suck sap from leaves and devitalize the plants (Fig. 2.8). In severe cases, fruit bearing capacity of the tree is also affected. Fruits are rendered insipid in taste and blackened due to sooty mold. Such fruits fetch low price in the market.

Fig. 2.8. Black fly on citrus

For successful fruit set, a minimum of 2.2% organic nitrogen in leaf is must. Five to ten black flies/cm^2 area or 50 to 100 nymphs/leaf are sufficient to reduce leaf nitrogen level below 2.2%.

(ii) Pre-disposing Factors for Citrus Black Fly Incidence: Grown up orchards on heavy clay soils had evergreen canopies intermingling with each other thus creating a microniche underneath. Further, poor drainage in such soils adds to the dampness which together helps in pest buildup. The tall evergreen border shrubs aggravate the pest problem further by sheltering pest population and in a way provide alternative to main host citrus plants.

(iii) Integrated Management

(a) Two Bioagents: Combined release of *M. boninensis* (30 larvae/tree) and *T. radiata* (40 adults/tree) resulted in 28-30 % population, respectively.

2.2.1.3. Green Scale, *Coccus viridis*

The green scale is a serious pest of citrus and coffee in Kodagu (Karnataka) and Palani and Shevroy hills of Tamil Nadu.

(i) Damage: The females breed parthenogenetically producing 500 nymphs or crawlers. The nymphs settle on all parts of the leaves preferring to settle on the under surface of the leaves along the midribs. The nymphs suck sap and excrete honeydew. The vigour of the infested plant is reduced and the black sooty mold fungus develops on the honeydew excreted. The insect passes through 3-4 generations. In cases of severe attack, the fruits also get smudged with black sooty mold and the market value of such fruits is lost.

(ii) Integrated Management

(a) Bioagents and Chemicals: In mixed planted orchards (citrus + coffee) with more shade and less light interception (900-1400 lux), spray of *Verticillium lecanii* at 10×10^6 spores/ml + 0.005% quinalphos + 0.05% teepol just before the onset of rainy season was highly effective against green scale (*C. viridis*) both in citrus and coffee. In pure citrus orchards, the combination was only effective during the rainy season (Singh, 1995).

2.2.1.4. Brown Scale, *Saissetia coffeae*

The brown scale is present in citrus and coffee plantations throughout the year. Its outbreaks is recorded which cause more concern than even *Coccus viridis*.

(i) Damage: By and large *S. coffeae* behaves similar to *C. viridis*. The eggs hatch inside the body of the female and the nymphs or crawlers start emerging from the underside of the hemispherical shell. Nymphs settle on the leaves and the damaging cycle begins.

(ii) Integrated Management

(a) Bioagents and Chemicals: In mixed planted orchards (citrus + coffee) with more shade and less light interception (900-1400 lux), spray of *Verticillium lecanii* at 10×10^6 spores/ml + 0.005% quinalphos + 0.05% teepol just before the onset of rainy season was highly effective against brown scale both in citrus and coffee. In pure citrus orchards, the combination was only effective during the rainy season. *V. lecanii* at 3×10^8 spores/ml is found effective in reducing the population of brown scale in humid areas.

2.2.2. Diseases

2.2.2.1. Damping-Off, *Phytophthora nicotianae, P. citrophthora, P. palmivora, Rhizoctonia solani, Pythium* spp.

Damping-off of seedlings in nursery bed is widespread problem in citrus industry. The disease frequently occurs in field nurseries where maintenance of sanitary measures is difficult. More than 20% seedling mortality has been observed in Central India due to this disease (Naqvi, 2000).

(i) Symptoms: Necrosis of tissue and typical damping-off of seedlings occur due to fungal infection just above the soil level. The seeds/soil infested with the pathogen results in pre-emergence rot of seeds and post-emergence damping-off of seedlings. In infested seed beds, the mortality of seedlings occurs in patches. The seedling mortality increases where excessive soil moisture accompanies the favourable temperature for the pathogen. Pathogens survive in soil either through saprophytic growth (*R. solani*) or production of resistant structures such as chlamydospores or oospores (*Phytophthora* spp.). Seedlings become tolerant to *R. solani* infection on maturity of first true leaf. Flood irrigation in flat bed system spreads the pathogen from one bed to other. However, infection of seedlings with *Phytophthora* spp. in primary nursery beds perpetuates and causes further losses to seedlings in secondary nursery beds. The budded plants show stunted, chlorotic growth with development of poor feeder roots.

(ii) Integrated Management

(a) Bioagents and Botanicals: Mixing 1 kg of *Trichoderma viride* in 40 kg of FYM and incubating the mixture for 24 hr. and application at 250 g mixture/m^2 is effective.

2.2.2.2. Foot Rot, Root Rot, Crown Rot, Gummosis, Leaf Fall and Fruit Rot, *Phytophthora palmivora, P. nicotianae* var. *parasitica*

The disease seems to occur especially in the high rainfall areas. Its prevalence has been reported in south India, Maharashtra, Gujarat, Punjab and Assam states. *P. nicotianae* var. *parasitica* is widespread in Assam, while *P. palmivora* is prevalent throughout India.

(i) Symptoms: Profuse gumming on the surface of the attacked bark is the main symptom. When gumming occurs on the stem, droplets of gum trickle down the stem (Fig. 2.9). The bark gradually turns brown to dark brown and develops longitudinal cracks. A thin layer of wood tissue is also affected. When gumming starts close to the soil, the disease spreads to the main roots and then around the base of the trunk. As a result of severe gumming, the bark becomes completely rotten and the tree dies owing to girdling effect. The trees usually blossom heavily and die after the fruits mature. In such cases, the disease is called as foot rot or collar rot. The pathogen produces symptoms of decline through rotting of the rootlets, girdling of the trunk and dropping of the blighted leaves. The fruits lying on the ground are liable to invasion by the pathogen and develop brown rot.

Fig. 2.9. Gummosis on main stem, foot rot and decline of Nagpur mandarin tree.

Leaf fall and fruit rot phase of the disease is severe on mandarin oranges in heavy rainfall areas of south India. Quick shedding of leaves is the earliest symptom. The infection starts as water-soaked lesions at the leaf base. By the time the lesions extend to the whole leaf, the leaf drop off. The infection may spread to young twigs and fruits of all stages. The affected fruits show water-soaked patches on rind and subsequently drop off and rot.

(ii) Epidemiology: Severe occurrence of the disease is noticed in sweet oranges, acid lime and lemon. Heavy soil, high water table, high soil moisture, soil pH of 5.4 to 7.5 and temperature of 25° to 28°C are conducive for disease development. Low grafting, deep planting and nearness of bud union to ground level increase the chances for soil-borne infection. The fungi survive on fallen fruits, twigs, leaves and in cracks of the tree and spread by irrigation water, rain splashes, wind and insects to stems, leaves and fruits.

(iii) Integrated Management

(a) Bioagents and Botanicals: Soil application of *Trichoderma* spp. along with organic matter in the ratio of 1: 40 at 2 kg/plant or FYM enriched with *Trichoderma* spp. at 2 kg/plant is effective against the pathogen.

Trichoderma harzianum has been advocated to have a potent antagonistic action against *Phytophthora* root rot of Coorg mandarin when applied along with coffee waste, poultry manure and FYM (Sawant and Sawant, 1989).

2.2.2.3. Pencillium Rot, *Pencillium digitatum, P. italicum*

Green mold (*P. digitatum*) and blue mold (*P. italicum*) decays are important post-harvest diseases which occur in all citrus growing areas and often constitute the predominant type of decay (Gardner *et al.*, 1986). Post-harvest losses of citrus fruits caused by these pathogens can account for more than 90% of all post-harvest losses in semi-arid production areas of the world (Eckert and Eaks, 1989).

(i) Symptoms: A soft water-soaked area is developed at the infection site in both the diseases. Coloured spore mass is developed at the centre of the lesion surrounded by broad band of white mycelial growth in green mold infection, whereas white mycelial growth around the spore mass of blue mold is usually not more than 2 mm wide. Both the pathogens occur frequently, but green mold grows faster at moderate temperature and contaminates the fruit. The spores of green mold are unable to infect healthy uninjured adjacent fruits while the blue mold develops nesting onto uninjured healthy fruits and may cause serious damage (Fig. 2.10).

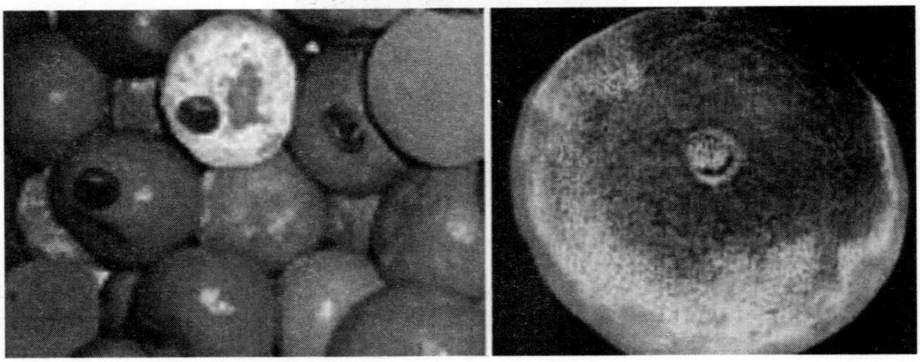

Fig. 2.10. Citrus green mold, *Pencillium digitatum*

(ii) Integrated Management

Bioagent and Chemical: Combining 0.2% glycolchitosan (antimicrobial substance) with the antagonist *Candida saitoana* was more effective in controlling green mold of oranges and lemons than either treatment alone.

2.2.2.4. Canker, *Xanthomonas campestris* pv. *citri*

Citrus canker was first reported from U.K. and the U.S.A. in 1933 on herbarium specimens of *Citrus medica* collected at Dehra Dun during 1827-1831. The disease is known to occur in almost all citrus growing areas. The disease is very severe on acid lime, lemon and grapefruit.

(i) Symptoms: The disease affects leaves, twigs, older branches, thorns and fruits (Fig. 2.11). Lesions first appear as small, round, watery, translucent spots on lower surface of the leaves and then on the upper surface. As the disease progresses, spots become white or grayish and give a rough corky crater appearance. The lesions are surrounded by yellowish halo. Elongated lesions are formed on twigs. On larger branches, the cankers are irregular, rough and more prominent. Cankers on fruits are similar to those on leaves except that the yellow halo is absent and crater-like depressions in the centre are more pronounced. The lesions on fruits remain confined to the rind, but sometimes, it causes cracks and fissures on the skin.

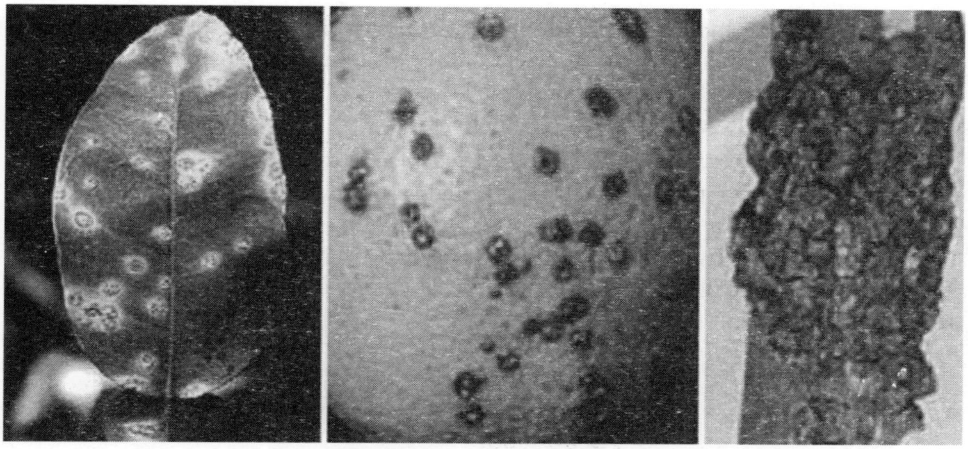

Fig. 2.11. Bacterial canker on acid lime leaf, fruit and bark (Courtesy: K. Srinivasan, TAFE, Chennai;V.K. Das, NRCC, Nagpur).

(ii) Epidemiology: The disease is carried from one season to another in the cankerous leaves, twigs and branches which form the main source of inoculum. The pathogen survives up to 5 months in the infected leaves (Rao and Hingorani, 1963). Chakravarti *et al.* (1966) reported that the bacterium can survive up to 76 months in infected twigs. The pathogen enters the host through stomata, lenticels and wounds. Temperatures between 20° and 30°C with evenly distributed rains are most suitable for the disease development. Presence of free moisture on the host surface for 20 min is essential for successful infection (Ramakrishnan, 1954). The pathogen from cankers is disseminated mainly by wind splashed rains. Dissemination through leaf miner (*Phyllocnistis citrella*) is reported by Nirvan (1961). The long distance dissemination takes place through infected planting material.

(iii) Integrated Management

(a) Cultural and Chemical: Pruning of infected twigs before monsoon and spraying 3 times with 500 ppm streptomycin sulphate + 0.2% copper oxychloride at 20 days interval was found effective to manage the disease (Ravi Kumar *et al.*, 2001).

2.2.3. Nematodes

2.2.3.1. Citrus Nematode, *Tylenchulus semipenetrans*

Siddiqi (1961) reported the citrus nematode for the first time from India and observed that about 80 per cent of the citrus trees at Aligarh, Uttar Pradesh, were infested with this nematode. The nematode causes "slow decline" and is considered to be one of the factors responsible for die-back of citrus trees. The disease is known to occur in almost all citrus growing areas. The disease is very severe on acid lime, lemon and grapefruit.

(i) Economic Importance and Losses: Yield increases of 40 to 200 per cent have been obtained following the control of the nematode over a range of growing conditions (Swarup and Seshadri, 1974). *T. semipenetrans* was responsible for 69%, 29% and 19% loss in fruit yield of sweet orange (Baghel and Bhatti, 1983a), lemon (Mukhopadhyaya and Suryanarayana, 1969) and sweet lime (Mukhopadhyaya and Dalal, 1971), respectively.

(ii) Symptoms: 'Slow decline' of citrus is a diseased condition of trees with symptoms similar to those caused by drought and malnutrition. Affected trees exhibit reduced vigour (Fig. 2.12), chlorosis and falling of leaves, twig dieback and consequently, reduced fruit production (Prasad and Chawla, 1965). This decline of the tree is gradual and persists until the crop is so small that tree maintenance may become uneconomical.

Fig. 2.12. Sweet orange plants (9 month-old) infected with *Tylenchulus semipenetrans*. Left – Healthy, Right – Infected.

As the nematode feeds on roots (Fig. 2.13) and reproduces, a large proportion of the feeder roots of citrus trees, particularly in the upper soil layers, is inactivated or destroyed, the uptake of water and minerals from the soil is reduced and the symptoms appear on the above ground tree parts. Heavily infested roots are darker in colour with branch rootlets shortened, swollen and irregular in appearance than in normal ones (Chona *et al.*, 1965).

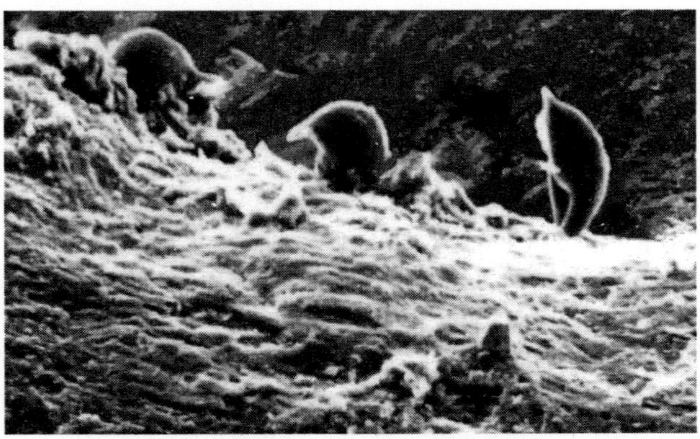

Fig. 2.13. Citrus root infected with *Tylenchulus semipenetrans*.

Soil particles usually cling tightly, even after washing, to the gelatinous egg masses which cover the protruding part of the nematode body (Fig. 2.14). In heavily infested roots, the cortex separates readily from the vesicular stele.

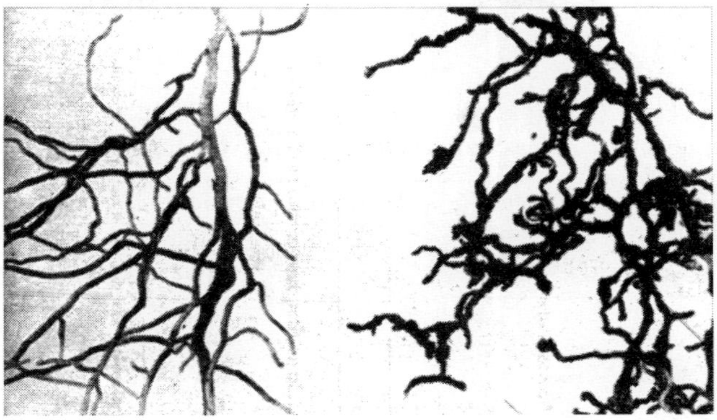

Fig. 2.14. Citrus root infected with *Tylenchulus semipenetrans*. Left – Healthy, Right – Infected

(iii) Integrated Management

(a) Bioagents and Botanicals: Management of the citrus nematode based on the application of neem cake and castor cake both at 10 kg per plant along with a parasitic fungus, *Paecilomyces lilacinus* at 250 g (grown on paddy seeds) per plant 3 times in a year at 15 cm depth and 50 cm away from the trunk was found to be extremely effective in reducing the citrus nematode population with a consequent increase in the growth of acid lime trees. The above treatment also gave highest parasitization of egg masses and eggs of *T. semipenetrans* and increased spore density of *P. lilacinus* in soil (Parvatha Reddy *et al*., 1991, 1993) (Table 2.5).

Table 2.5. Effect of oil cakes and *Paecilomyces lilacinus* on *Tylenchulus semipenetrans* on acid lime

Treatment	Dose (g/plant)	Nematode popn./plant		% egg masses parasitized	% eggs infected	Spore density (cfu/g soil)
		Soil	Root			
Castor cake	20	3980	2342	---	---	---
Karanj cake	20	5040	2932	---	---	---
Neem cake	20	3840	2320	---	---	---
P. lilacinus	8	4680	2074	52.8	56.3	5556
Castor cake + *P. lilacinus*	20 + 4	2500	1606	58.0	65.0	6471
Karanj cake + *P. lilacinus*	20 + 4	4240	1650	57.0	59.0	5997
Neem cake + *P. lilacinus*	20 + 4	2120	1330	67.5	72.4	7123
Control	---	16080	4498	---	---	---
CD (P = 0.05)	**---**	**995.3**	**449.5**	**5.58**	**5.5**	**576**

Trichoderma harzianum in combination with neem oil cake was effective in increasing the growth of acid lime trees and reducing the citrus nematode population both in soil and roots. The parasitization of citrus nematode females with *T. harzianum* increased in the presence of oil cakes (Parvatha Reddy *et al*., 1996b) (Table 2.6)

Table 2.6. Effect of *Trichoderma harzianum* and oil cakes on plant growth and population of *Tylenchulus semipenetrans* on acid lime.

Treatment	Dose (g)/ plant	Dry shoot wt. (g)	Nematode popn.		Parasitization of females (%)
			Roots (10g)	Soil (200 ml)	
T. harzianum	4	4.7	228	184	22
Castor cake	40	5.5	246	170	---
Karanj cake	40	5.3	274	180	---
Neem cake	40	5.8	196	166	---
T. harzianum + Castor cake	2 + 20	6.8	140	132	34
T. harzianum + Karanj cake	2 + 20	6.5	184	142	30
T. harzianum + Neem cake	2 + 20	8.0	128	108	42
Control	---	3.5	1520	1180	---
CD (P = 0.05)	**---**	**0.88**	**56.3**	**53.8**	**2.3**

Incorporation *Verticillium lecanii* with neem cake facilitated the effective management of *T. semipenetrans* on acid lime (Parvatha Reddy *et al*., 1996a).

Integration of neem cake with *Pseudomonas fluorescens* gave maximum reduction in citrus nematode population both in soil and roots and increased plant growth of acid lime (Parvatha Reddy *et al*., 2000) (Table 2.7)

Table 2.7. Effect of integrated use of *Pseudomonas fluorescens* and oil cakes on plant growth and population of *Tylenchulus semipenetrans* on acid lime

Treatment	Dose (g)/ plant	Plant wt. (g)	CFU/g root	Nematode population	
				Soil (100 ml)	Root (10 g)
P. fluorescens	4 x 10^9 spores	21.14	2.0 x 10^7	9116	2418
Castor cake	50 g	23.00	---	9012	2264
Karanj cake	50 g	22.09	---	9228	2426
Neem cake	50 g	23.40	---	8804	2310
P. fluorescens + Castor cake	½ dose each	25.97	12.4 x 10^7	6398	1218
P. fluorescens + Karanj cake	½ dose each	24.42	10.2 x 10^7	6548	1298
P. fluorescens + Neem cake	½ dose each	28.97	18.6 x 10^7	6034	1010
Control	---	17.60	---	13456	6142
CD(P = 0.05)	**---**	**1.76**	**---**	**234.64**	**212.34**

(b) AMF and Botanicals: Inoculation of endomycorrhiza, *Glomus fasciculatum* in the soils amended with neem cake was found effective for the management of *T. semipenetrans* on acid lime. This strategy can help in combating the menace of citrus nematode at nursery stage and also provide highly mycorrhizal seedlings of acid lime for transplanting in the main field for the management of *T. semipenetrans* under field conditions (Parvatha Reddy *et al.*, 1995).

(c) Bioagents and AMF: Integration of *P. lilacinus* (4 g/kg soil), *T. harzianum* (4 g/kg soil) and *G. fasciculatum* (500 spores/kg soil) was effective in increasing plant growth parameters and in reducing the citrus nematode population both in soil and roots of Rough lemon (Walode *et al.*, 2008).

(d) Two Bioagents: Application of bacterial bioagent, *Pasteuria penetrans* (at 2 x 10^9 spores/plant) and fungal bioagent, *P. lilacinus* (at 50 g/plant with 4 x 10^7 spores/g) was effective in reducing the *T. semipenetrans* population and in increasing the parasitization of larvae by *P. penetrans* and eggs by *P. lilacinus* (Parvatha Reddy and Nagesh, 2000) (Table 2.8).

Table 2.8. Effect of integration of *Pasteuria penetrans* and *Paecilomyces lilacinus* on population of *Tylenchulus semipenetrans* and parasitization by bioagents on acid lime

Treatment*	Dose/plant	Nema popn.		Spores (cfu)		% parasitization	
		Root (5 g)	Soil (250 ml)	*Pp* (100 ml soil	*Pl* (5 g roots)	*Pp* (Female)	*Pl* (Egg masses)
Pp_1	2 x 10^6 spores	1668	460	64.6	---	16	---
Pl_1	2 x 10^6 spores	2298	588	---	1698	---	24
$Pp_1 + Pl_1$	½ dose each	1268	386	62.6	1726	15	22
$Pp_1 + Pl_2$	½ dose each	994	228	59.9	1122	28	24
$Pp_2 + Pl_1$	½ dose each	1694	612	36.2	1456	10	22
$Pp_2 + Pl_2$	½ dose each	1832	644	39.4	1232	12	20
Control	---	2724	698	---	---	---	---
CD (P = 0.05)	---	**22.2**	**12.4**	**3.6**	**99.8**	**2.8**	**1.2**

* Pp - *Pasteuria penetrans*, Pl - *Paecilomyces lilacinus*

(e) Bioagents and Chemicals: Integration of *P. lilacinus* (at 4 g/plant) with carbofuran (at 30 mg a.i./plant) was found effective in increasing the plant growth parameters of acid lime and in reducing the citrus nematode population both in soil and roots (Parvatha Reddy *et al.*, 1996c) (Table 2.9).

(f) Botanicals and Chemicals: Neem cake at 1 kg/plant combined with carbofuran at 2 kg/ha reduced 47.1% nematode population and increased the yield by 41.1% (Baghel, 1995).

Table 2.9. Effect of *Paecilomyces lilacinus* and pesticides on plant growth and population of *Tylenchulus semipenetrans* on acid lime

Treatment	Dose/plant	Plant weight (g)	Nematode population	
			Soil	Root
P. lilacinus	8 g	19.98	6888	1811
Carbofuran	60 mg a.i.	26.60	9794	1948
Phorate	60 mg a.i.	15.86	3391	853
P. lilacinus + Carbofuran	4 g + 30 mg a.i.	27.20	2813	643
P. lilacinus + Phorate	4 g + 30 mg a.i.	16.87	3176	676
Control	---	13.86	15057	4440
CD (P = 0.05)	**---**	**3.14**	**431.8**	**221.4**

2.2.3.2. Root-knot Nematodes, *Meloidogyne javanica, M. indica*

In India, Thirumala Rao (1956) reported that *Meloidogyne* sp. caused considerable damage to citrus in Andhra Pradesh when a susceptible crop like tobacco or okra is grown as an intercrop. This is the first report of nematode damage to citrus in India.

Mani (1986) reported that *M. javanica* caused severe crop loss to acid lime in Andhra Pradesh and was pathogenic to acid lime and sweet orange. It is confined to only coastal Andhra Pradesh.

Chitwood and Toung (1960) observed the root-knot nematode resembling *M. incognita* infecting *C. sinensis* from Delhi and proved its pathogenicity to citrus.

Whitehead (1968) reported *M. indica* on citrus from India (Fig. 2.15).

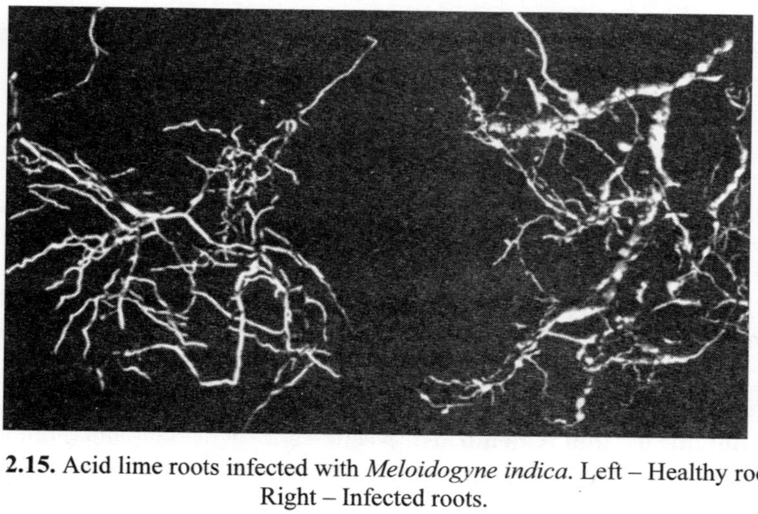

Fig. 2.15. Acid lime roots infected with *Meloidogyne indica*. Left – Healthy roots; Right – Infected roots.

(i) Symptoms: *M. javanica* infected trees are poor in vigour, unthrifty in appearance and show severe stunted growth. They fail to flower and produce fruits even after several years of planting. The roots have conspicuous galls on pioneer and fibrous roots (Fig. 2.16). In advanced stage, large cavities can be observed in place of galls. Egg masses can be seen as thin films spread over the root surface. Nematode infestation gets aggravated if vegetables like okra, brinjal, cucurbits, tomato and tobacco are grown as intercrops in orchards or as rotational crops in nurseries (Mani, 1986).

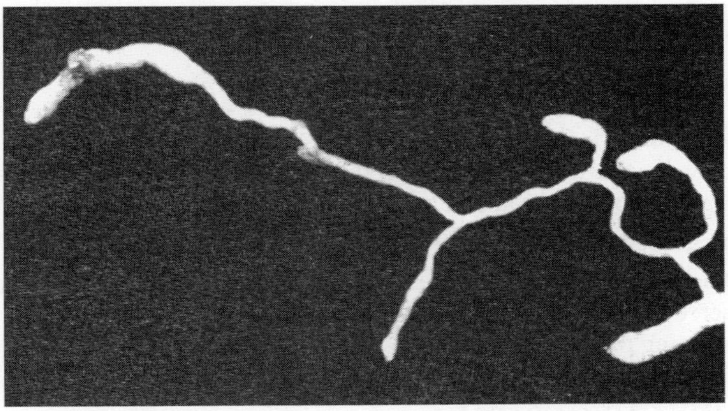

Fig. 2.16. Citrus roots infected with *Meloidogyne javanica.*

(ii) Integrated Management

(a) Two Bioagents: The combined use of *P. lilacinus* and *P. chlamydosporia* each at 5 g/kg soil significantly reduced root galling, nematode population (*M. javanica*) both in soil and roots of acid lime, number of eggs per egg mass, and increased egg parasitization, root colonization and spore density of bioagents in soil (Rao, 2005) (Table 2.10).

Table 2.10. Effect of integration of bioagents for the management of *Meloidogyne javanica* infecting acid lime

Treatment/ Dose/kg soil	Root-knot index	% egg parasitization		Root coloniz-ation (cfu/g)		Spore density (cfu/g soil)	
		Pl	*Pc*	*Pl*	*Pc*	*Pl*	*Pc*
P. lilacinus – 5 g	5.5	43.9	---	31678	---	27975	---
P. lilacinus – 10 g	5.1	57.4	---	34578	---	29874	---
P. chlamydosporia – 5 g	6.2	---	51.9	---	28765	---	25439
P. chlamydosporia – 10 g	5.9	---	58.9	---	32674	---	27896
P. lilacinus – 5 g + *P. chlamydosporia* – 5 g	3.2	40.7	40.8	30785	27347	28753	25873
P. lilacinus – 10 g + *P. chlamydosporia* – 10 g	3.8	56.9	30.8	35687	23879	28796	21784
Control (Untreated)	7.9	---	---	---	---	---	---
CD (P = 0.05)	**1.34**	**9.6**	**7.6**	**2598.4**	**2145.9**	**2566.6**	**1987.5**

(b) Bioagents and Botanicals: Combined application of *Aspergillus niger* and *P. lilacinus* at transplantation, 10 days prior to which mustard cake was introduced, significantly increased the plant vigour and reduced the root-knot nematode population.

2.3. SAPOTA, *Manilkara achras*

2.3.1. Diseases

2.3.1.1. Dry Root Rot/Wilt, *Fusarium solani*

(i) Symptoms: Infected plants initially show yellowing of leaves. When the disease is severe, the infected stems produce brown to black colour lesions along with disintegration of tissues at soil level resulting in death of affected plants.

(ii) Integrated Management

(a) Bioagents and Chemicals: Soil drenching with 0.1% carbendazim plus 0.3% copper oxychloride followed by soil application of *Trichoderma harzianum* helps in managing the disease.

2.4. PAPAYA, *CARICA PAPAYA*

2.4.1. Diseases

2.4.1.1. Damping-off, *Pythium aphanidermatum, Phytophthora parasitica*

(i) Symptoms: The typical symptoms caused are the pre- and post-emergence damping-off. The post-emergence damping-off is characterized by dull green to pale seedlings showing water-soaked lesions on the cotyledons which become weak and the seedlings collapse. There is also rotting of roots where *P. parasitica* is involved.

(ii) Integrated Management

(a) Bioagents and Botanicals: Soil treatment with neem cake + *Trichoderma harzianum* gave good germination and seedling stand by controlling damping-off disease.

(b) AMF and Botanicals: Plant mortality was least in *G. fasciculatum* + vermicompost (25%) followed by *G. fasciculatum* (31.35%). The least disease incidence was recorded in *G. fasciculatum* + vermicompost (7.01%) which was at par with *Sclerocystis dussii* (7.74%) and *S. dussii* + vermicompost (8.27%).

2.4.2. Nematodes

2.4.2.1. Root-knot Nematodes, *Meloidogyne* spp.

M. incognita and *M. javanica* have been reported to be the major nematode pests of papaya in India.

(i) Economic Importance: Ponte (1980) and Taylor *et al.* (1982) reported 10 to 20 per cent reduction in papaya fruit yield due to root-knot nematodes.

(ii) Symptoms: Papaya orchards infected with *Meloidogyne* spp. show patches of poor growth with many plants missing in the rows. General symptoms visible in field include poor growth, yellowing of foliage, dropping of leaves, reduction in leaf production, weak vigour and premature dropping of fruits. Roots exhibit typical below-ground symptoms *i.e.* galls of varying sizes (Fig. 2.17). The lateral branching of roots is limited. In heavily infected old roots, adjacent galls join together and form large galls. In mild infestation, root tips become swollen and root growth inhibition is distinctly seen.

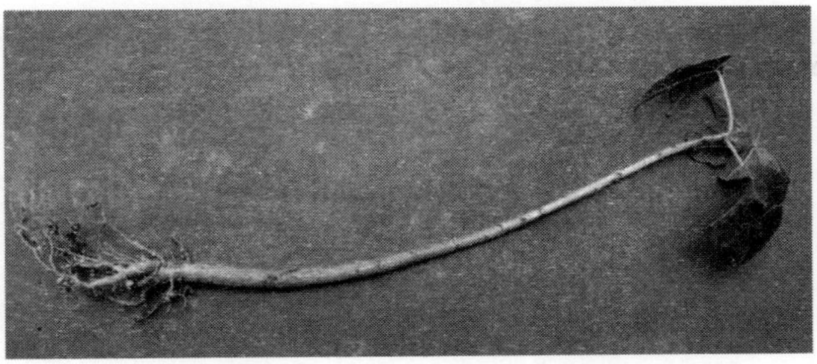

Fig. 2.17. Papaya plant infected with *Meloidogyne incognita.*

(iii) Integrated Management

(a) Two Bioagents: Nursery bed treatment with *T. harzianum* and *P. lilacinus* each at 5 or 10 g/kg soil resulted in production of highly vigorous papaya seedlings whose roots were colonized with both the bioagents. There was significant reduction in root galling in the combination treatments (2.9 to 3.2) compared to control (8.9) (Rao and Naik, 2003) (Table 2.11).

Table 2.11. Effect of integration of bioagents for the management of *Meloidogyne incognita* infecting papaya

Treatment*	Root-knot index	Root colonization (cfu/g)		Propagule density (cfu/g soil)		% egg masses parasitized by *Th* and/or *Pl*
		Th	*Pl*	*Th*	*Pl*	
T1	5.6	---	26434	---	22097	59.36
T2	5.2	---	29157	---	23134	57.44
T3	6.4	35765	---	29347	---	58.42
T4	5.2	37464	---	30956	---	62.46
T5	3.0	35124	23301	27437	17104	80.62
T6	2.8	35624	24211	30147	16122	73.84
T7	8.5	---	---	---	---	---
CD (P=0.05)	**0.65**	**2613.46**	**1905.27**	**2116.82**	**1817.43**	**8.75**

* T1 – Nursery soil mixed with *P. lilacinus* (5 g/kg), T2 - Nursery soil mixed with *P. lilacinus* (10 g/kg), T3 - Nursery soil mixed with *T. harzianum* (5 g/kg), T4 - Nursery soil mixed with *T. harzianum* (10 g/kg), T5 – T1 + T3, T6 – T2 + T4, T7 – Control.

Seed treatment with *P. fluorescens* (10^8 spores/g) combined with soil application of *T. harzianum* (10^6 spores/g) and *P. fluorescens* (10^8 spores/g) at 5 g/kg soil gave significant reduction in *M. incognita* population both in soil and roots, number of eggs per egg mass and hatching of eggs and increased root colonization by both the bioagents in papaya under nursery conditions (Rao, 2007a) (Table 2.12).

Table 2.12. Effect of integration of bioagents for the management of *Meloidogyne incognita* infecting papaya

Treatment*	No. of nemas/10 g roots	No. of J_2/ 100 ml soil	No. of eggs/egg mass	% egg hatch suppression	Root colonization (cfu/g)	
					T. harzianum	*P. fluorescens*
T1	47	78	389	32	---	8769
T2	42	70	364	45	---	18457
T3	50	65	375	43	22649	---
T4	35	52	358	65	21895	18249
T5	38	56	326	55	---	24531
T6	45	63	347	54	22412	8566
T7	31	45	310	67	20874	24124
T8	68	126	412	---	---	---
CD at 5%	**6.6**	**8.5**	**33.2**	**7.4**	**2365.7**	**2487.2**

* T1 - Seed treatment with *P. fluorescens*, T2 – Nursery soil mixed with *P. fluorescens* (5 g/kg), T3 - Nursery soil mixed with *T. harzianum* (5 g/kg), T4 - Nursery soil mixed with *T. harzianum* (5 g/kg) + *P. fluorescens* (5 g/kg), T5 - Seed treatment with *P. fluorescens* + Nursery soil mixed with *P. fluorescens* (5 g/kg), T6 - Seed treatment with *P. fluorescens* + Nursery soil mixed with *T. harzianum* (5 g/kg), T7 - Seed treatment with *P. fluorescens* + Nursery soil mixed with *T. harzianum* (5 g/kg) and *P. fluorescens* (5 g/kg), T8 – Control.

Simultaneous application of *P. lilacinus* and *T. harzianum* both at 1 g per plant gave maximum increase in plant growth parameters and highest reduction in reproduction factor and root galling in papaya (Khan, 1991).

Integrated management strategy for root-knot nematodes (*M. incognita*) infecting papaya was standardized by nursery bed treatment with neem based formulation of *P. lilacinus* (1 x 10^6 cfu/g) + *T. harzianum* (1 x 10^9 cfu/g) (each at 5 or 10 g/kg soil) and application of 35 g each of *P. lilacinus* + *T. harzianum*/pit while transplanting. The above treatment increased plant height, plant weight and reduced root galling, number of egg masses/plant and number of eggs/egg mass in nursery (Table 2.13). The integration of both the bioagents also increased root colonization of bioagents, propagule density in soil and parasitization of egg masses under field conditions (Table 2.14).

Table 2.13. Effect of neem based formulations of bioagents on plant growth and management of *Meloidogyne incognita* infecting papaya in nursery

Treatment/Dose	**Plant height (cm)**	**Plant weight (g)**	**Root-knot index (1-10 scale)**	**No. of egg masses/ seedling**	**No. of eggs/ egg mass**
P. lilacinus - 5 g/kg soil	25.35	5.58	5.8	15.45	356
P. lilacinus - 10 g/kg soil	28.87	6.21	5.2	12.86	325
T. harzianum - 5 g/kg soil	23.69	5.76	6.5	17.89	421
T. harzianum - 10 g/kg soil	29.72	6.89	5.4	14.59	398
P. lilacinus - 5 g/kg soil+ *T. harzianum*- 5 g/kg soil	32.89	7.99	3.1	10.33	270
P. lilacinus - 10 g/kg soil + *T. harzianum* - 10 g/kg soil	33.72	8.65	2.8	9.65	262
Control	18.64	4.12	8.7	26.27	412
CD (P = 0.05)	**3.67**	**1.25**	**0.65**	**2.38**	**34.87**

Table 2.14. Effect of neem based formulations of bioagents on plant growth and management of *Meloidogyne incognita* infecting papaya in main field.

Treatment/Dose	**Root colonization (cfu/g)**		**Propagule density in soil (cfu/g)**		**Parasitization of egg masses (%)**	
	Th	***Pl***	***Th***	***Pl***	***Th***	***Pl***
P. lilacinus - 5 g/kg soil	---	25,876	---	21,765	---	55.89
P. lilacinus - 10 g/kg soil	---	28231	---	23,867	---	61.98
T. harzianum - 5 g/kg soil	34,789	---	28,765	---	56.76	---
T. harzianum - 10 g/kg soil	37,895	---	31,194	---	66.89	---
P. lilacinus - 5 g/kg soil+ *T. harzianum*- 5 g/kg soil	34,853	24,129	26,986	17,459	41.68	35.98
P. lilacinus - 10 g/kg soil + *T. harzianum* - 10 g/kg soil	36,432	23,679	29,457	16,934	40.21	32.49
Control	---	---	---	---	---	---
CD (P = 0.05)	**2655**	**2106**	**2317**	**1567**	**7.89**	**7.95**

(b) Bioagents and Botanicals: Application of 2 kg FYM enriched with *P. fluorescens* (10^9 spores/g) and *P. lilacinus* (1 x 10^6 cfu/g) per plant at the time of planting and at an interval of 6 months significantly reduced reniform and root-knot nematodes on roots by 66 and 70%, respectively. Significant increase in fruit yield (28%) was also observed. Benefit cost ratio (calculated for marginal cost of biopesticides and returns accrued by application of biopesticides) was 3.2 (Anon, 2012).

2.4.2.2. Root-knot, *Meloidogyne incognita* and Wilt, *Fusarium solani* Disease Complex

(i) Symptoms: Increasing inoculum of *M. incognita*, whether present alone or together with *F. solani*, decreased seedling emergence of papaya. The combination of highest inoculum of both pathogens (3000 nematodes and 3g culture of fungus) caused maximum inhibition of seedling emergence and also increased post-emergence damping-off of papaya seedlings (Khan and Hussain, 1990).

Significant reduction was observed in plant height, root length, shoot and root weight in plants inoculated with nematode and fungus simultaneously, and prior inoculation of nematode followed by fungus 12 days later. Root galling was highest in case of nematode alone followed by nematode inoculation 12 days prior to fungus, and simultaneous inoculation of nematode and fungus (Table 2.15) (Kishore *et al.*, 2005).

Table 2.15. Effect of *Meloidogne incognita* and *Fusarium solani* on disease complex in papaya

Treatment	Root rot (%)	Plant height (cm)	Shoot weight (g)	Root weight (g)	Galls/g roots
M. incognita	12	98.7	58.2	10.8	74
F. solani	48	106.3	63.3	13.3	0
M. incognita + *F. solani* (simultaneously)	64	65.3	53.6	9.2	43
M. incognita (prior) + *F. solani* (later)	94	73.2	60.3	9.7	65
F. solani (prior) + *M. incognita* (later)	52	83.7	76.6	15.6	35
Uninoculated Control	0	125.7	85.8	70.6	0
CD (P = 0.05)	**7.15**	**12.66**	**5.12**	**2.37**	---

(ii) Integrated Management

(a) Two Bioagents: Khan *et al.* (1997) have reported that *P. lilacinus* gave better over all protection of papaya plants against *M. incognita* – *F. solani* disease complex than *T. harzianum.* They reported that application of both the biocontrol agents further limited the damage caused by *M. incognita* and *F. solani* and gave a 35% increase in plant growth compared to individual applications.

2.5. JACKFRUIT, *ARTOCARPUS HETEROPHYLLUS*

2.5.1. Diseases

2.5.1.1. Die-back, *Botryodiplodia theobromae*

(i) Symptoms: This is the most destructive disease of the jackfruit. The onset of die-back symptoms becomes evident by discolouration and darkening of the bark at some distance from the tip. The dark area spreads and young green twigs start withering first at the base and then extending outwards along the veins of leaf edges. The affected leaves turn brown and their margins roll upwards. At this stage, the twig or branch dies, shrivels and drops. There may be exudation of gum from affected branches. Such branches have also been found to be affected by shoot borers. Infected twigs show internal discolouration. In early stages, epidermal and sub-epidermal cells of twigs appear slightly shriveled.

(ii) Integrated Management

(a) Cultural and Chemical: Pruning of infected twigs followed by spraying of 0.1% carbendazim or Topsin M or 0.2% chlorothalanil has been recommended for the management of the disease.

2.6. STRAWBERRY, *FRAGARIA* spp.

2.6.1. Diseases

2.6.1.1. Fruit Rot, *Phytophthora cactorum*

(i) Integrated Management

(a) Physical, Cultural and Chemical: Maximum disease control was achieved in plots which were mulched with polythene and sprayed with ridomil MZ (74.47%) followed by paddy straw mulch and ridomil MZ sprays (64.92%). The fruit yield of strawberry also increased significantly in combined treatment of polythene mulch and ridomil MZ (97.57%) followed by paddy straw mulch and ridomil MZ (90.04%) (Table 2.16) (Bharadwaj and Sharma, 2000).

Table 2.16. Effect of fungicide alone and in combination with mulch on incidence of fruit rot and yield of strawberry

Treatment	Disease incidence (%)	Fruit rot control (%)	Fruit yield (kg/plot)	% increase in fruit yield
Polythene mulch	18.43	54.36	1.600	71.99
Ridomil MZ	24.67	38.92	1.258	35.23
Paddy straw mulch	28.02	30.63	1.056	13.51
Polythene mulch + Ridomil MZ	10.31	74.47	1.838	97.57
Paddy straw mulch + Ridomil MZ	14.17	64.92	1.768	90.04
Control	40.39	---	0.930	---
CD (P = 0.05)	**5.025**	**14.263**	---	---

(b) Physical and Bioagents: The combination of soil solarization and application of *Trichoderma* spp. reduced *P. cactorum* soil population to the maximum extent (88.9% in January 2001, 97.6% in 2002, and 99.0% in 2003). The very promising effect of *Trichoderma* spp. and solarization against *P. cactorum* indicates that there may be future alternatives to traditional chemicals for disease control (Porras *et al.*, 2007).

2.6.1.2. Root Rot, *Phytophthora cinnamoni*

(i) Symptoms: Infected roots turn dark and lose their feeder roots (Fig. 2.18). Black root rot causes poor plant vigour. This problem is associated with soils of high clay content, excessive irrigation, and soil compaction. Avoid these soils and provide adequate soil aeration for vigorous root growth.

Fig. 2.18. Root rot symptoms on strawberry

(ii) Integrated Management

(a) Biofumigation and Solarization: Soil solarization and biofumigation with *Brassica carinata* (B+S) increased plant growth (foliar surface), fruit weight, and strawberry yield the most each year. Plant growth differences were observed relative to soil solarization (S) and untreated control (C), with foliar surface (cm^2) of B + S, S and C of 502, 414 and 351, and 435, 346 and 228 in January 2006 and 2007, respectively. Furthermore, S increased plant growth and strawberry yield relative to C. Fruit weight (g/fruit) of B + S, S and C was 25, 22 and 17 and 23, 20 and 17 in 2006 and 2007, respectively. In addition, both treatments reduced *Phytophthora* soil population relative to C. The current work contributes to the development and optimization of biofumigation with *Brassica* and soil solarization as alternatives to the traditional use of chemicals in strawberry production (Porras *et al.*, 2009).

2.6.1.3. Fusarium Wilt

(i) Symptoms: Symptoms consisted of wilting of foliage, drying and withering of older leaves, stunting of plants, and reduced fruit production. Plants eventually collapsed and died. Internal vascular and cortical tissues of plant crowns showed a brown to orange brown discolouration (Fig. 2.19).

Fig. 2.19. Fusarium wilt affected crown with brown discolouration of water conducting tissues

(ii) Integrated Management

(a) Botanicals and AMF: Combined use of AMF *Gigaspora margarita* and 3-15% charcoal compost (which contained antagonistic microorganisms such as *Bacillus subtilis, Thermomonospora* sp. and *Thermoactinomyces* sp.) drastically reduced Fusarium wilt in strawberry. Moreover, AMF and charcoal compost stimulated rooting and increased the root volume, and hence plant growth (Kabayashi, 1989a).

2.6.2. Nematodes

2.6.2.1. Lesion Nematode, *Pratylenchus penetrans* and Wilt, *Verticillium dahliae* Disease Complex

(i) Symptoms: Infested plants are stunted, pale green, have reduced yields and eventually die. Root symptoms appear as distinct brown lesions in early or light infestation. The entire root system eventually becomes black and necrotic as populations increase and secondary organisms invade (Fig. 2.20).

Fig. 2.20. The lesion nematode (*Pratylenchus penetrans*) on strawberry. A-Left 2 plants, infected; Right, healthy. B-Top, Lesion nematodes inside the root. C- Bottom, Lesions on the root.

(ii) Integrated Management

(a) Crop Rotation and Biofumigation: Nematode population densities were significantly lower after biofumigation [canola was ploughed under at late bloom (end of July) followed by white mustard which was ploughed at full bloom (October)] along with crop rotation with both pearl millet cultivars (CFPM 101 and Tifleaf) than after corn and oats. In 2006, both pearl millets and biofumigation significantly reduced by an average of 21% of the incidence of Verticillium wilt and allowed a 54% average increase in the development of strawberry plants cv. Jewel.

(b) Biofumigation and Solarization: Plastic mulch significantly improved biofumigation (by incorporating poultry manure at 15 t/ha at canola-crop ploughing time) with a 95% reduction in the number of *P. penetrans* densities when compared to canola ploughed with poultry manure but no plastic mulch (Bélair and Coulombe, 2008).

2.6.2.2. Root-knot Nematodes, *Meloidogyne* spp.

(i) Symptoms: Symptoms associated with root-knot nematodes include: stunting, yellowing of leaves, reduced berry yields, reduced production of runner plants, wilting, and general loss of vigour.

This nematode causes tiny galls (about 3 mm in diameter) on feeder roots. Several short branch roots stick out from each tiny swelling or gall (Fig. 2.21). Injured plants appear stunted, take on an "off" colour and produce little fruit. Weakened plants are more subject to drought damage and make fewer runners.

Fig. 2.21. L - Strawberry plants infested with nematodes appear stunted (left rows) and produce fewer berries than healthy plants not infested with nematodes (right rows). R - Root swelling or knot symptoms on strawberry roots caused by root-knot nematodes.

(ii) Integrated Management

(a) Botanicals and Chemicals: Application of neem cake in combination with carbofuran at the time of planting increased the yield by 35.2% over control (Laqman Khan, 2001).

(b) Solarization and Chemicals: Solarization + MB/Telopic or MB, were effective in controlling root-knot nematodes compared to solarization and biofumigation (Table 2.17)

Table 2.17. Effect of different soil fumigation methods for the management of root-knot nematodes

Treatment	Location: Cartaya[1]			Location: Moguer[1]		
	Incidence[2]	Severity[3]	♀/g[4]	Incidence[2]	Severity[3]	♀/g[4]
Control	54.2 a[5]	2.01 a	13.5 a	15.0 a	0.99 a	13.1 a
Solarization	36.7 a	1.48 ab	5.8 ab	8.3 ab	1.20 a	10.3 a
Biofumigation	32.5 ab	1.10 bc	5.0 ab	0.8 b	0.25 b	0.2 a
Solarization + Metham sodium	26.7 b	0.63 bcd	3.6 bc	0.0 b	0.00 b	0.0 a
Solarization + MB/Telopic	12.5 b	0.72 cd	1.7 bc	0.0 b	0.00 b	0.0 a
Methyl Bromide	0.8 c	0.08 d	0.0 c	0.0 b	0.00 b	0.0 a

1: Mean of 10 plants/plot, 3 blocks/year and 4 years; 2: Percentage of disease plants; 3: Severity (0=no symptoms; 4=more than 90% of roots affected); 4: Number of females/g of roots; 5: Means followed by the same letter are not different under a MDS test, ($P<0.05$).

2.7. MULBERRY, *MORUS* spp.

2.7.1. Diseases

2.7.1.1. Nursery Diseases: Stem Canker and Dieback, *Botryodiplodia theobromae;* Cutting Rot, *Fusarium solani;* Collar Rot, *Phoma mororum, P. sorghina*

(i) Symptoms: These diseases cause more than 30% mortality of mulberry stem cuttings and death of saplings. Initial establishment of mulberry is greatly affected resulting in poor plant survivability, growth and leaf yield.

(ii) Integrated Management

(a) Bioagents, Botanicals and Chemicals: Integration of 'Nursery-Guard' (*Trichoderma pseudokoningii*) applied to soil as pre-planting treatment after mixing with FYM (1 : 60 ratio) with soaking of stem cuttings in 0.1% mancozeb resulted in reducing the severity of nursery diseases by 82% and increasing the plant stand by 40%. The integration help to improve the overall establishment and growth of mulberry plantation, and thus provide better yield (Gupta, 2000, 2001; Gupta *et al.*, 1998, 1999).

2.7.1.2. Root Rot, *Fusarium oxysporum, F. solani*

***(i)* Symptoms:** The disease symptoms include decaying of roots resulting in sudden withering of leaves followed by defoliation and death of plants (Fig. 2.22) (Gupta, 2001). Occurrence is mostly seen in summer. In initial stage, leaf blade turn to wilt and then spread to entire plant. At later stage, black fungus appear on branches and stem. The disease spread through soil and water.

Fig. 2.22. Root rot symptoms on mulberry

(ii) Integrated Management

(a) Bioagents, Botanicals and Chemicals: Soil application of enriched FYM with *T. harzianum* (commercial product 'Raksha' 10 g + 500 g FYM) at 500 g/pit before replanting in infested area and *T. harzianum* integrated with mancozeb (dressing of stem cutting or roots of sapling before planting) gave about 80% control of root rot disease in mulberry (Philip *et al*., 1996; Gupta, 2001).

2.7.2. Nematodes

2.7.2.1. Root-knot Nematodes, *Meloidogyne* spp.

Govindaiah *et al*. (1991) reported 11.8% loss in herbage yield of mulberry due to *M. incognita*.

(i) Symptoms: Above-ground symptoms include stunted growth, poor and delayed sprouting, reduced leaf size and yield, chlorosis and marginal necrosis of leaves, yellowing and wilting of leaves in spite of adequate soil moisture availability, and death of plants in severe cases. These symptoms first appear as isolated patches, slowly spreading over the entire garden.

Below-ground symptoms include formation of galls or knots on roots (Fig. 2.23), reduced and stubby root system, and necrotic lesions on the root surfaces and death of roots.

(ii) Integrated Management

(a) Bioagents and AMF: Combined inoculation of AMF and *Pseudomonas fluorescens* had positive effect on root-knot nematode control on mulberry (Kumutha, 2001).

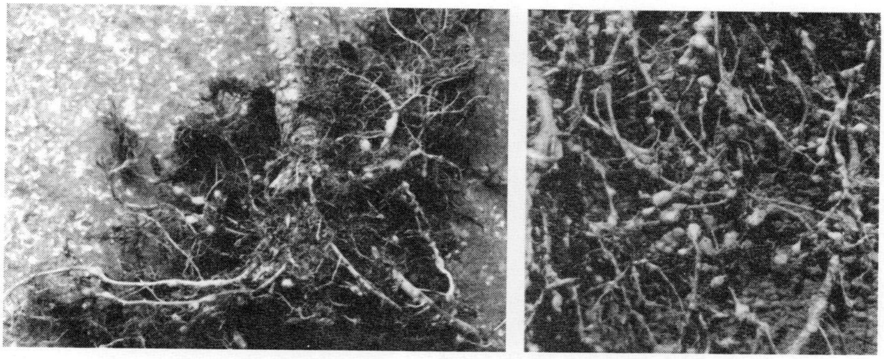

Fig. 2.23. Root-knot galls on mulberry roots

(b) Cultural and Botanicals: Intercropping of mulberry with marigold at wider spacing of 30 to 45 cm controlled *M. incognita*, increased leaf moisture and leaf yield of mulberry (Govindaiah *et al*., 1990).

(c) Two Bioagents: Combined soil application of *Pseudomonas fluorescens* and *Trichoderma viride* each at 10 g/plant was effective to check the nematode population and improve growth of mulberry with increased leaf yield (Muthulakshmi *et al*., 2010).

Chapter - 3

SUB-TROPICAL FRUIT CROPS

3.1. MANGO, *MANGIFERA INDICA*

3.1.1. Insect Pests

3.1.1.1. Fruit fly, *Dacus dorsalis*

This is one of the most serious pests of mango in India which has created problem in the export of fresh fruits. Its infestation is more in southern states than in northern states. The pest is active throughout the year in south India, whereas in northern part it hibernates during winter (November to March) in pupal stage (Kapoor *et al.*, 1993).

(i) Symptoms: The female punctures the outer wall of the mature fruits with the help of its pointed ovipositor and inserts eggs just below the fruit epidermis (1-4 mm deep) in small clusters inside the mesocarp. After hatching the maggots feed on pulp of these fruits (Fig. 3.1) and the infested fruits start rotting and fall down. As a result the brown patch appears around the place of oviposition and the infested fruits start rotting. The maggots come out of affected fruits to pupate in the soil. The flies breed on fruits that are mature and population increases rapidly during summer (May to July). The population declines slowly from August to September. In methyl eugenol traps, it was found that peak trap catch was between 16 and 17 hr (Jayanthi and Verghese, 1998).

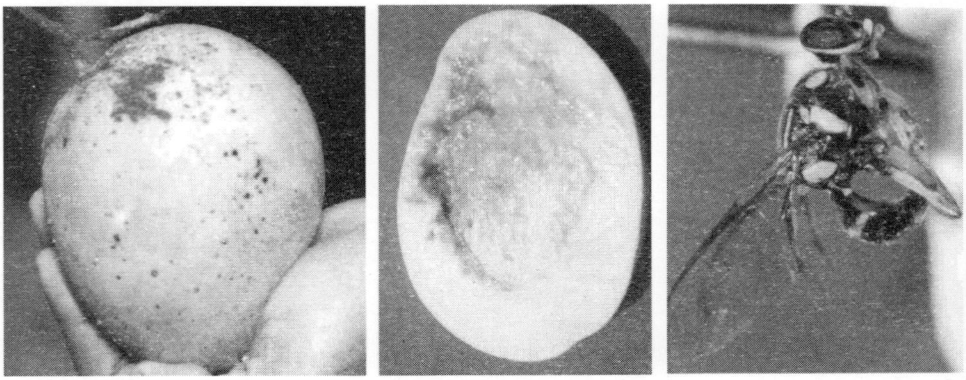

Fig. 3.1. Mango fruit fly and its damage to the fruit (Courtesy: A. Verghese, IIHR, Bangalore).

High temperature coupled with high humidity prevailing during May-July months are the favourable environmental factors for the development and reproduction of fruit fly.

(ii) Integrated Management

(a) Cultural, Chemical and Physical: The following pre- and post-harvest IPM combining sanitation + intercultural ploughing + 3 insecticidal sprays at 15 days interval prior to harvest + hot water treatment of fruits gives 95 -100% control (Verghese *et al.*, 2000).

Pre-harvest Management

- 45 days before harvest the following precautions need to be taken:
  - Destroy all fallen fruits at weekly intervals.
  - Install at least 10 methyl eugenol bottle traps (0.1 %) per ha.
  - Plough/rake the soil at the tree basin at frequent intervals (1 or 2 times between flowering and harvest) to expose pupae to sun.
  - Avoid delay in harvesting.
- Three weeks before the harvest, spray decamethrin 2.8 EC at 0.5 ml/l + Azadirachtin (0.3%) at 2 ml/l and take up timely harvest.

Post-harvest Management

Within 24 hr after harvest, hot water treatment of fruits at 48°C maintained by thermostat for 1 hr, gave 100% fruit fly infestation-free fruits in cvs. Banganapalli and Totapuri. Hot water treatment in 5% salt solution at 55°C for 30 min without thermostat gave 100% control in cvs. Banganapalli, Totapuri and Alphanso.

3.1.2. Validation of Mango IPM (Uttarakhand)

- Spraying of 0.3% copper oxychloride for control of die-back, anthracnose and red rust diseases wherever they appeared during September-October.
- Ploughing of orchard in November-December to expose pupae of fruit flies, midges, leaf hoppers and eggs of mealy bugs to natural enemies.
- Polythene banding of tree trunk in December-January and application of 5% NSKE and *Beauveria bassiana* in January.
- Spraying of 0.2% sulfex for the control of powdery mildew disease.
- Spraying of *Verticillium lecanii* in orchards for control of hoppers.
- Fixing methyl eugenol traps (wooden blocks impregnated with methyl eugenol) to control fruit flies from April to August.
- Mechanical removal of mango leaf webber larvae and webs by leaf web removing device (developed by the Central Institute of Sub-tropical Horticulture, Lucknow) from April to September-October.

The IPM package was successfully validated in 16.8 ha of mango orchards in Gulabkhera, Habibpur, Budhadia, Pathakganj, Rehmankhera and Kanar villages in Malihabad and Kakori belt of mango near Lucknow on Dashehari variety during 2000-04. As a result of adoption of IPM, yield of mango increased from 6.0-9.0 MT/ha as it was 3.5-7.0 MT/ha earlier in non-IPM orchards. By adopting IPM, the mango growers in that area earned a net profit of Rs. 30,000/- to Rs. 55,000/- while the farmers who did not adopt IPM, earned a net profit of Rs.17,000/- to 35,000/- per ha only (Table 3.1) (Trivedi *et al.*, 2004).

Table 3.1. Economics of IPM in Mango

Parameters	IPM plots	Non - IPM plots	% increase
Yield MT/ha	6.0-9.0	3.5-7.0	28.57-71.43
Net profit (Rs/ha)	30,000-55,000	17,000-35,000	57.14-76.47

3.2. GRAPEVINE, *VITIS VINIFERA*

3.2.1. Insect Pests

3.2.1.1. Lepidopterous Caterpillars, *Helicoverpa armigera, Spodoptera litura*

(i) Symptoms: These polyphagous caterpillar pests cause severe damage to the leaves and berries of grapes.

(ii) Integrated Management

(a) Bioagents, Cultural and Chemicals: A combination of pheromone traps (2/acre) + NPV at 250 LE/acre + sprays of endosulfan 35 EC at 2ml/l or cypermethrin 25 EC at 0.5ml/l (depending on severity) + hand picking and destroying was found effective.

3.2.2. Diseases

3.2.2.1. Grey Mold Rot, *Botrytis cinerea*

The disease occurs in the entire grape growing regions of the world. It is one of the principal causes of post-harvest spoilage in storage. The pathogen is capable of growing at low temperatures.

(i) Symptoms: In early stage, tissue just beneath the surface of fruit is infected loosening the skin from the flesh. The affected area turns light brown. The fungal infection advances into the inner flesh resulting in a soft watery mass of decayed tissue. Under moist atmosphere, the fungus sporulates on the surface of the fruit and the typical powdery grey mold stage becomes evident (Fig. 3.2). Infected fruits shrivel and turn dark brown. The disease starts in midseason and continues to develop until harvest in the absence of rain. The fungus infects stigma and style and becomes latent in the necrotic stigma and style tissues at the stylar end of the berry. In a compact infected bunch, fruits inside may split during growth. Infection of such fruits results in bunch rot. Even a single field infected berry may cause 'nest rot' in transit and storage.

Fig. 3.2. Grey mold rot of grapevine

(ii) Integrated Management

(a) Bioagents and Chemicals: Application of *T. harzianum* 1293-22 at bloom and early fruit development, followed by a tank-mix application of the antagonist and half rates of iprodione, suppressed the *Botrytis* bunch rot by 98% (Harman *et al.*, 1989).

3.2.3. Nematodes

3.2.3.1. Root-knot Nematodes, *Meloidogyne* spp.

Two species of root-knot nematodes, *M. javanica* and *M. incognita* are recognized as the major pests of grapes causing economic damage. In India, *M. javanica* is most prevalent in northern part of the country (Baghel *et al.*, 1980) and *M. incognita* in southern part (Darekar and Patil, 1985). *M. incognita* is also reported from some parts of Haryana. *M. incognita* was responsible for 55% loss in fruit yield of grapes (Rajagopalan and Naganathan, 1977a), while *M. javanica* caused 53% loss in yield (Baghel and Bhatti, 1983b).

(i) Symptoms: The root-knot nematode infestation is not manifested by any typical above-ground symptoms. Patches of poorly branched vines with scant foliage, pale and small leaves, and poor bearing are the indications of root-knot nematode damage (Fig. 3.3). In young plants, premature decline, weak vegetative growth are commonly associated with nematode attack. The visibly unthrifty growth is generally attributed to moisture stress, low fertility, nutritional deficiency and other adverse conditions. However, the confirmation of nematode attack is possible by assaying soil and root samples. The root system shows typical localized swellings particularly on feeder roots and young secondary roots (Fig. 3.3) and females may be found on internodal trunk just below the ground level. *Meloidogyne incognita* has been reported to stimulate the production of many new fine rootlets above the site of nematode infection resulting in "hairy root" condition. Depending upon the variety of grape, *M. javanica* forms galls of varying size and shape and distorts the normal appearance of roots.

Fig. 3.3. Root-knot nematodes on grapevine

(ii) Integrated Management

(a) Cultural and Bioagents: Pruning (during July) and soil application of 4g talc formulation of *P. fluorescens* (containing 15 x 10^8 cfu/g)/vine around root-knot infested grapevine at 15 cm depth in the basin significantly reduced root galling due to *M. incognita* (39%), number of egg masses (250%) and increased fruit yield (166%) (Shanthi *et al.*, 1998) (Table 3.2).

Table 3.2. Field evaluation of biocontrol potential of *Pseudomonas fluorescens* against *Meloidogyne incognita* infecting grapevine.

Treatment	No. of galls/ 5 g roots	No. of egg masses/5 g roots	Root colonization (10^8 cfu/g)	Yield (MT/ha)
P. fluorescens – 1 g/vine	428c	94c	24	12.07b
P. fluorescens – 2 g/vine	390b	85b	36	15.41c
P. fluorescens – 4 g/vine	326a	72a	58	22.07d
Carbofuran 3G – 60 g/vine	300a	67a	---	31.65e
Control (Untreated)	535d	180d	---	8.33a

Figures with different letters are significantly different from each other at 5% level by analysis of variance test.

(b) Two Bioagents: The combined application of two bioagents *P. lilacinus* and *P. penetrans* was most effective in increasing the fruit yield of grapevine besides checking *M. incognita* with benefit: cost ratio of 3.3 (Sivakumar and Vadivelu, 1999).

(c) Botanicals and Chemicals: Application of neem cake (200 g/plant) + carbofuran (10 g/plant) reduced the nematode populations in the soil and roots. Combination of garlic and carbofuran at 0.1 g a.i./m^2 reduced nematode population

(44.2-47.6%) and increased yield (88.3-105.6%) over control with benefit: cost ratio of 4-5 (Anon, 1993).

(d) Bioagents and Botanicals: Application of neem cake at 200 g and *Paecilomyces lilacinus* at 50 g/vine helps in improving vine growth and yield. Application of FYM enriched with *Trichoderma harzianum* and *P. lilacinus* to the field at the rate of 2 kg per plant for 5-6 times at an interval of 2 months reduced the nematode problems significantly and improved the yield levels of grapes (one kg of neem based formulation of *T. harzianum* and *P. lilacinus* was applied to 1 ton of farm yard manure. One hundred kg of neem cake was applied to the farm yard manure to enhance the rate of enrichment by bio-agents. The farm yard manure was kept moist for 15 days under shade with thorough mixing of farm yard manure at 5 days interval).

(e) Botanical, Cultural and Chemicals: Combined application of neem cake and nematicide at reduced doses, along with intercropping with onion and garlic is highly profitable.

3.2.3.2. Root-knot Nematode, *Meloidogyne incognita* and Wilt, *Fusarium moniliforme* Disease Complex

(i) Integrated Management

(a) Bioagents and Botanicals: Soil application of *Pseudomonas fluorescens* at 100 g and FYM at 20 kg/vine gave effective management of disease complex and improved the plant stand by reducing the final soil nematode population (56.9%), root gall index (1.8) and % disease incidence (15.67%). This treatment also increased the number and weight of fruit bunches (17.83 and 155.40%, respectively) and fruit quality (more TSS-13.53 Brix, TSS: acid ratio-14.87, lower acidity-0.91%). The bunch weight of grapevine increased by 155.4% compared to untreated control (Senthil Kumar and Rajendran, 2004).

Chapter - 4

TEMPERATE FRUIT CROPS

4.1. APPLE, *PYRUS MALUS*

4.1.1. Insect Pests

4.1.1.1. Codling Moth, *Cydia pomonella*

The codling moth is a serious pest of apple and other fruits in Leh and Kargil districts of Ladakh region (Khaltse, Nurla, Nemo, Sapol, Basgo, Leh, Saboo, Kargil, etc.) of Jammu and Kashmir.

(i) Damage: About 30-70% of the apple fruits are rendered unmarketable by this pest. The females lay eggs on fruit or leaves and the black-headed yellow larvae attack the fruit immediately upon hatching. Each larva burrows into the fruit, eats for around three weeks (Fig.4.1), then leaves the fruit to overwinter and pupate elsewhere. Most nourishment is obtained by feeding on the proteinaceous seeds.

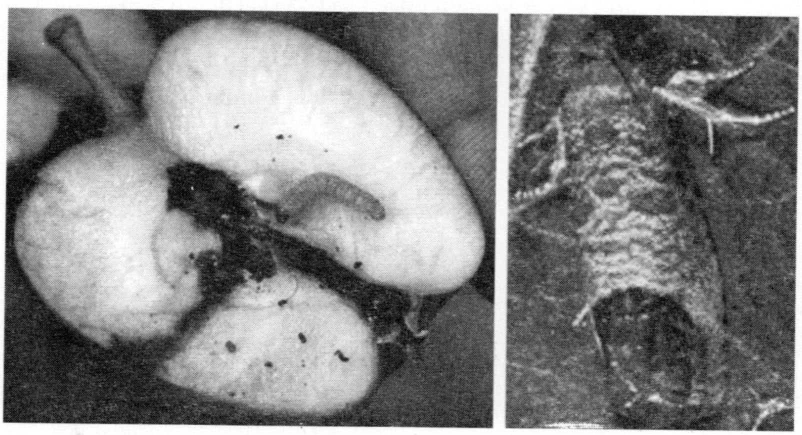

Fig.4.1. Codling moth damage on apple and adult.

(ii) Integrated Management

(a) Bioagents and Chemicals: Synchronizing the release of egg parasitoid *Trichogramma embryophagum* at 2000 adults/tree with the first appearance of the moth, along with use of pheromone traps (E, E-10, 12-dodecadien-1-01) increased the efficiency of controlling codling moth.

4.1.2. Diseases

4.1.2.1. Scab, *Venturia inequalis*

Scab is the most destructive disease of apple and present in all the countries of the world where apples are grown. The disease is particularly severe in high rainfall and humid areas. In India, it is present in all the apple growing areas of Jammu and Kashmir, Himachal Pradesh, Uttar Pradesh, Sikkim, Arunachal Pradesh and Nilgiris. Losses from scab result from mid-season defoliation of trees, reduction in fruit production, devaluation in fruit grade, weakening of the trees, failure of fruit bud formation and increased expenses on spray operations. The first epidemic of scab in Kashmir valley in 1973 completely ruined the apple crop worth US$ 540,000. During 1983, it rendered 10% apple crop (30,000 MT) unfit for consumption in Himachal Pradesh which has to be destroyed and as a result the state suffered a loss of Rs. 1.5 crores (Gupta, 1985).

(i) Symptoms: Scab may appear on leaves, fruits, petioles and green twigs. The most striking symptoms of scab are commonly observed on leaves and fruits. Fruits may show small, rough, black, circular lesions on their skin (Fig. 4.2) while on the tree or after keeping in cold storage. Fruits picked from infected trees appear apparently healthy but are too sticky to keep even in cold storage. Such fruits soon develop scab symptoms even at low temperature and may not last long in storage. The affected fruits rot due to secondary infection of the lesions. Secondary infections on leaves are so numerous that the entire leaf surface appears covered with scab, commonly referred as sheet scab. Lesions on young fruits resemble those on leaves but turn dark brown to black and become corky or scab-like with time. Infections are often limited to one or two spots per fruit. Secondary infections are clumped together.

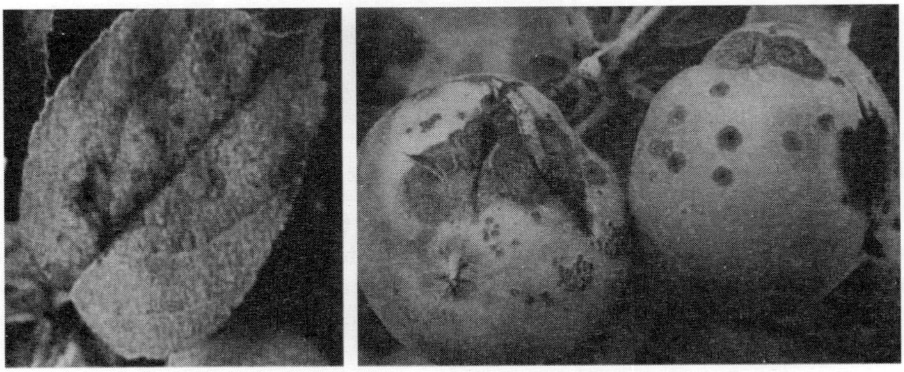

Fig. 4.2. Apple leaf and fruits infected with scab.

(ii) Integrated Management

(a) Bioagents and Chemicals: Integrated control of apple scab by modifying the nutrient status of overwintering leaves is well established. Treatment of senescent apple leaves on trees shortly before leaf fall or of fallen leaves on the orchard floor with a solution of 2% urea greatly reduced ascospore production in the spring. Urea treatment of fallen apple leaves greatly increased the bacterial populations found on decaying leaves. Fluorescent Pseudomonads increased greatly in number on urea treated leaves. Pseudothecial development may be reduced directly by antibiotics produced by fluorescent Pseudomonads on decaying leaves, or indirectly by degeneration of leaf structure due to enhanced degradation of leaf material in the presence of higher bacterial populations induced by treatment with a good nitrogen source (Burchill and Cook, 1971).

4.1.2.2. White Root Rot, *Dematophora necatrix*

(i) Symptoms: This is a soil-borne disease which is most important as the fungus can cause death of plants. The disease appears on underground plant parts and causes complete rotting of the roots. The fine roots are attacked first which are completely devoured and infection spreads to main root through secondary root system. Lateral roots turn dark brown and become infected with white flocculent fungus during monsoon months. Bronzing of leaves, stunted growth and size are important above-ground symptoms (Fig. 4.3). Root rot affected trees are usually associated with a heavy blossom and fruiting next year. However, in succeeding years, few leaves emerge and much of the immature fruits fail to reach maturity. Infected trees often persist for 2-3 years depending upon infestation of fungus.

Fig. 4.3. The tree with the bronzed leaves is in the process of being killed by the disease

(ii) Epidemiology: Small black sclerotia may be formed over the dead bark of wood. Infection of new roots is brought about by the mycelial growth in the soil or contact of plant with old dead roots left in the soil from previous infected plants. The disease incidence is severe when soil moisture is excessive during June to September and 100-140 cm rainfall. Soil type and soil pH between 6.0-6.5 are more favourable for survival of the pathogen. High textured soil is more favourable for development of the disease.

(iii) Integrated Management

(a) Physical and Bioagents: Use of antagonists like *T. viride*, *T. harzianum, Enterobacter aerogenes* and *Bacillus subtilis* along with soil solarization have been found to protect the plants from root rot infection.

(b) Bioagents and Chemicals: Application of *Enterobacter aerogenes* + carbendazim (0.1 or 0.05%) showed more than 92% disease control when applied as pre-inoculation to the pathogen. In simultaneous inoculation, 0.1% carbendazim in combination with *E. aerogenes*, completely prevented the appearance of the disease, however, 0.05% in combination with *E. aerogenes, T. viride* and *G. virens* gave 90.4, 86.9 and 81.0% disease control, respectively (Gupta and Sharma, 2004).

(c) Physical, Botanicals and Bioagents: Soil solarization in combination with organic amendments and biocontrol agents in general gave good control of the disease in nursery. However, cent percent control of the disease was achieved in plots where *T. viride* or *T. harzianum* or *Bacillus* sp. was incorporated in combination with organic amendments (deodar needles, neem cake) (Table 4.1).

Table 4.1. Effect of soil solarization, organic amendments and biocontrol agents against soil-borne diseases in apple nursery.

Treatment	White root rot (%)	Collar rot (%)	Hairy root (%)	Crown gall (%)
Soil solarization (SS)	1.15	0.00	2.25	4.00
SS + Neem cake (NC)	0.00	0.00	4.15	1.50
SS + Deodar needles (DN)	3.75	0.00	2.00	0.50
SS + *Trichoderma viride* (Tv)	0.00	0.00	0.00	2.50
SS + *Trichoderma harzianum* (Th)	0.00	0.00	0.00	0.00
SS + *Bacillus* sp. (Ba)	0.00	0.00	0.00	0.00
SS + NC + Tv	0.00	0.00	0.00	0.00
SS + NC + Th	0.00	0.00	0.00	0.00
SS + DN + Tv	2.38	0.00	1.75	0.00
SS + DN + Th	0.00	0.00	0.00	0.00
SS + NC + DN + Tv	0.00	0.00	0.00	0.00
SS + NC + DN + Th	0.00	0.00	0.00	0.00
US = Unsterilized soil (US) + NC	4.50	1.00	7.00	3.00
US + DN	7.75	0.75	9.45	6.55
US+ Tv	1.50	0.00	6.75	3.95
US+ Th	1.43	0.00	5.85	4.15
US + Ba	2.01	0.00	3.15	5.25
US + NC + Tv	0.00	0.00	0.50	0.00
US + NC + Th	0.00	0.00	0.00	1.50
US + DN + Tv	2.36	0.00	3.75	1.75
US + DN + Th	2.27	0.00	4.25	2.15
US + NC + DN + Tv	0.00	0.00	2.05	0.05
US + NC + DN + Th	0.00	0.00	1.00	0.00
US	22.82	2.05	16.00	8.00

4.1.2.3. Collar/Crown Rot, *Phytophthora cactorum*

The disease is also known as crown rot and trunk canker is universally present in all the apple growing regions of the world including USA, UK, Canada, New Zealand, The Netherlands, Germany, Australia and India. On susceptible varieties, it causes extensive losses even resulting in death of apple trees within a few years.

(i) Symptoms: The above-ground symptoms are often confused with white root rot. The infection starts from the collar region and spreads to the under-ground parts. Bark at the soil level becomes slimy and rots resulting in cankered areas (Fig. 4.4) . The infected trees are recognized by chlorotic foliage with red colouration of veins and margins.

Fig. 4.4. Crown rot on apple

(ii) Epidemiology: The fungus is known to survive in most of the orchard soils but a soil temperature of 12°-20°C associated with pH 5-6 is found to be the best for its survival. Majority of infections in apple occur through mycelial penetration of stems near the ground line but zoospore infection may occur since these are liberated in the soil. Pieces of bark containing the oospores may also serve as the source of infection, which usually occurs during damp and cool weather in the spring. Temperature of 20°-25°C along with high soil moisture is favourable for sporangial production but oospores are produced at low moisture level at the same temperature.

(iii) Integrated Management

(a) Soil Solarization and Bioagents: Use of antagonists like *T. viride*, *T. harzianum, Enterobacter aerogenes* and *Bacillus subtilis* along with soil solarization have been found to protect the plants from collar rot infection.

(b) Soil Solarization, Organic Amendments and Bioagents: Soil solarization in combination with organic amendments and biocontrol agents in general gave good control of collar rot disease in nursery. However, cent percent control of the disease was achieved in plots where *T. viride* or *T. harzianum* or *Bacillus* sp. was incorporated in combination with organic amendments (deodar needles, neem cake). Collar rot pathogen was highly sensitive to the biocontrol agents and organic amendments both in solarized and unsolarized soil.

4.1.2.4. Hairy Root, *Agrobacterium rhizogenes*

(i) Symptoms: At the union between scion and root piece, an enlargement somewhat resembling a newly formed crown gall appears. From this arise numerous roots, fleshy or fibrous in texture, with many containing numerous branches (Fig. 4.5). The surface of these enlargements bears numerous convolutions with fissures extending deep into the interior of the enlargements.

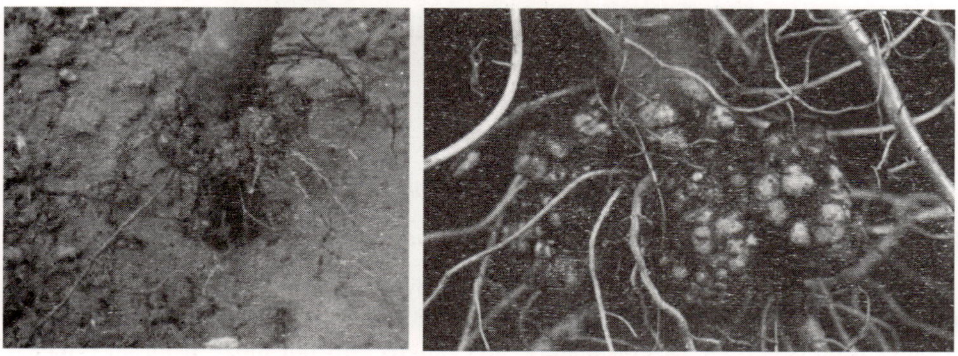

Fig. 4.5. Hairy root symptoms on apple

(ii) Integrated Management

(a) Soil Solarization, Organic Amendments and Bioagents: Soil solarization in combination with organic amendments and biocontrol agents in general gave good control of hairy root disease in nursery. However, cent percent control of the disease was achieved in plots where *T. viride* or *T. harzianum* or *Bacillus* sp. was incorporated in combination with organic amendments (deodar needles, neem cake).

4.1.2.5. Crown Gall, *Agrobacterium tumefaciens*

(i) Symptoms: Crown gall bacterium enters the plant through wounds in roots or stems and stimulates the plant tissues to grow in a disorganized way, producing swollen galls. Galls are present all year. Crown gall is identified by overgrowths appearing as galls on roots and at the base or "crown" of apple (Fig. 4.6).

(ii) Integrated Management

(a) Physical, Botanicals and Bioagents: Soil solarization in combination with organic amendments and biocontrol agents in general gave good control of crown gall disease in nursery. However, cent percent control of the disease was achieved in plots where *T. viride* or *T. harzianum* or *Bacillus* sp. was incorporated in combination with organic amendments (deodar needles, neem cake).

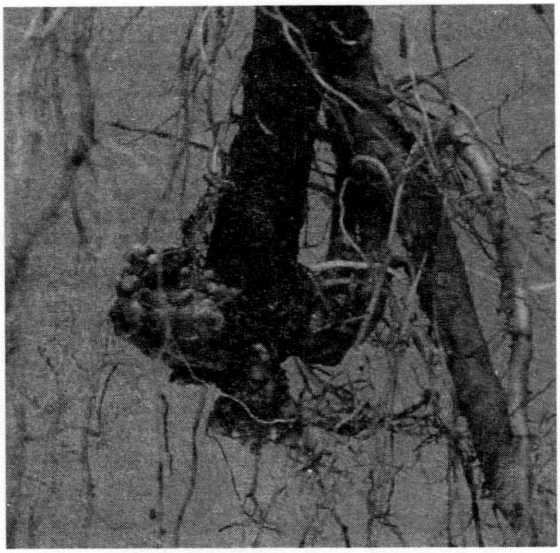

Fig. 4.6. *Agrobacterium tumefaciens* induced galls on apple roots (crown gall)

4.1.2.6. Fruit Rot, Blue Mold, *Pencillium expansum;* Gray Mold, *Botrytis cinerea*

(i) Symptoms: When apples are infected by blue mold, the rotted areas on the fruit are soft, watery and light brown in colour. Bluish-green spores cover the surface of older lesions (Fig. 4.7). Fortunately, this fungus will not move from an infected fruit to an adjacent fruit unless the fruit is bruised or punctured in some way.

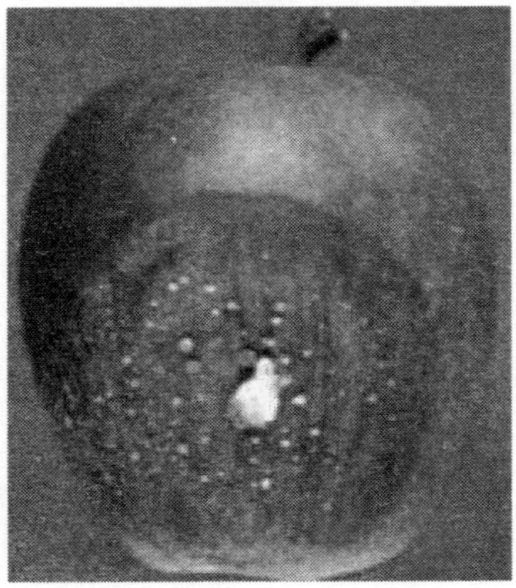

Fig. 4.7. Blue mold on apple fruit

When apples are infected by grey mold, the texture of the decay on the fruit is firm and the skin is tight and tough. The lesions are pale tan to brown and the surfaces of older lesions are covered with grayish mycelium and dark brown spores (Fig. 4.8). Unfortunately, grey mold can move from fruit to fruit and whole bins or boxes of fruit can become infected.

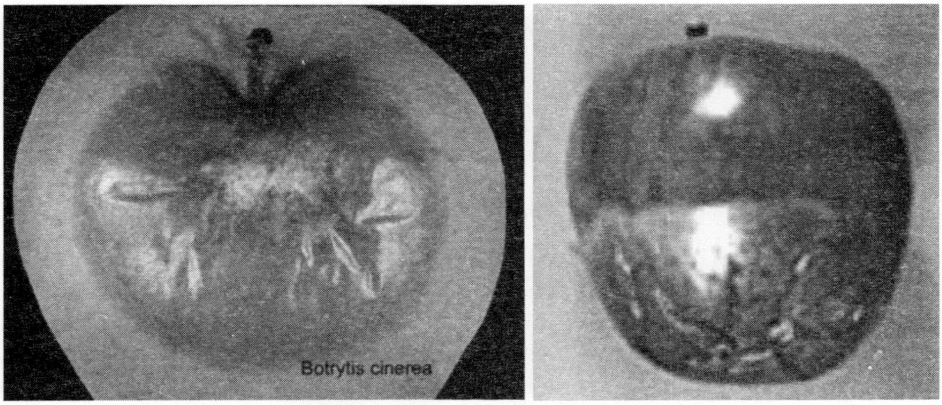

Fig. 4.8. Gray mold on apple fruits

(ii) Integrated Management

(a) Two Bioagents: Combined application of *Pseudomonas* sp. and *Acromonium breve* gave complete control of *P. expansum* and *B. cinerea* on apple (Janisiewicz, 1988).

A co-application involving the bacterial antagonist *P. syringe* and the yeast Sporobolomyces *roseus* applied in equal biomass provided control of blue mold that was superior to that obtained by treatment with the individual agents applied separately (Janisiewicz and Bors, 1995).

Pre-harvest combined application of two strains of *Aureobasidium pullulans* and an isolate of *Rhodotorula glutinis* was superior to any of the strains applied individually in controlling decay caused by *P. expansum, P. malicorticis* and *B. cinerea* (Leibinger *et al*., 1997).

Mixtures of *R. glutinis* SL1 with *Candida albidus* SL43 and *R. glutinis* SL30 with *C. albidus* SL43 showed synergistic effect against *P. expansum*, but not against *B. cinerea* (Calvo *et al*., 2003).

(b) Bioagents and Chemicals: Combining 0.2% glycolchitosan (antimicrobial substance) with the antagonist *Candida saitoana* was more effective in controlling gray and blue molds of apples than either treatment alone (El-Ghaouth *et al*., 2000).

Chand-Goyal and Spotts (1996a, b) showed that the control of apple blue mold by *Candida laurentii* HRA5 was increased by combining it with TBZ fungicide. Combination of *Pseudomonas syringae* MA-4 at 1-3 x 10^7 cfu/ml with cyprodinil at 5 to 10 µg/ml controlled both blue and gray mold by more than 90% on apple, demonstrating that the integration could not only improve disease control efficacy but also extended the degree of control to more than one important disease (Zhou *et al*., 2002).

McLaughlin *et al*. (1990) showed that addition of 2% $CaCl_2$ to a cell suspension of yeast *Candida guilliermondii* significantly increased its efficacy in control of apple post-harvest diseases, compared with either yeast or calcium used alone. In another report (Wisniewski *et al*., 1995), addition of 90 or 180 mM $CaCl_2$ enhanced biocontrol activity of *Candida oleophila* isolate 182 against *B. cinerea* and *P. expansum*.

Aspire, when combined with 2% sodium bicarbonate, showed enhanced efficacy in the control of *B. cinerea* and *P. expansum* rot in apple, compared with that used alone (Droby *et al*., 2003).

4.1.2.7. Fire Blight, *Erwinia amylovora*

(i) Symptoms: The intensity of blossom infection with consequent loss of crop and the possible loss of some branches as a result of cankering would be very serious (Fig. 4.9). Primary infection frequently occurs as a result of transfer of the bacteria by pollinating insects to open blossom. At that time (April) in North India, temperature above 24°C and plentiful rains occur which favour infection and rapid spread of the disease. Presence of naturally occurring hosts (pear) in the vicinity of apple orchards constitute a permanent reservoir of infection.

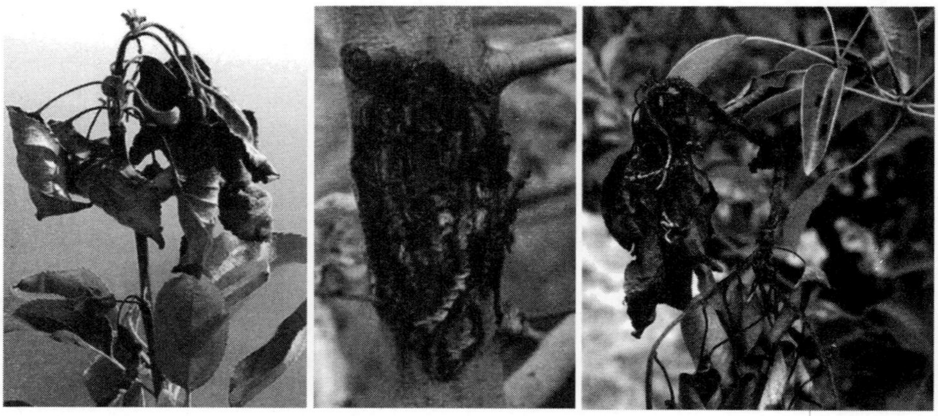

Fig. 4.9. Fire blight of apple

(ii) Integrated Management

(a) Bioagents and Chemicals: Studies conducted with *P. fluorescens* A 506 in combination with antibiotic (streptomycin/oxytetracycline) applications (7 days after application of the antagonist) suggest that the control achieved is likely to be additive in nature (Lindow *et al*., 1996).

4.1.3. Validation of Apple IPM (Himachal Pradesh)

4.1.3.1. Winter Module (October-November)

- Use of 5% urea at leaf shedding stage for early decomposition of the infested leaves and to encourage the population of antagonists in the plant rhizosphere.
- Use of Bordeaux paint during autumn on the naked plant stem to overcome the direct effect of UV rays on the plant skin to reduce sun burn and canker disease complex.

- Overwinter spray of Bordeaux mixture as eradicative action to pathogens and total disinfection of plant surface.

4.1.3.2. Spring Module (April-June)

- Use of Neemarin at pink bud stage i.e. pre-bloom stage to manage blossom thrips population.

4.1.3.3. Summer Module (July-September)

- Use of *Bacillus thuringiensis* at fruit development stage for the management of fruit scrapper insect pests.
- Use of *Trichoderma viride* (Bioderma) for the control of root rot fungus.
- Use of Bordeaux mixture for the control of root and collar rots.

During the period 2001-04, IPM for apple crop was validated and promoted in 30 ha of orchards in Kotkhai, Jhubbal, Theneder and Rohroo villages of Himachal Pradesh. By adopting IPM package, farmers were able to harvest 580 boxes (4.99 MT/ha) of apple in IPM plots as compared to 380 boxes (4.12 MT/ha) in case of non-IPM farmers. Apple growers who adopted IPM earned a profit of Rs. 145733/- while the farmers who did not adopt IPM earned a profit of Rs. 110889/-. Average benefit cost ratio of IPM to non-IPM was 4.07 to 3.01 (Table 4.2) (Trivedi *et al.*, 2004).

Table 4.2. Economics of apple IPM

Parameters	IPM plots	Non-IPM plots	% Increase
Yield (MT/ha)	4.99	4.12	21.11
Net profit (Rs/ha)	145733	110889	31.42
Benefit: cost ratio	4.07	3.01	35.21

4.2. PEACH, *Prunus persica* AND PLUM, *Prunus salicina*

4.2.1. Diseases

4.2.1.1. Brown Rot of Fruits, *Monilinia fructicola, M. fructigena, M. laxa, M. laxa* f. sp. *mali*

Among the fruit rots, brown rot is the most important in stone fruits including peach.

(i) Symptoms: Symptoms of the disease are blighting of blossom and leaves, canker production on woody tissues and rotting of fruits (Fig. 4.10). Blossom blight is the first symptom during spring and attacked parts turn grey to dark brown. The fungus spreads through peduncle and reach branches causing twig blight. Stem cankers usually develop from blighted twigs or fruit spurs. Fruit rot is the most destructive phase of the disease. Rotted fruits either fall down or hang as firm mummies. Conidia and ascospores produced on the mummified fruits and canker spots serve as a source of primary infection.

Fig. 4.10. Brown rot on peach and plum fruits

(ii) Integrated Management

(a) Bioagents and Chemicals: Biocontrol agents can be used in combination with fungicides. Such an approach was successful in controlling rots on peach, in cases where *B. subtilis* (B3), effective against brown rot (incited by *M. fructicola*), was combined with dicloran used for the control of *Rhizopus* rot (Pusey *et al*., 1986).

Zhou *et al*. (1999) reported that addition of 0.5% calcium to cell suspension of *Pseudomonas syringae* MA-4 resulted in a greater reduction of peach brown rot incidence when sprayed on peaches naturally infected with *Monilinia fructicola*. Preharvest application of *P. syringae* MA-4 with a foliar calcium fertilizer also significantly increased biocontrol efficacy against peach brown rot (Zhou and Schneider, 1998).

Aspire, when combined with 2% sodium bicarbonate, showed enhanced efficacy in the control of *Monilinia* and *Rhizopus* rot in peach, compared with that used alone (Droby *et al*., 2003).

4.2.2. Nematodes

4.2.2.1. Peach-Tree Short-Life, *Mesocriconema xenoplax*

(i) Symptoms: It causes pruning and necrosis of fine feeder roots, especially on young plants, but also feeds on older parts of the root. It predisposes some *Prunus* spp. and *Malus* spp. to infection by *Pseudomonas syringae* pv *syringae,* resulting in tree mortality due to bacterial canker (BC) and to winter frost damage. The combined effect of the nematode, bacterium and cold injury result in enhanced tree mortality (Fig. 4.11).

(ii) Integrated Management

(a) Biofumigation and Solarization: Four months (January 1999) after establishing the sorghum biofumigant and methyl bromide plots and prior to planting peach trees, ring nematode (*M. xenoplax*) populations were greatest ($P < 0.05$) in the unfumigated soil than in sorghum + plastic, sorghum without plastic, and methyl bromide fumigated plots. In September 1999 (12 months after incorporating the sorghum as a green manure), no differences in nematode populations were detected among the unfumigated and two sorghum treatment plots.

Fig. 4.11. Peach tree short life symptoms.

However, nematode populations were still suppressed ($P < 0.05$) in the methyl bromide plots. At 24 months after methyl bromide application (September 2000), the nematode population density in fumigated soil did not differ from the other treatment plots. Ring nematode populations continued to increase in subsequent sampling dates. Peach trees developed typical peach-tree short-life (PTSL) symptoms and died during this experiment. In 2001 and 2002, percentage of PTSL tree death was greater in unfumigated w/o plastic (29% and 54%) followed by sorghum + plastic (12% and 46%), sorghum w/o plastic (8% and 50%), unfumigated soil + plastic (4% and 29%), and methyl bromide (4% and 29%), respectively. No differences in trunk diameter were detected among any of the treatments. Sorghum as a green manure with and without plastic did suppress the population of *M. xenoplax* in the early stages of this experiment, but suppression did not last as long as preplant methyl bromide fumigation (i.e., 12 vs. 24 months, respectively) (Nyczepir and Rodriguez-Kabana, 2004).

4.3. PEAR, *Pyrus communis*

4.3.1. Diseases

4.3.1.1. Gray Mold, *Botrytis cinerea*

(a) **Symptoms:** Gray mold and its many strains cause death of flower parts, leaves, buds, shoots, seedlings and fruits (Fig. 4.12). The disease needs moisture as one of its criteria for infection. The wetter the plant is, the more likely the gray mold will show up on plants. Not only are the numbers of infected areas increased but also so are the numbers of plants attacked as well as the severity of the infections (quicker growth of the disease and death of tissue).

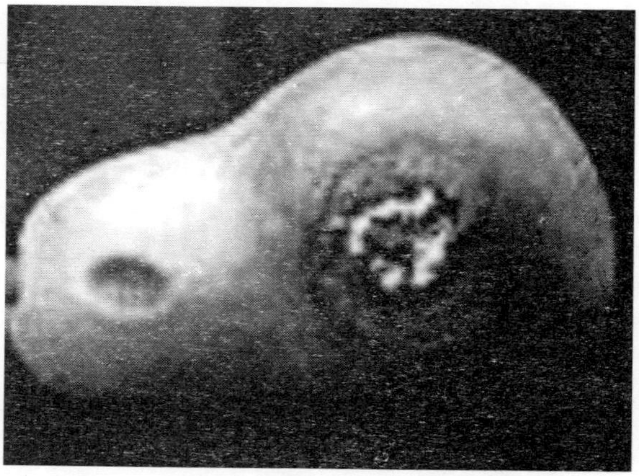

Fig. 4.12. Gray mold on pear fruit

(ii) Integrated Management

(a) Bioagents and Chemicals: In a packing house trial, combination of Bio-Save 110 or Aspire with TBZ at 100 μg/ml (about 17.6% of the label rate) provided control of blue mold and gray mold of pears, similar to that of TBZ alone used at the label rate (569 μg/ml) (Sugar and Spotts, 1999).

4.3.1.2. Blue Mold, *Pencillium expansum*

(i) Symptoms: The rotted areas are soft, watery and light brown in colour. The surface of older lesions may be covered by bluish-green spores that initially are nearly snow white in colour (Fig. 4.13). The lesions are of varying shades of brown, being lighter on the yellow or green varieties. Two characteristics are of importance in the recognition of *P. expansum*, the most common species, namely the musty odor and the formation of conidial tufts or coremia on the surface of well developed lesions.

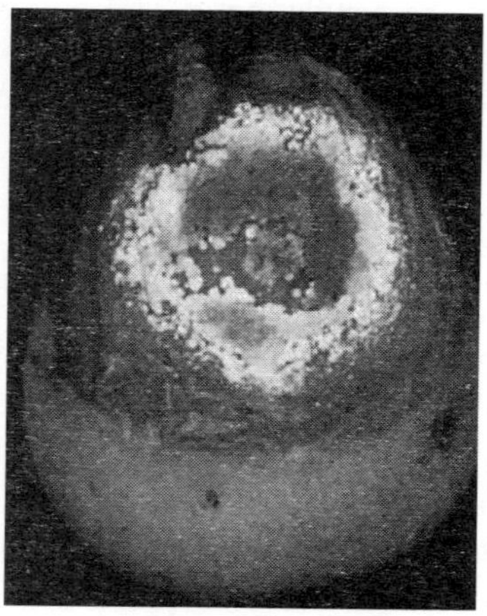

Fig. 4.13. Blue mold rot on pear fruit

(ii) Integrated Management

(a) Bioagents and Chemicals: In a packing house trial, combination of Bio-Save 110 or Aspire with TBZ at 100 μg/ml (about 17.6% of the label rate) provided control of blue mold and gray mold of pears, similar to that of TBZ alone used at the label rate (569 μg/ml) (Sugar and Spotts, 1999).

4.3.1.3. Fire Blight, *Erwinia amylovora*

(i) Symptoms: The term "fire blight" describes the appearance of the disease, which can make affected areas appear blackened, shrunken and cracked, as though scorched by fire (Fig. 4.14). Observe blighted limbs and shoots for removal during the normal pruning operation.

Fig. 4.14. Fire blight infection on tree and fruit

(ii) Integrated Management

(a) Bioagents and Chemicals: Lindow *et al.* (1996) reported that use of *P. fluorescens* strain A 506 in combination with streptomycin and oxytetracycline could reduce pear fire blight by 40 to 50%.

Chapter - 5

SEMI-ARID FRUIT CROPS

5.1. POMEGRANATE, *Punica granatum*

5.1.1. Insect Pests

5.1.1.1. Anar Butterfly, *Deudorix isocrates, D. epijarbas*

(i) Damage: The larvae bore into the fruits of pomegranate and can destroy up to 50% of fruits (Fig. 5.1). The female butterfly lays eggs singly on calyx of flowers or small fruits. On hatching, the caterpillars bore inside the developing fruits and feed on pulp and seeds. This will allow the entry of fungi and bacteria causing fruit rot. The conspicuous symptoms of damage are offensive smell and excreta of caterpillars coming out of the entry holes. The affected fruits rot, get dried and ultimately fall down. Besides pomegranate, the pest also damages guava, amla, annona, apple, ber, citrus, litchi, loquat, sapota, mulberry, peach, pear, plum *etc*.

Fig. 5.1. Anar butterfly damage on pomegranate fruit

(ii) Integrated Management

Botanical, Bioagent and Chemical: An IPM technology has been developed against the fruit borer with the following components:

- Spraying 3% neem oil at the time of butterfly activity.
- One week later, release of *T. chilonis* at 2.5 lakhs/ha.
- Five days later, spraying of 0.05% monocrotophos as an ovicide.

- Spraying of 0.07% endosulfan twice, first 15 days after ovicide spray and the second 15 days thereafter reduced the incidence of the pest to a greater extent.

5.1.2. Diseases

5.1.2.1. Bacterial Blight, *Xanthomonas axonopodis* pv. *punicae*

(i) Symptoms: Small irregular, water-soaked spots 2-5 mm in diameter with necrotic centre of pinhead size appear on the leaves. Spots are translucent which later turn to light brown to dark brown and are surrounded by prominent water-soaked margins (Fig. 5.2). Spots coalesce to form larger patches. Severely infected leaves fall off. The bacterium attacks stems, branches and fruits also. On the stem, the disease starts as brown to black spots around the nodes resulting in girdling and cracking of nodes. Finally, the branches break down. Brown to black spots appears on the pericarp with L- or Y-shaped cracks. The spots on fruits are raised and oily in appearance. In severe cases, there will be extensive cracking of fruits (Fig. 5.2).

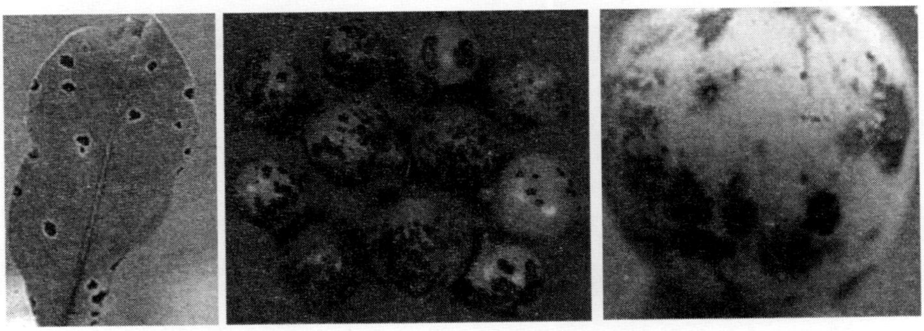

Fig. 5.2. Bacterial blight on leaf and fruits of pomegranate

(ii) Epidemiology: High temperature and low humidity favour the disease. The bacterium survives on the tree. It can survive for 120 days on fallen leaves during off-season. The primary infection is through infected cuttings. The disease spreads through wind-splashed rains.

(iii) Integrated Management

(a) Cultural and Chemicals

Before Planting

- Use disease-free seedlings for planting.
- Application of FYM/compost/vermicompost helps in building up resistance in plants to bacterial blight.

After Planting

- Practice field sanitation (collection and burning of diseased leaves, stem and fruits) to prevent the spread of the disease.

- Before pruning, spray 1% Bordeaux mixture on diseased leaves. Then spray with ethrel solution (2.0-2.5 ml/litre) to defoliate the diseased leaves. Collect and destroy the fallen leaves.
- Dust the tree basins with bleaching powder at 20-25 kg/ha to kill the bacteria on left over leaves.
- After pruning, paste the diseased stems with bromopal (Bacterinashak or Bitrenetol) at 0.5 g/litre or copper oxychloride (3 g/litre) mixed with red sandy loam soil.
- At the beginning of disease incidence stage, give 5 to 6 sprays with bromopal (0.5 g/litre) or copper oxychloride (2 g/litre) at 10 days interval.
- After each spraying of bactericides, give mineral spray (1 g each of Cu SO_4, Mg SO_4, Ca SO_4 and Boron in 1 litre of water) to reduce the disease severity in plants.

Section - III

BIOINTENSIVE INTEGRATED PEST MANAGEMENT IN VEGETABLE CROPS

Chapter - 6

SOLANACEOUS VEGETABLE CROPS

6.1. POTATO, *Solanum tuberosum*

6.1.1. Diseases

6.1.1.1. Black Scurf, *Rhizoctonia solani*

(i) Symptoms: Lesions characteristic of *Rhizoctonia* on stems and stolons are brown to black and sunken. These cankers can continue to expand and are capable of girdling stems and stolons of young developing plants (Fig. 6.1). *Rhizoctonia* infection of older plants very seldom lead to girdled stems that die. However, the health of these plants can be severely compromised and they can frequently become more susceptible to other diseases, particularly early blight.

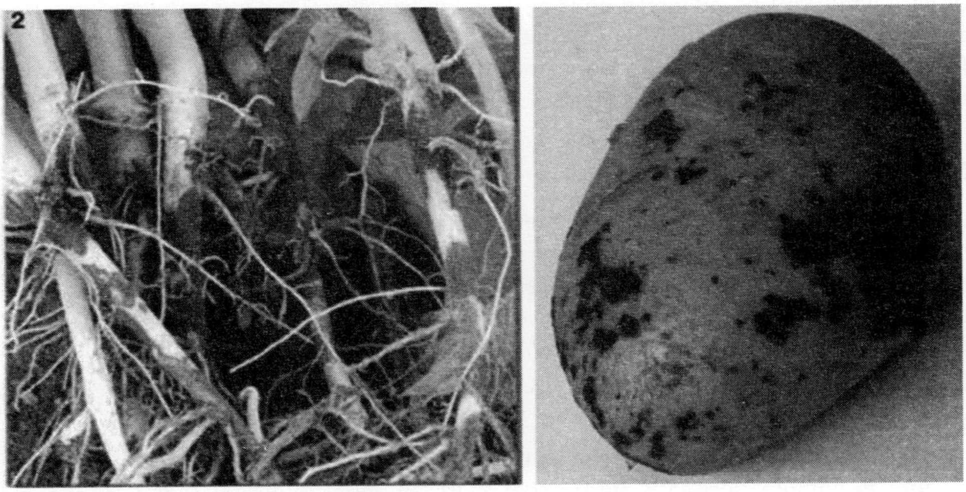

Fig. 6.1. Black scurf on collar region of potato.

Perhaps the most readily apparent phase of *Rhizoctonia* disease is the black scurf or sclerotia present on tuber surfaces. These sclerotia can vary in size from very small, flat, superficial black specks to large, raised, irregularly shaped masses that can cover a major portion of the tuber (Fig. 6.1). While black scurf can, under extreme conditions, affect the marketability of table stock/fresh pack potatoes, it is an extremely important seed-borne phase of the pathogen that deserves attention by all potato producers.

Potato plants are most severely affected in the spring when underground sprouts can be killed prior to emergence. The secondary sprouts that develop are generally less vigorous and emerge much later causing irregular, uneven stands. Sprouts of severely affected plants that do not die are frequently stunted and remain so for the rest of the growing season. Infections of stolons that occur early in the growing season frequently result in pruning to tuber formation or abortion of tubers early in their development.

Mid-season *Rhizoctonia* infections of potato plants will result in deep, sunken cankers on the main stem. The above ground portion of the plant will appear yellow with some purpling and upward curling of the foliage.

(ii) Epidemiology: *R. solani* can be either a soil-borne or seed-borne pathogen. The fungus survives in soil as mycelium in decomposing plant tissues. It also survives as sclerotia on tuber surfaces (seed-borne) or in the soil for extended periods. Populations of *R. solani* decline in the absence of a susceptible host although the rate of decline is affected by soil type, rotational crops and possibly the amount of organic matter present in the soil.

Disease development on emerging sprouts is favoured by cold, wet soil conditions. Although these conditions may not increase disease incidence, disease severity is generally much greater. These conditions slow sprout development but favour germination of sclerotia and infection causing cankers to develop on young, under-developed tissues. Tuber-borne inoculum is very important in this phase of the disease while soil-borne inoculum is believed to be generally more important in stem and stolon infections. Cool, moist conditions, with moisture being the most critical factor, also favour disease development on stems and stolons. Disease development is optimum at around 65°F and decreases as soil temperatures increase.

(iii) Integrated Management

(a) Physical/Chemical and Bioagents: Although solar heating or methyl bromide fumigation significantly reduced the disease, the combination of either solar heating or methyl bromide soil fumigation followed by *T. harzianum* application improved control over solarization or fumigation alone and additionally controlled *Verticillium dahliae* and *Sclerotium rolfsii* (Elad *et al*., 1980).

(b) Physical and Bioagents: Combination of soil solarization of black scurf infested fields and seed treatment with *T. viride* further improved the disease control.

(c) Biofumigation and Solarization: Application of mustard chopped green manure in combination with plastic coat (with an appropriate polymer formulation that forms continuous plastic coats) gave maximum reduction in black scurf followed by millet chopped green manure in combination with plastic coat. The maximum yield was obtained in corn chopped green manure in combination with plastic coat followed by mustard chopped green manure in combination with plastic coat (Table 6.1).

Table 6.1. Effect of biofumigation and solarization on black scurf and yield of potato

Treatment	Soil solarization	Yield (kg/m^2)	Black scurf
Millet	Ecotex	2.208	4.4%
	None	1.943	21.5%
Cabbage	Ecotex	2.187	15.1%
	None	2.101	15.9%
Mustard	Ecotex	2.279	2.0%
	None	2.090	22.0%
Corn	Ecotex	2.401	13.1%
	None	2.067	17.2%
Control	Ecotex	2.135	14.3%
	None	1.996	35.1%

6.1.1.2. Verticillium Wilt, *Verticillium dahliae, V. albo-atrum*

(i) Symptoms: Foliar symptoms first appear as chlorosis and necrosis beginning in the lower leaves. On warm, sunny days, leaves may appear limp and flaccid. Sometimes symptoms occur on only one side of the leaf or the plant (Fig. 6.2). In severely diseased plants, medium-tan discolouration of the vascular tissue is evident (Fig. 6.2), and the plants may be stunted. Tubers of some cultivars may develop a light brown discolouration of the vascular ring, although other factors may cause this symptom. Tuber yield is reduced because of the decreased rate of photosynthesis and premature death of foliage. The optimum temperature range for potato growth is 18-20°C. When the temperature rises above 20°C, plant stress increases and symptoms of Verticillium wilt are more severe.

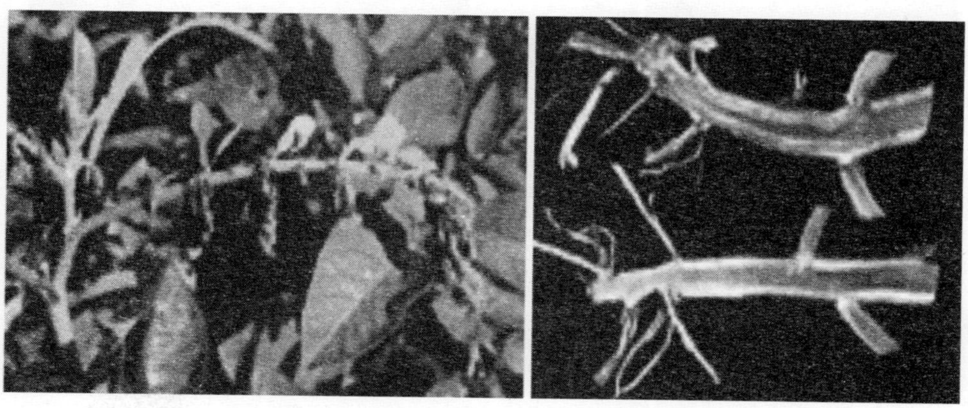

Fig. 6.2. Verticillium wilt affected potato plant

(ii) Integrated Management

(a) Physical/Chemical and Bioagents: Although solar heating or methyl bromide fumigation significantly reduced the disease, the combination of either solar heating or methyl bromide soil fumigation followed by *T. harzianum* application improved control over solarization or fumigation alone and additionally controlled *R. solani* and *Sclerotium rolfsii* (Elad *et al.*, 1980).

(b) Bioagents and Chemicals: Ordentlich *et al.* (1990) integrated *T. harzianum* with captan to protect potato tubers against *V. dahliae* in order to reduce the disease incidence and to increase potato yield by 15.7% under field conditions.

6.1.2. Nematodes

6.1.2.1. Root-knot Nematode, *Meloidogyne incognita*

The root-knot nematodes are by far the most important nematode pests of potatoes in India. Severe infestation of root-knot nematodes leading to crop failure has been noticed in Mahasu district of Himachal Pradesh and Hassan in Karnataka where farmers did not get even as much as planted by them. An initial level of 2 root-knot larvae per gram of soil resulted in 100% tuber infection with an overall yield reduction of 42.5% (Prasad, 1989).

(i) Symptoms: Plants affected by root-knot nematodes are generally stunted, slightly yellow and may wilt during hot weather. The diagnostic symptoms of a *Meloidogyne* attack on potatoes are the galls found on both roots and tubers. Tuber surface becomes uneven and warty because of numerous blister-like galls (Fig. 6.3). When an infested tuber is transversely cut, the white glistening swollen females of the size of pin head can be easily seen embedded in the potato tissue (Fig. 6.3). Such tubers are invariably small, unfit for marketing, rot quickly and cause considerable yield losses. Galling on tubers may render them unsalable.

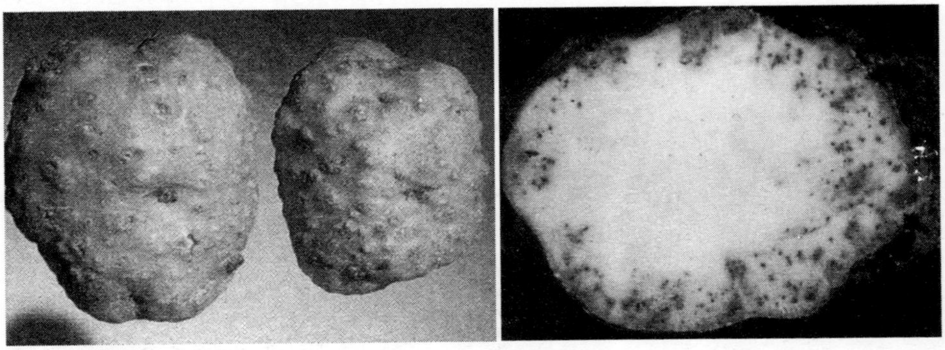

Fig. 6.3. Potato tubers with blisters incited by *Meloidogyne incognita*

(ii) Integrated Management

(a) Bioagents and Botanicals: Application of neem cake/FYM/compost enriched with *Trichoderma harzianum/Paecilomyces lilacinus* gave effective control of root-knot nematodes.

(b) Physical and Cultural: Following 2 years crop rotational sequence of maize-wheat-potato-wheat coupled with summer fallow after 2 or 3 deep ploughings in North Western Plains reduced root-knot damage significantly.

(c) Bioagents, Cultural and Chemical: Aldicarb at 2.5 kg a.i./ha and aldicarb + *P. lilacinus* combined with crop rotation with maize reduced root-knot nematodes and increased tuber quality in *P. lilacinus* and *P. lilacinus* + aldicarb plots. Yield of maize, a rotation crop, in *P. lilacinus* and *P. lilacinus* + aldicarb-treated plots was double than that in control plots. Yield of potato in *P. lilacinus* treated plots was higher than in the plots treated with only aldicarb (Jatala, 1985).

(d) Physical, Botanical and Chemical: Several chemical and non-chemical alternatives to standardized cold or hot diffusion of MB (50 g/m^2) are able to maintain adequate sanitary (against *M. incognita*) and productivity levels on potato cultivation (Table 6.2).

Table 6.2. Effect of integration of physical, chemical and organic amendment on yield of potato tubers

Treatment	Potato yield (kg/m^2)	Relative yield
Soil solarization + MB	2.80 a	125
Soil solarization + Telopic	2.53 ab	113
Soil solarization + fresh chicken manure	2.45 ab	109
Soil solarization alone	2.40 ab	107
MB VIF	2.24 bc	100
Control (no disinfestation)	1.81 c	81

6.1.2.2. Columbia Root-knot Nematode, *Meloidogyne chitwoodi*

(i) Symptoms: Pimple-like galls galls are produced, they appear as small, raised lumps above the developing nematodes, giving the skin a rough appearance (Fig. 6.4). Galls may be grouped in a single area or scattered near the tuber eyes. Infestations are difficult to detect in freshly harvested tubers, but after a few months the egg sacs turn from translucent to brown and can be seen as brown spots in the cortex of cut tubers. Brown spots only become evident when the females begin egg production. Internally, brown spots are usually within 5-6 mm of the tuber surface. There are no symptoms on potato roots and above ground symptoms are generally lacking.

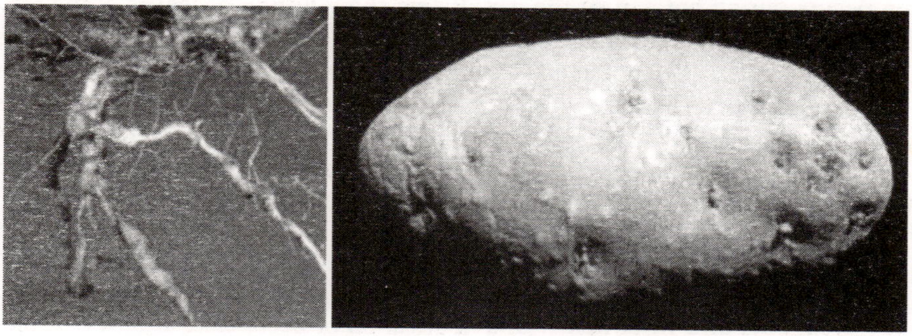

Fig. 6.4. Columbia root-knot nematode infestation on potato

(ii) Integrated Management

(a) Biofumigation and Chemicals: Riga *et al.* (2006) reported a strategy in which a *Brassica* crop arugula (*Eruca sativa* var. Nemat) cover crop was combined with lower rates of a synthetic nematicide to manage nematodes and reducing pest management costs by 50%. Results are promising, with nematodes reduced up to 80%. The current recommendation is the use of rapeseed or mustard cover crop plus the application of nematicide (mocap). This regimen costs about the same as fumigation (2006 prices).

Arugula in combination with half the recommended rate of Telone/Temik reduced root knot nematode populations, *M. chitwoodi*. In addition, Arugula did not reduce the beneficial free-living nematode populations and the non-pathogenic *Pseudomonas*. The cost of growing and incorporating Arugula and combining it with ½ rate of Telone and one fungicide was approximately half of the present commercial cost of Telone and fungicide applications.

6.1.2.3. Cyst Nematode, *Globodera rostochiensis*

The potato cyst nematode is established as one of the major crop protection problems of the world. The ability of this nematode to build up to damageable levels in a short span of 5-6 years, substantial yield reductions in the crop, lack of inexpensive nematicides for soil treatment capable of providing adequate level of control under field conditions, the relative ease with which the cysts are dispersed with soil adhering to the seed tubers and the long persistence of eggs within the cyst in the absence of the host makes this nematode as probably the most important pest problem on potatoes.

The avoidable yield loss in susceptible potato due to cyst nematodes was 99.5 to 99.8% in summer and autumn crops. Total failure of the crop has been reported under severe infestation conditions. An average loss of about 9% of global potato is accounted to the cyst nematodes amounting to about 45 million tonnes (Prasad, 1989).

(i) **Symptoms:** The infested plants exhibit typical symptoms of patchy growth of weak and stunted plants. The patches increase in time with continuous potato cropping. Under conditions of severe infestation, the plant growth is stunted and wilting occurs during hot part of the day. Plants show tufting of leaves at the top as the outer leaves turn yellow and die. Root system is infected with adults and cysts (Fig. 6.5). Root system is smothered, secondary roots are induced at the collar region and the plants can be easily pulled out. Tubers formed are less in number and reduced in size. In extreme cases, tuber formation is arrested. The total photosynthesis per plant is also significantly reduced as a result of reduced leaf area and this is reflected in reduced potato yield.

(ii) Integrated Management

(a) Bioagents and Botanicals: Soil application of neem cake at 5 MT/ha along with *Trichoderma viride* at 0.5 kg/ha recorded maximum tuber yield (23.14 MT/ha) and reduction in potato cyst nematode multiplication (Reproduction factor-1.09).

Fig. 6.5. Potato plants and roots infected with *Globodera rostochiensis*.

(b) Cultural, Chemical and Host Resistance: The combination of disease escape (by planting early maturing varieties) and the use of nematicides gave good control of the potato cyst nematode. Hygiene in the form of seed certification, combined with crop rotation in seed growing areas is effective. Use of a resistant variety followed by a nematicide would kill 99% of the nematode population. The potato cyst nematode problem in Nilgiris is being managed by chemical treatments, crop rotations and utilizing the available sources of resistance (cv. Kufri Swarna).

The most effective management combines crop rotation, use of nematicides and resistant varieties to keep the nematode at an economically acceptable level (Jones, 1969) (Table 6.3).

Table 6.3. Integrated management of the potato cyst nematode.

Sl. No.	Management method (s)	Resulting kill popn. (% initial popn)	Kill (%)	Popn. after growing & harvesting a susc. cv., calculated at 2 assumed nema multpln. rates (% initial popn)	
				30 x	70 x
1	4 years without potato	3	97	90	< 100
2	1 year with resistant Potato	20	80	> 100	> 100
3	Nematicide treatment (Fumigant)	25	75	> 100	> 100
4	1 and 2	0.6	99.4	18	42
5	1 and 3	0.75	99.25	22.5	52.5
6	2 and 3	5	95	> 100	> 100
7	All 3 methods	0.15	99.85	4.5	10.5

A well known integration of methods can control *G. rostochiensis* in potato in a 4-year rotation combining soil fumigation (causes 30% reduction in nematode population), cultivation of non-host (50% reduction), cultivation of resistant potato variety (30% reduction), cultivation of non-host (50% reduction) and then cultivation of susceptible variety of potato (Oostenbrink, 1972).

(c) Solarization, Organic Amendments and Bioagents: Better control of *G. rostochiensis* by *Pasteuria penetrans* was possible with soil solarization combined with application of FYM or karanj cake which resulted in greater kill of nematodes (Sitaramaiah and Naidu, 2003).

6.1.2.4. Potato Rot Nematode, *Ditylenchus destructor*

D. destructor is the most important pest of potato tubers and is responsible for dry rot of tubers. High yield losses occur in the areas where climatological conditions favour establishment of the potato rot nematode. The effect of nematodes will manifest itself at harvest or storage when infected tubers will rot.

(i) Symptoms: *D. destructor* enters potato tubers through lenticels and initially causes small white mealy spots just below the surface that are only visible if the skin is removed. Infested areas enlarge and coalesce and light brown lesions, consisting of dry granular tissue, may be visible beneath the skin. As the infestation progresses, the tissues dry and shrink and the skin becomes cracked and papery (Fig. 6.6). Internal tissues gradually darken and there are often secondary invasions of fungi, bacteria, mites, etc. If stored in moist conditions, a general rot may ensue and spread to neighbouring tubers.

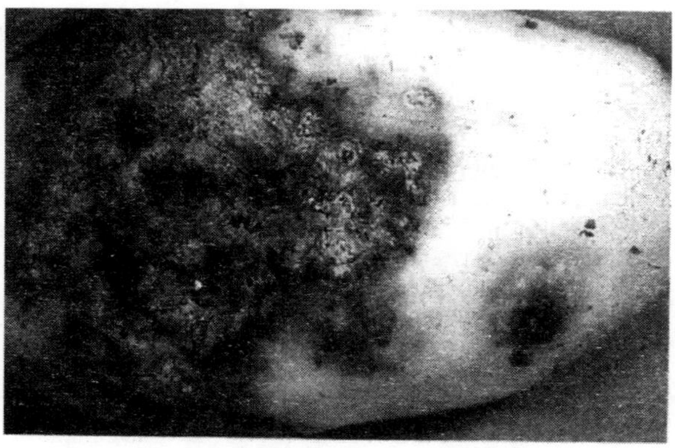

Fig. 6.6. Dry rot of potato tuber incited by *Ditylenchus destructor*

(ii) Survival and Spread: *D. destructor* has a wide host range, can survive on weeds and on a wide range of soil-inhabiting fungi. It can also survive on infected tubers left in the field. Spread occurs by introduction of infected tubers and in soil adhering to seed pieces. Irrigation water and cultivation by infested farm tools and machinery are other sources of inoculum dissemination.

(iii) Integrated Management

The control of potato rot nematode was achieved by the combination of disease escape, hygiene, in the form of seed certification and crop rotation as follows (Winslow and Willis, 1972):

- Healthy seed, planted late, harvested early and stored as cool and dry as possible.

- Proper rotation of potatoes with non-host crops and growing of potatoes not more frequently than once in 3 or 4 years.
- Field hygiene in the form of removal of old infested tubers and weed control.

6.1.2.5. Early Dying/Lesion Nematode, *Pratylenchus* penetrans and Wilt, *Verticillium dahliae* Disease Complex

(i) Symptoms: The potato early dying disease results in premature vine senescence (Fig. 6.7) and can limit potato tuber yield by as much as 30 to 50%. Early dying is primarily caused by *Verticillium dahliae*, a fungal vascular wilt pathogen, but co-infection of potato by both *V. dahliae* and the lesion nematode, Pratylenchus penetrans, can greatly increase the severity of disease.

Fig. 6.7. Early dying of potato plant

(ii) Integrated Management

(a) Physical/Chemical and Bioagents: Although solar heating or methyl bromide fumigation significantly reduced the disease, the combination of either solar heating or methyl bromide soil fumigation followed by *T. harzianum* application improved control over solarization or fumigation alone and additionally controlled *R. solani* and *Sclerotium rolfsii* (Elad *et al*., 1980).

6.2. TOMATO, *LYCOPERSICON ESCULENTUM*

6.2.1. Insect Pests

6.2.1.1. Fruit Borer, *Helicoverpa armigera*

Tomato fruit borer is a polyphagous pest infesting tomato, reducing marketable yield and market value of the crop. This pest has already developed resistance to a number of insecticides in different crops, and is thus difficult to manage. Losses up to 50% in Tamil Nadu (Srinivasan, 1958); 65% in Punjab (Singh and Singh, 1975) and 22 to 38% in Karnataka (Tewari and Krishna Murthy, 1984) have been reported due to this pest attack.

(i) **Damage:** The first instar larvae initially feed on the leaves and later migrate to the developing green fruits. Later the larvae bore into the fruits with the posterior end outside the hole (Fig. 6.8). Pupation takes place in the soil. Eight or more eggs/30 trifoliate leaves below the top most flower cluster was observed to cause economic loss.

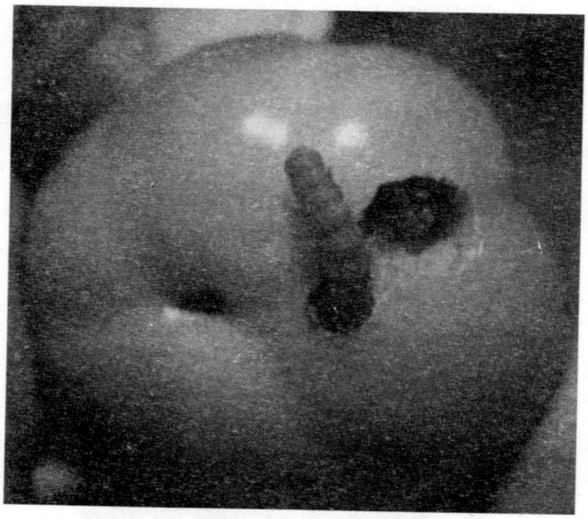

Fig. 6.8. Fruit borer damage on tomato.

(ii) Integrated Management

(a) Bioagents/Botanicals and Cultural: Use of African marigold (*Tagetes erecta*) as a trap crop for the management of fruit borer on tomato involves planting one row of 45 day-old marigold seedlings after every 16 rows of 25 day-old tomato seedlings and spraying of *Ha* NPV at 250 LE/ha or 4% NSKE or 4% pulverized NSPE, 28 and 45 DAP coinciding with peak flowering (Srinivasan *et al.*, 1994) (Fig. 6.9).

Fig. 6.9. Integrated management of tomato fruit borer using African marigold as a trap crop.

(b) **Two Bioagents:** For effective control of *H. armigera*, the egg parasitoid, *T. pretiosum* and *Ha* NPV could be integrated to tackle different stages of the pest. *T. pretiosum* was released at 2.5 lakh/ha and *Ha* NPV was sprayed 2 times (28 and 35 DAP) at 250 LE/ha for effective control of the pest (4.18% fruit borer damage as compared to 21.79% fruit borer damage in control). The increase in yield was 65.5% over control and the reduction in fruit borer damage was 80.8% over control (Krishnamoorthy *et al.*, 1999).

Three releases of *T. pretiosum* + 3 sprays of *B. thuringiensis* at 1 kg/ha was found highly effective against fruit borer in Himachal Pradesh (Gupta and Rajaram Mohan Babu, 1998).

(c) Bioagents and Chemicals: An endosulfan tolerant strain of *T. chilonis* has been developed at the Project Directorate of Biological Control at Bangalore, and transferred to the private industry, which is now marketed under the trade name of "Endogram". This strain has been further developed for multiple tolerance to monocrotophos and fenvalerate. This strain of parasitoid can be utilized in places where both the releases of egg parasitoid *T. chilonis* and spraying of endosulfan/monocrotophos/fenvalerate for the control of other pests are warranted (Singh, 2000).

*Ha*NPV with endosulfan, both at reduced doses, is recommended for minimizing the borer damage effectively (Ganguli *et al.*, 1997). The reduction in larval population of *H. armigera* was maximum (69.3%) in the treatments where the 3 components (Bt + *Ha*NPV + endosulfan-half dose) were applied 55 and 75 DAP. This was followed by *Ha*NPV + endosulfan (55.28%) and Bt + endosulfan (52.68%) (Mahalingam and Saminathan, 2003). The observations on the fruit damage revealed that combination of Bt + *Ha*NPV + endosulfan-half dose recorded minimum fruit damage (7.82%).

(d) Cultural, Bioagents and Botanicals: Use of nylon nets (40 gauge) to avoid insect vectors, spraying of *Ha* NPV twice at 28 and 35 DAP, spraying of pongamia soap at 1% given at 40 DAP was found effective for the management of fruit borer and leaf miner. The bored fruits were removed mechanically once at 40 DAP. The fruit borer and leaf miner incidence were 2% and 0.7-2.7 mines/leaf, respectively.

6.2.2. Diseases

6.2.2.1. Damping-off, *Pythium aphanidermatum, P. ultimum*

(i) Symptoms: In the pre-emergence phase of the disease, the young seedlings are killed even before they emerge out of the soil surface. However, this disease mostly occurs at post-emergence stage which is characterized by the toppling over of the infected seedlings anytime after they emerge from the soil (Fig. 6.10) until the stem has hardened sufficiently to resist invasion by the pathogen. Infestation usually occurs at ground level or through roots. The infected tissue appears soft and water-soaked. As the disease advances, the stem becomes constricted at the base of the plants that collapse later. Seedlings that are healthy looking one day may have collapsed the next morning.

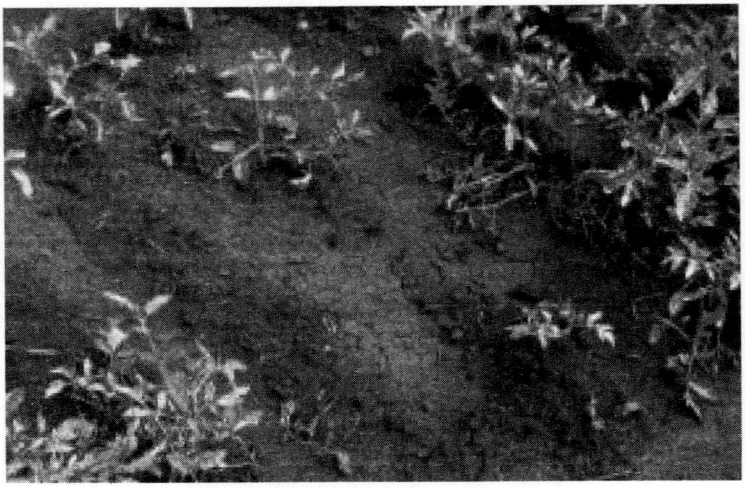

Fig. 6.10. Damping-off of tomato seedlings

(ii) Integrated Management

(a) Bioagents and Botanicals: Seed treatment with *T. viride* and *Pseudomonas fluorescens* and addition of neem cake to the nursery beds enhanced seed germination and seedling stand.

Seed treatment with *Trichoderma viride* + soil application of FYM enriched with *T. viride* gave effective control of damping-off (Table 6.4) (Rahman *et al.*, 2002).

Table 6.4. Comparative efficacy of bioagents and organic amendments against damping-off of tomato

Treatment	% Disease index
Seed treatment with *Trichoderma viride* at 4g/kg + soil application of 50 kg FYM enriched with 500g of *T. viride*	5.70 (13.71)
Seed treatment with *Pseudomonas fluorescens* at 10g/kg + soil application of 50 kg FYM enriched with 2.5 kg of *P. fluorescens*	7.93 (16.06)
Seed treatment with *Trichoderma harzianum* at 4g/kg + soil application of 50 kg FYM enriched with 500g of *T. harzianum*	7.83 (16.03)
Seed treatment with *Azotobacter croococcum* at 16g/kg + soil application of 50 kg FYM enriched with 500g of *A. croococcum*	8.46 (16.68)
Control (Check)	30.64 (33.55)
CD at 5%	**(3.23)**

Figures in parentheses indicate the arc sin transformed values

(b) Bioagents and Chemicals: Dey and Mukhopadhyay (1994) reported effective control of damping-off of tomato by integration of *T. virens* with thiram apron.

(c) Physical and Bioagents: Combination of the seed/root application of *T. harzianum* or *P. fluorescens* with soil solarization was very effective in management of damping-off of tomato in nursery at farmers' field.

(d) Physical and Botanicals: Composted chicken manure applied to solarized soil reduced numbers of *P. ultimum* (Table 6.5). Fungal numbers and galling index generally decreased with increasing fertilizer dosage. Soil solarization reduced cfu of *P. ultimum* in all cases.

Table 6.5. *In vitro* effect of amending soil with composted chicken manure[1] and/or soil heating[2] on numbers (cfu) of *Pythium* sp.

Treatment manure dosage	*Pythium* sp. (cfu/g soil)
Control	14.4a[3]
Solarization only	12.0b
Chicken manure at 2690 kg/ha	9.6b
Chicken manure at 5381 kg/ha	1.1c
Chicken manure at 2690 kg/ha + Solarization	1.8c
Chicken manure at 5381 kg/ha + Solarization	0.0 c

1 Containing 3280 mg NH_4-N/kg.
2 Three day incubation. Diurnal healing regime: 42°C high; 18°C low.
3 Values within columns followed by different letters are different at $P<0.05$ according to Duncan's Multiple Range Test.

6.2.2.2. Wilt, *Fusarium oxysporum* f. sp. *lycopersici*

(i) Symptoms: The disease is characterized by yellowing and wilting of leaves and finally the entire plant wilts and dies prematurely (Fig. 6.11). Stem tissue often is discoloured throughout the plant. Vascular browning takes place in the root (Fig. 6.11). The fungus can persist in the soil for many years and is virulent at moderate temperature (26-28°C).

Fig. 6.11. Tomato plant infected by *Fusarium* wilt.

(ii) Integrated Management

(a) Two Bioagents: Mao *et al.* (1998) found that combined inoculation of *T. virens* and *Burkholderia cepacia* resulted in increased plant stand and greater yield than those obtained with either biocontrol agent alone.

6.2.2.3. Southern Blight, *Sclerotium rolfsii*

(i) Symptoms: The first symptom of the disease is observed as soft tissue necrosis of bark of the stem near soil line. White growth of cottony mycelium is clearly visible on the affected portion just below the soil surface (Fig. 6.12). Later, dense silvery fungus growth along with white to light brown mustard-like sclerotia is observed on the same portion. Progressive drooping, yellowing or wilting of the entire plant is observed (Fig. 6.12). Sometimes the plant is collapsed soon after infection. The disease is soil-borne where the pathogen survives in the form of sclerotia.

Fig. 6.12. Southern blight symptoms on tomato

(ii) Integrated Management

(a) Solarization and Bioagents: Studies by Elad *et al.* (1980) have demonstrated that 2-4 week soil solarization during the summer was able to reduce sclerotial numbers and limit disease. Additionally, combining solar heating with applications of *Trichoderma harzianum* was more effective than either treatment alone. The growth of the biological control agent *T. harzianum* is enhanced with solar heating (Jenkins and Averre, 1986).

Studies in North Carolina showed that soil solarization combined with *T. virens* reduced the disease incidence by 49% during the first season after solarization and 60% during the second season (Ristaino *et al.*, 1991).

(b) Biofumigation, Solarization and Chemicals: Pre-plant soil treatments utilizing composted amendments, biofumigation, solarization, and low dosage dazomet (Basamid) produced higher yields and resulted in a lower incidence of the soil-borne disease, southern blight, than did controls (Fig. 6.13). The compost amended plots produced yields 59% higher than controls. MITC and biofumigation plots had 39 and 40% higher yields than controls. However, the combined MITC +

compost plots produced the highest yields - 82% above control. These results suggest that the most effective treatments for prevention of tomato diseases and enhanced yields may be those that integrate two or more control mechanisms while also favouring the growth of beneficial soil organisms.

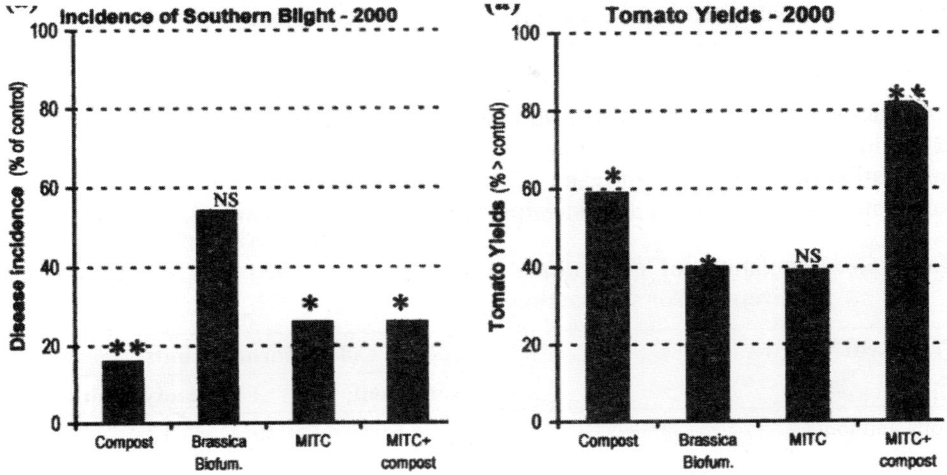

Fig. 6.13. Incidence of southern blight and tomato yields in different treatments

6.2.2.4. Wilt, *Verticillium dahliae*

(i) Symptoms: Verticillium wilt is a soil-borne fungal disease that results in the yellowing, and eventual browning and death of foliage, particularly in branches closest to the soil (Fig. 6.14). The wilt starts as yellow, V-shaped areas that narrow at the leaf margins. These yellow areas grow over time, turn brown, and then the leaf dies. Often, entire branches are infected.

Fig. 6.14. Verticillium wilt of tomato

(ii) Integrated Management

(a) Physical and Chemical: In Morocco, solar heating of the soil proved to be efficient in controlling *V. dahliae* on tomato (Besri and Drame, 1982). The mean maximum temperature at 15 cm depth in the solarized plot was 42°C (9°C higher than non-covered control soil), and the mean minimum was 34°C (7°C higher than the control). Maximum temperatures at 15 cm depth in the fine sandy loam plot were 44°C and 36°C in solarized and control plots, respectively.

Factorial ANOVA revealed significant activity of fertilizers against *V. dahliae* in the loam soil, but not in the fine sandy loam (Table 6.6). Fungicidal activity of solarization was highly significant in both soils. A highly significant fertilizer-solarization interaction was found against *V. dahliae* in the loam soil.

Table 6.6. Populations of *Verticillium dahliae* in soil treated with ammonia-based fertilizers and/or solarization[1]

Treatment	Colony-forming units[2]	
	Loam soil	Fine sandy loam soil
Aqua ammonia	5.5	17.4
Aqua ammonia + Solarization	0.1	5.2
Urea	1.6	23.0
Urea + Solarization	0	5.7
Ammonium sulphate	3.0	0.5
Ammonium sulphate + Solarization	0	0
Ammonium phosphate	8.9	14.0
Ammonium phosphate + Solarization	0	0.4
Non-treated Control	9.5	19.2
Solarization	0.1	0
LSD	3.6	13.1
Factorial analysis for variance		
Fertilizers	$P<0.01$	NS
Solarization	$P<0.01$	$P<0.01$
Fertilizers[3] x Solarization[4]	$P<0.01$	NS

1 Fertilizers were applied at 305 kg N/ha.

2 Populations are colony-forming units (cfu)/g of oven-dried soil at 0.15 cm soil depth

3 $P<0.05$ or 0.01; NS = no significant differences.

4 Interaction of treatments.

There were no differences in number of surviving plants in either soil, and no symptoms of *Verticillium* wilt were observed. Factorial ANOVA showed both fertilizers and solarization to be highly significant in increasing all growth parameters in the loam soil; however, no fertilizer-solarization interactions were found (Table 6.7). No significant differences in tomato growth due to fertilizers were found in the sandy loam plot. Solarization, however, was highly significant in increasing numbers of fruit and fruit fresh weight.

Table 6.7. Fresh market tomato ("Early Pak 7") growth in soil treated with ammonia-based fertilizers and/or solarization

Soil Treatment[1]	**Loam soil**			**Fine sandy loam soil**		
	No. of fruits[2]	**Fruit fresh wt.(g)**	**Vegetative fresh wt. (g)**	**Number of fruits**	**Fruit fresh wt. (g)**	**Vegetative fresh wt.(g)**
Aqua ammonia	5.9	90.8	604	22.7	398.3	1386
Aqua ammonia + Solarization	14.0	210.4	1102	30.5	493.6	1384
Urea	8.7	151.8	538	26.0	490.6	1130
Urea + Solarization	14.5	206.4	1 080	30.7	605.0	1246
Ammonium Sulphate	12.3	175.1	690	22.6	404.8	1160
Ammonium Sulphate + Solarization	21.4	382.6	1 582	32.3	552.0	1660
Ammonium Phosphate	18.5	321.5	1 162	27.9	359.7	1754
Ammonium Phosphate + Solarization	26.1	502.7	1 668	28.3	533.0	1766
Non-treated Control	11.7	132.1	638	22.4	360.3	1180
Solarization	15.0	215.7	1 078	33.5	574.4	1448
LSD	8.0	171.9	437	11.9	247.6	670
Factorial analysis of variance	P<0.01	P<0.01	P<0.01	NS	NS	NS
Fertilizers						
Solarization	P<0.01	P<0.01	P<0.01	P<0.01	P<0.01	NS
Fertilizers[3] x Solarization[4]	NS	NS	NS	NS	NS	NS

1 All fertilizers were applied at 305 kg N/ha.
2 All values are given on a per-plant basis.
3 P = 0.05 or 0.01; NS = no significant differences.
4 Interaction of treatments

6.2.2.5. Corky Root, *Pyrenochaeta lycopersici*

(i) Symptoms: A key diagnostic symptom seen in problem fields is the presence of brown, corky bands on the roots that may develop into dark, rotted roots (Fig. 6.15); and loss of small feeder roots. Other symptoms include stunting, slow growth and premature defoliation. On severely affected roots, the outer layer (cortex) can be easily pulled off the root core (stele).

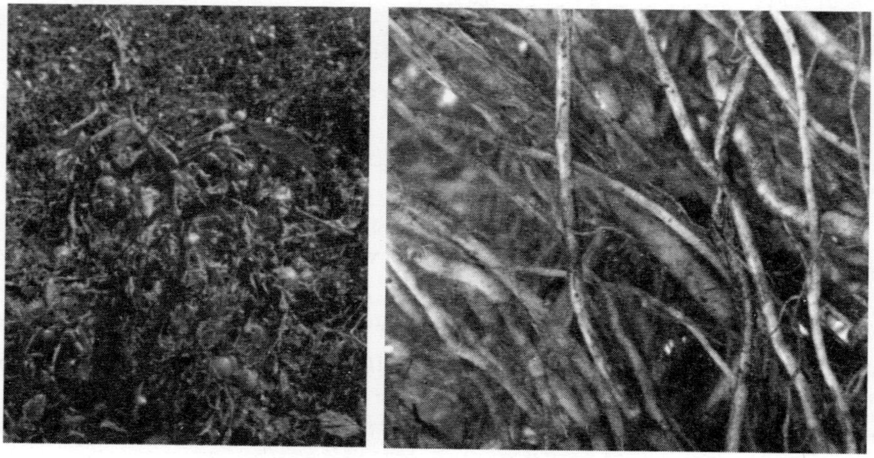

Fig. 6.15. Corky root symptoms on tomato

(ii) Integrated Management

(a) Biofumigation and Solarization: Soil solarization alone or in combination with *Brassica* green manures resulted in a significant ($P < 0.05$) decrease of corky root severity of tomato in all three years of trial (1999 through 2002). *Brassica oleracea* ("Senshi") green manure treatment reduced significantly ($P<0.05$) the disease in two years out of three. *B. rapa* local ecotype caused a significant ($P<0.05$) reduction of corky root, and tomato yield increased in one year of testing out of three. Soil manuring with *B. oleracea* ("Senshi") resulted in a significant ($P < 0.05$) increase of tomato yield in two years out of three.

Immediately after its application, soil-solarization alone or in combination with green manures significantly ($P<0.05$) reduced fluorescent *Pseudomonas* sp. population. However, *Pseudomonas* population increased at the end of tomato cropping in most cases. At the end of tomato cultivation, *Bacillus* sp. population was higher in solarized plots than in non-solarized ones. Corky root severity was negatively correlated with *Bacillus* sp. population density in all trials (Amenduni *et al*., 2004).

6.2.2.6. Bacterial Wilt, *Ralstonia solanacearum*

(i) Symptoms: This is the most serious bacterial disease of solanaceous crops. Characteristic symptoms of the disease are: wilting without leaf yellowing and collapse of the entire plant (Fig. 6.16). The vascular system shows browning. If a segment of the stem is cut and squeezed, bacterial ooze is visible (Fig. 6.16). It causes sudden wilt and ultimately, the plant is killed resulting in total loss.

(ii) Integrated Management

(a) Bioagents, AMF and Botanicals: Even though soil application of *Trichoderma viride* at 5 kg/ha mixed in FYM and *Glomus mosseae* at 500 g/m^2 gave maximum fruit yield (142.1% and 141.1% increase over control), karanj cake along with bleaching powder was more effective in reducing the apparent bacterial growth rate at flowering stage, which is the most critical stage of bacterial wilt, in which

132.2% more yield was obtained which was at par with that of *T. viride* and *G. mosseae* (Table 6.8).

Fig. 6.16. Bacterial wilt affected tomato plant.

Table 6.8. Effect ofbioagents, plant extracts, oil cakes and chemicals on bacterial wilt and yield of tomato cv. Pusa Ruby

Treatment	% reduction of initial inoculum	% decrease over control	% increase in yield over control
Glomus mosseae	41.4	39.41	141.1
Trichoderma viride	29.0	27.52	142.1
Azotobacter + Phosphobacteria	37.3	27.62	89.9
Asafoetida + Turmeric extract	39.3	44.01	71.2
Onion extract	32.7	38.09	71.2
Garlic extract	24.4	9.80	64.4
Karanj cake + Bleaching powder	32.5	34.87	123.4
Bleaching powder + lime	36.2	37.37	107.5
Urea + lime	30.8	22.72	52.6
Control	33.6	---	---

(b) Botanicals and Bioagents: Integration of growing and incorporation of sunnhemp (green manure crop) into the soil and seed treatment with *Pseudomonas fluorescens* (10^{10} cfu/ml) was highly effective in reducing the bacterial wilt incidence to 6.75% and significantly increasing the fruit yield to 25.5 MT/ha as compared to high wilt incidence (55.6%) and low yield (7.8 MT/ha) in control. Thus, the efficacy of *Pseudomonas fluorescens* is increased under green manure incorporated soil against bacterial wilt of tomato (Gopalakrishnan and Ajit Kumar, 2006).

6.2.2.7. Bacterial Speck, *Pseudomonas syringae* pv. *tomato*

(i) Symptoms: The disease starts as small, round, water-soaked spots, which gradually become yellowish brown and finally blackish brown with sunken ashy centre and pale yellowish green halo. Later on, spots become irregular to circular, measuring 1 mm in diameter, distributed over the entire leaf lamina and in severe cases forming large blotches. Spots on petioles are more or less oval.

(ii) Integrated Management

(a) Bioagents and Chemicals: Application of copper fungicide (Kocide), preceding application of the non-pathogenic, copper-resistant mutant of *P. syringae* pv. *tomato*, resulted in greater reduction in disease than either treatment alone (Cooksey, 1988) (Table 6.9).

Table 6.9. Effect of copper-sensitive strain (PT12) and copper-resistant strains (PT23.200 and PT23.201) of *Pseudomonas syringae* pv. *tomato* with and without Kocide on the incidence of bacterial speck on tomato.

Pseudomonas syringae pv. *tomato* strain	Preinoculation treatment with Kocide 101	Lesions/leaflet
PT12 (2 x 10^7/ml)	–	47.9 a
	+	5.6 c
PT12 (2 x 10^7)+ PT23.200 (5 x 10^8/ml)	–	16.3 b
	+	5.6 c
PT12 (2 x 10^7) + PT23.201 (5 x 10^8/ml)	–	7.9 c
	+	2.2 d

6.2.3. Nematodes

6.2.3.1. Root-knot Nematodes, *Meloidogyne* spp.

(i) Economic Importance: *M. incognita* was responsible for 30.57 to 46.92% loss in fruit yield of tomato (Bhatti and Jain, 1977; Reddy, 1985; Darekar and Mahse, 1988), while *M. javanica* caused 77.5% loss in yield (Anon, 1993).

(ii) Symptoms: Above ground symptoms are stunting, yellowing, wilting, reduced yield, and premature death of plants. Below ground symptoms are swollen or knotted roots (root galls) or a stubby root system (Fig. 6.17). Root galls vary in size and shape depending on the nematode population levels, and species of root-knot nematode present in the soil.

(iii) Integrated Management

(a) Botanicals and Bioagents: Application of *Aspergillus niger* and *P. lilacinus* along with mustard cake gave maximum reduction in nematode population both in root and soil with enhanced plant vigour. *A. niger* being toxic agent, kills the second stage juveniles present in the rhizosphere; while *P. lilacinus* being the egg parasite, invade the eggs of *M. incognita* which escaped from the toxicity of *A. niger*. As a result, there is an overall reduction in root and soil population. Further, addition of mustard cake also helps to maintain the general plant health in addition to possessing nematicidal properties (Goswami *et al*., 1998) (Table 6.10).

Fig. 6.17. Heavy galling of tomato roots with *Meloidogyne incognita*.

Table 6.10. Effect of *Aspergillus niger, Paecilomyces lilacinus* and mustard cake on biomass and multiplication of *Meloidogyne incognita* infecting tomato

Treatment	Biomass	No. of galls/plant	No. of egg masses /plant	No. of eggs/egg mass	Nematode popn./500 g soil
Aspergillus niger	9.8	34	18	272	470
Paecilomyces lilacinus	10.2	42	23	262	580
A. niger + *P. lilacinus*	12.2	26	16	264	270
A. niger + *P. lilacinus* + Mustard cake	12.6	22	10	255	290
Control	9.0	98	72	320	2880
CD at 5%	**2.33**	**7.24**	**3.24**	**27.18**	**162.80**

Combined application of *P. penetrans* + *T. viride* + neem/castor cake each at 1/3 doses was significantly superior compared to their individual applications in terms of increased tomato plant growth and reduced root galling, egg mass production and final populations of *M. incognita*. Similarly, total parasitization by bioagents was higher in plants treated with these bioagents in terms of number of juveniles encumbered and adult females infected with *P. penetrans* and egg masses parasitized by *T. viride*, compared to their individual applications (Rangaswamy *et al*., 1999).

The integration of castor leaves with *P. lilacinus* increased the efficiency of nematode control (*M. javanica*) in tomato. The residual effect on the second crop of tomato revealed the reduction in root-knot index (29.6 to 356.8%) and J_2 population in soil (26.4 to 57.9%) (Zaki, 1998). *P. lilacinus* in combination with castor leaves reduced the nematode population up to 89% and increased plant growth and yield in tomato (Zaki and Bhatti, 1991).

In nursery, integration of *Pasteuria penetrans* (at 28 x 10^4 spores/m^2), *P. lilacinus* (at 10 g/m^2 with 19 x 10^9 spores/g) and neem cake (at 0.5 kg/m^2) gave maximum increase in plant growth and number of seedlings/bed. Parasitization of *M. incognita* females was highest when neem cake was integrated with *P. penetrans*, while parasitization of eggs was highest when neem cake was integrated with *P. lilacinus*. In field, planting of tomato seedlings (raised in nursery beds amended with neem cake + *P. penetrans*) in pits incorporated with *P. lilacinus* (at 0.5 g/plant) gave least root galling and nematode multiplication rate and increased fruit weight and yield of tomato (Table 6.11) (Parvatha Reddy *et al*., 1997).

Table 6.11. Effect of neem cake, *Pasteuria penetrans* and *Paecilomyces lilacinus* on root galling and yield of tomato

Treatment		Root-knot index	Yield (kg)/ 6 m^2
Nursery (m^2)	**Main field (per plant)**		
Neem cake -1 kg	*P. lilacinus* – 0.5 g	3.4	9.168
Neem cake -1 kg	*P. penetrans* (28 x 10^4 spores)	3.2	9.312
P. lilacinus – 20 g	*P. penetrans* (28 x 10^4 spores)	3.0	9.504
P. penetrans (28 x 10^7 spores)	*P. lilacinus* – 0.5 g	2.9	9.624
Neem cake – 0.5 kg + *P. lilacinus* – 10 g	*P. penetrans* (28 x 10^4 spores)	2.5	9.672
Neem cake – 0.5 kg + *P. penetrans* (28 x 10^4 spores)	*P. lilacinus* – 0.5 g	2.0	9.984
Neem cake – 0.5 kg + *P. lilacinus* – 10 g + *P. penetrans* (28 x 10^4 spores)	---	2.6	9.600
Control	---	4.6	8.352
CD (P = 0.05)		**0.14**	**0.100**

Soil drenching with 5% Wellgro solution (at 200 ml/seed pan) along with *P. lilacinus* (at 2 x 10^4 spores/ml) gave maximum increase in seedling weight and highest reduction in root galling. Roots of tomato seedlings raised in the above treatment dipped in 5% Wellgro solution mixed with *P. lilacinus* spores (at 2 x 10^4 spores/ml) for 20 min when transplanted in pots gave maximum increase in plant growth, root colonization and parasitization of egg masses of *M. incognita* by the bioagent and highest reduction in root galling and final nematode population both in soil and roots (Rao and Parvatha Reddy, 1993a).

Bare root-dip treatment of tomato seedlings in 10% castor leaf extract mixed with *P. lilacinus* spores (at 1.5 x 10^8 spores /ml) for 20 min significantly increased the plant growth and reduced root galling and final nematode population. The above treatment also gave significant increase in parasitization of eggs and egg masses and propagule density of *P. lilacinus* in roots (Rao *et al.*, 1999). Similarly, root-dip treatment of tomato seedlings in 5 and 10% neem leaf extract mixed with *P. lilacinus* spores (at 6 x 10^4 spores /ml) for 30 min gave significant reduction in root galling and final nematode population. Significant increases in root colonization, parasitization of eggs and egg masses and propagule density of the bioagent in soil were also noticed in the above treatment (Parvatha Reddy *et al.*, 1998).

Application of *Pochonia chlamydosporia* (100 ml/seed pan containing 4 x 10^5 spores/ml) with neem leaves (150 g/seed pan) increased seedling weight and colonization of roots with the bioagent. Tomato seedlings raised in the above treatment transplanted in pots resulted in maximum increase in plant growth, least root galling, nematode population both in soil and roots and highest parasitization of eggs, egg masses and propagule density of *P. chlamydosporia* in roots (Parvatha Reddy *et al.*, 1999).

Incorporation of neem cake (20 g/pot) along with *P. chlamydosporia* (10 ml/pot containing 4 x 10^5 spores/ml) gave maximum increase in plant growth and significant reduction in root galling and nematode population both in soil and roots. The above treatment also gave highest parasitization of eggs and egg masses of *M. incognita* and maximum propagule density of the bioagent in soil and roots (Rao *et al.*, 1998b).

Integration of neem cake (40 g/plant) with *T. harzianum* (at 5 g/plant with 4 x 10^8 spores/g) was effective in increasing the plant growth and reducing root galling and final population of *M. incognita* infecting tomato. The above treatment also gave maximum reduction in number of eggs/egg mass and increased root colonization, spore density in soil and parasitization of adult females with *T. harzianum* (Rao *et al.*, 1997c).

Nursery bed treatment with *P. fluorescens* (with 1 x 10^9 spores/g) and *P. chlamydosporia* (10 ml/pot containing 4 x 10^6 spores/ml) each at 20 g/m^2 and field application of 5 tonnes of FYM enriched with the above bioagents each at 5 kg, significantly reduced root-knot nematodes in tomato by 76% over untreated check. The yield increase was up to 21.7% with benefit: cost ratio of 4.9.

Integration of *P. lilacinus* and *T. viride* with mustard cake gave significant reduction in root galling, egg mass production, fecundity, reproduction factor and nematode population both in soil and roots coupled with increase in plant growth parameters (Tripathi and Singh, 2006).

(b) Botanicals and AMF: Application of calotropis leaves at 400 g/m^2 along with *G. fasciculatum* (250 g/m^2 containing 25-30 chlamydospores/g) in nursery beds gave significant reduction in root galling and fecundity of *M. incognita*. The above treatment also gave maximum increase in plant growth and root colonization with *G. fasciculatum* (Rao *et al.*, 1996). Organic amendments (oil cakes, calotropis leaves) in combination with AMF enhance the colonization of AMF on tomato roots which further increased plant growth and reduced gall index.

Inoculation of *G. mosseae* or *G. fasciculatum* in the nursery beds amended with neem cake/castor cake/neem leaf/calotropis leaf reduced the infestation of root-knot and reniform nematodes to the maximum extent on tomato. Amendment of botanicals in the nursery beds increased the multiplication of endomycorrhizae and colonization of tomato root which in turn could protect the crop from these nematodes in the main field resulting in increased yields (Rao *et al.*, 1996; 1997).

In nursery, integration of neem cake (at 500 g/m^2) with *Glomus mosseae* (at 250 g/m^2 containing 16 chlamydospores/g) significantly reduced *M. incognita* population in soil, root galling, egg mass production, fecundity and produced vigorous tomato seedlings with increased root colonization with *G. mosseae*. The seedlings raised in the above treatment when planted in the main field significantly reduced root galling and increased fruit yield, root colonization with *G. mosseae* and chlamydospore population in soil (Parvatha Reddy *et al.*, 1998; Rao *et al.*, 1995) (Table 6.12).

Table 6.12. Effect of integration of *Glomus mosseae* with oil cakes on root galling and yield of tomato infected with *Meloidogyne incognita*

Treatment	**Dose/plant**	**Root gall index**	**Yield (kg/plant)**
Control	---	4.4	1.44
Castor cake	50 g	3.9	2.00
Karanj cake	50 g	4.1	2.06
Neem cake	50 g	3.6	2.23
G. mosseae	500 cm^3 soil	3.4	2.19
Castor cake + *G. mosseae*	50 g + 500 cm^3 soil/m^3	3.4	2.53
Karanj cake + *G. mosseae*	50 g + 500 cm^3 soil/m^3	3.0	2.69
Neem cake + *G. mosseae*	50 g + 500 cm^3 soil/m^3	2.8	2.98
Carbofuran	0.1 g	3.0	2.33
CD (P = 0.05)	---	**0.27**	**0.43**

(c) Bioagents and AMF: Interestingly, integration of bio-agent (*P. lilacinus*) and endomycorrhizae (*G. mosseae/G. fasciculatum*) has culminated in the successful management of *M. incognita* infecting tomato. This phenomenon facilitated standardization of a strategy wherein inoculation of mycorrhizae and bio-agent in the nursery beds protected the seedlings of tomato from the attack of *M. incognita*. Further, these mycorrhizal seedlings (colonized either with *G. mosseae* or *G. fasciculatum*) can be given a root-dip treatment with spore suspension of *P. lilacinus* for 5-10 minutes for the effective management of nematodes in the main field after transplanting (Rao *et al.*, 1993a).

Integration of *G. deserticola* with *P. chlamydosporia* gave effective management of *M. incognita* in tomato, increased seedling weight, root colonization with *G. deserticola* and *P. chlamydosporia* and parasitization of eggs of *M. incognita* with *P. chlamydosporia* and reduced root galling, egg mass production and fecundity of root-knot nematodes (Rao *et al.*, 1997).

(d) Bioagents, AMF and Botanicals: Inoculation of *G. mosseae* in neem leaf/ neem cake amended nursery beds followed by the root-dip treatment of mycorrhizal seedlings of tomato in spore suspension of *P. lilacinus* gave effective management of *M. incognita* under field conditions (Rao *et al.*, 1995).

In nursery, integration of neem cake (at 500 g/m^2) with *G. fasciculatum* (200 g/m^2 containing 15 spores/g) gave maximum increase in plant growth and highest root colonization with *G. fasciculatum* and least root galling. Integration of neem cake (at 500 g/m^2) with *T. harzianum* (at 100 g/ m^2 with 4 x 10^8 spores/g) gave maximum parasitization of eggs with *T. harzianum*. In field, planting of tomato seedlings raised in nursery beds treated with neem cake + *G. fasciculatum* in pits incorporated with *T. harzianum* (at 0.5 g/ plant with 4 x 10^8 spores/g) was effective in increasing tomato fruit yield, root colonization with *G. fasciculatum* and egg parasitization with *T. harzianum* and significant reduction in root galling and multiplication rate of *M. incognita* (Parvatha Reddy *et al.*, 1998) (Table 6.13).

Table 6.13. Effect of integration of neem cake with *Trichoderma harzianum/Glomus fasciculatum* on root galling and yield of tomato infected with *Meloidogyne incognita*

Treatment	Dose/m^2	Root-knot index	Yield (kg/plant)
Neem cake	1 kg	3.1	1.57
T. harzianum	100 g	3.2	1.43
G. fasciculatum	6000 spores	2.6	1.72
Neem cake + *T. harzianum*	500 g + 50 g	2.4	2.22
Neem cake + *G. fasciculatum*	500 g + 3000 spores	2.0	2.64
Control	---	4.2	1.10
CD (P = 0.05)	---	**0.21**	**0.38**

(e) Bioagents and Chemicals: Maheshwari *et al.* (1987) reported that application of *P. penetrans* in combination with carbofuran, aldicarb, miral, sebufos and phorate significantly improved plant growth of tomato by greatly reducing galling due to *M. javanica*.

Integration of *P. chlamydosopria* with carbofuran recorded maximum plant growth parameters and minimum gall index and nematode population. The above treatment also recorded maximum number of fruits/plant and yield/plant. Maximum parasitization of *M. incognita* with *P. chlamydosopria* was also noticed when the bioagent and the chemical were integrated (Gopinatha *et al.*, 2002).

(f) Two Bioagents: Integration of two bio-agents, *P. lilacinus* and *P. chlamydosopria* resulted in combined and complimentary effects for the successful management of *M. incognita* infecting tomato (Rao and Parvatha Reddy, 1992).

Tomato seedlings raised in nursery beds treated with *P. chlamydosopria* and *P. penetrans* when transplanted in the main field had significantly lower root-knot index, number of eggs per egg mass, nematode population in roots and soil. The above treatment also increased root colonization and egg parasitization with *P. chlamydosporia,* infection of *M. incognita* females with *P. penetrans* and increased tomato fruit yield. This method of integrated nematode management resulted in significant reduction in the amount of inoculum of the bioagents required for the treatment of soil only in the nursery beds (Rao *et al.*, 1998a) (Table 6.14).

Table 6.14. Effect of integration of *Pochonia chlamydosopria* with *Pasteuria penetrans* on root galling, fruit yield and egg parasitization by *P. chlamydosopria* on tomato infected with *Meloidogyne incognita*

Treatment	Root-knot index	Yield (kg) /6 m^2	% eggs parasitized by *P. chlamydosopria*
P. chlamydosopria	5.8	7.4	43
P. penetrans	6.6	6.3	---
P. chlamydosopria + *P. penetrans*	4.7	8.2	57
Control	8.0	5.1	---
CD (P = 0.05)	**0.78**	**0.88**	**4.72**

Maheshwari and Mani (1988) also observed that population of *M. incognita* and *M. javanica* were suppressed effectively and yields of tomato were greater when *P. penetrans* and *P. lilacinus* were applied together.

Combined soil application of *P. lilacinus* and *Aspergillus niger* at the time of transplanting tomato is very effective in reducing root-knot nematodes.

The combination treatment with *T. harzianum* + *T. viride* each at 50 g (4 x 10^8 cfu/g) considerably increased the plant growth, yield and reduced the root galls and soil nematode population (Hassan and Sobita Devi, 2004).

Consortial formulation of biocontrol agents *viz.*, *Pseudomonas fluorescens* Pf 128 and *Bacillus subtilis* Bbv 57 recorded the highest defense enzymatic activity (peroxidase, polyphenol oxidase and phenylalanine ammonia lyase) and lowest nematode population in tomato roots compared to other strains either alone or in combination (Sankarimeena *et al.*, 2012).

(g) Cultural and Chemical Methods: Treatment combination of 3 deep summer ploughings at 10 days interval (20 cm deep) during June (atmospheric temperature between 42-47°C) (92.3 % reduction in *M. javanica* population) + aldicarb treated nursery (0.4 g a.i./m^2) + spot application of aldicarb at 1 kg a.i./ha at transplanting of healthy tomato seedlings of 8-weeks-old gave maximum yield (7.8 kg/plot compared to 3.8 kg/plot in check) (102.6% higher yield over control) with minimum root-knot index (1.3) and final nematode population (85.5 to 93.7% reduction in nematode population) (Jain and Bhatti, 1985).

(h) Solarization and Biofumigation: Tsror *et al.* (2006) reported much improved control of *Meloidogyne* spp. on tomato by combining solarization with biofumigation compared to solarization alone.

Integration of soil solarization of nursery beds (with LLDPE transparent film of 25 μ for 15 days) with soil incorporation of calotropis leaves at 4 kg/1.44 m^2 resulted in increased plant height (29.9 cm compared to 20 cm in control), fresh shoot weight (176.49g compared to 100g in control), transplantable seedlings (546/m^2 compared to

297 in control) with minimum root-knot index (1.83 compared to 3.85 in control) (Patel *et al*., 2006). Soil solarization with clear LLDPE film (25μ) for 15 days in hot summer in combination with poultry manure at 2 t/ha proved effective in the management of nematodes and higher production of transplants in tomato nursery with benefit: cost ratio of 3.10.

(i) Botanicals and Cultural: In commercial greenhouse trials in Spain an integrated management system was developed, including biofumigation with sheep manure and mushroom residue and the cultivation of short-cycle vegetables acting as trap crops. Using this strategy, initial very high levels of *M. incognita* were reduced to near zero in the main susceptible tomato crop (Bello, 1998).

(j) Chemicals, Cultural and Physical: Significant nematode control (75 to 88%) was observed when soil solarization was introduced to the cropping system (a resistant crop preceding tomato), ranging from 40-51 percent. Percent nematode control increased when the soil was chemically treated, and particularly when solarization was included in the system (Table 6.15) (Heald and Robinson, 1987).

Table 6.15. The effect of short-term soil solarization and cropping system on the integrated control of *Meloidogyne* spp. under laizemeter conditions

Cropping system[1]	Egg masses/5g roots (% of control)	Females/5g roots (% of control)	Nematode control (%)
S-T	100	100	0
M-T	41	35	62
R-T	0	0	100
Z-S-T	64	55	40
Z-M-T	29	26	73
Z-R-T	0	0	100
S-Z-T	60	55	43
M-Z-T	23	21	78
R-Z-T	0	0	100
Z-S-Z-T	50	48	51
Z-M-Z-T	18	17	83
Z-R-Z-T	0	0	100
S-Tr	20	30	75
M-Tr	9	17	87
R-Tr	0	0	100
Z-S-Tr	11	21	84
Z-M-Tr	4	6	95
Z-R-Tr	0	0	100
S-Z-Tr	15	25	80
M-Z-Tr	1	10	95

Cropping system[1]	Egg masses/5g roots (% of control)	Females/5g roots (% of control)	Nematode control (%)
R-Z-Tr	0	0	100
Z-S-Z-Tr	10	15	88
Z-M-Z-Tr	3	6	96
Z-R-Z-Tr	0	0	100

1 S = Susceptible host; M = Moderately susceptible host; R = Resistant host; Z = Solarization; T = Tomato; and Tr = Tomato + Chemical treatment.

(k) Botanicals and Chemicals: Integration of chopped castor leaves (40 to 60 g/kg soil) 1 or 2 weeks before transplanting with application of aldicarb or carbofuran each at 2 kg a.i./ha at transplanting, significantly reduced number of galls due to *M. javanica* and enhanced the growth of tomato (Dutt and Bhatti, 1986).

In tomato, application of aldicarb and carbofuran each at 1 kg a.i. per ha in combination with neem cake and urea each at 10 kg N per ha at transplanting, produced maximum yield with lowest gall index (2.5) and nematode population, 90 days after planting (Routaray and Sahoo, 1985).

Integration of soil application of castor leaves with inorganic fertilizer (75 kg N/ha) enhanced plant growth of tomato and reduced *M. javanica* infestation (95% and 78% reduction in root galls and egg mass production, respectively) (Zaki and Bhatti, 1989).

Application of carbofuran at 1 kg a.i./ha in the nursery beds followed by neem cake at 400 kg/ha in the main field increased yield and reduced the gall index (Singh and Gill, 1998) (Table 6.16).

Table 6.16. Integrated management of root-knot nematodes in tomato using botanicals and chemicals

Treatment/Dose	% Reduction in root galling	% increase in yield
Carbofuran (1 kg a.i./ha) + Neem cake at 400 kg/ha	77.00	61.50
Carbofuran (1 kg a.i./ha) + Urea (23.8 kg/ha) + Neem cake at 200 kg/ha	67.90	49.20
Phenamiphos (1 kg a.i./ha) + Neem cake at 400 kg/ha	74.90	39.70
Phenamiphos (1 kg a.i./ha) + Urea (23.8 kg/ha) + Neem cake at 200 kg/ha	67.90	31.43

(l) Physical and Chemical: Addition of ammonium phosphate fertilizer to loam soil significantly reduced galling of tomato roots by *M. incognita*. Galling index generally decreased with increasing fertilizer dosage. Soil solarization along with ammonium phosphate fertilizer further reduced nematode galling in all cases (Table 6.17).

Factorial ANOVA showed both fertilizers and solarization to be highly significant in increasing all growth parameters in the loam soil; however, no fertilizer-solarization interactions were found (Tables 6.17).

Table 6.17. Effect of ammonium phosphate fertilization and soil solarization on galling of tomato roots by *Meloidogyne incognita*

Treatment and dosage	Root galling[1]	Number of fruits[2]	Fruit fresh wt. (g)	Vegetative fresh wt. (g)
Ammonium phosphate-100 mg/1	2.0	5.9	90.8	604
Ammonium phosphate-100 mg/1 + Soil solarization	0.8	14.0	210.4	1102
Ammonium phosphate-200 mg/1	1.0	8.7	151.8	538
Ammonium phosphate-200 mg/1 + Soil solarization	0.3	14.5	206.4	1080
Ammonium phosphate-300 mg/1	0.9	12.3	175.1	690
Ammonium phosphate-300 mg/1 + Soil solarization	0.2	21.4	382.6	1582
Ammonium phosphate-400 mg/1	0.7	18.5	321.5	1162
Ammonium phosphate-400 mg/1 + Soil solarization	0.0	26.1	502.7	1668
Control	3.5	11.7	132.1	638
Soil solarization only	1.8	15.0	215.7	1078
LSD	0.7	8.0	171.9	437
Factorial ANOVA Fertilizer	0.01[2]	0.01[3]	0.01	0.01
Soil solarization	0.01	0.01	0.01	0.01
Fertilizer x Soil solarization	NS[3]	NS	NS	NS

1 Based upon 0-4 rating scale.
2 Significance level.
3 NS = no significant difference.

The mean maximum temperature at 15 cm depth in the solarized plot was 42°C (9°C higher than non-covered control soil), and the mean minimum was 34°C (7°C higher than the control). Maximum temperatures al 15 cm depth in the fine sandy loam plot were 44°C and 36°C in solarized and control plots, respectively.

6.2.3.2. Reniform Nematode, *Rotylenchulus reniformis*

R. reniformis was responsible for 42.25 to 49.02% loss in fruit yield of tomato (Subramanyam *et al*., 1990).

(i) Symptoms: General symptoms include reduced root systems, leaf chlorosis, overall stunting of host plants, and reduced yields and plant longevity. Female nematodes and their eggs are often visible when plant roots are viewed under a dissecting microscope (Fig. 6.18).

Fig. 6.18. Tomato root infected with *Rotylenchulus reniformis*.

(ii) Integrated Management

(a) Botanicals and Chemicals: Application of neem cake in the nursery at 100 g/m^2 followed by carbofuran at 1 kg a.i./ha in the main field significantly reduced the soil population of *R. reniformis* and enhanced the fruit yield of tomato by 67% (Anitha and Subramanian, 1998).

(b) Bioagents and Chemicals: Integration of a bio-agent, *P. lilacinus* with carbofuran at 1 kg a.i./ha was found effective in the management of reniform nematode, *R. reniformis* infecting tomato (Parvatha Reddy and Khan, 1988) (Table 6.18).

Table 6.18. Effect of *Paecilomyces lilacinus* and carbofuran on *Rotylenchulus reniformis* infecting tomato

Treatment	Dose	Mature females in root/plant	Total nema population/ plant	Reproduction factor	% males
P. lilacinus	0.5 g/plant	29	3893	3.8	45.2
P. lilacinus	1.0 g/plant	9	3257	3.3	53.1
P. lilacinus	2.0 g/plant	14	2232	2.2	49.2
P. lilacinus + Carbofuran	2.0 g/plant + 2 kg a.i./ha	9	1537	1.5	56.1
Carbofuran	2 kg a.i./ha	8	1160	1.2	65.3
Control	---	57	6669	6.7	42.8
CD (P = 0.05)	---	**5.8**	**2011**	---	**15.9**

(c) Bioagents and Botanicals: Nursery bed treatment with *P. fluorescens* (with 1×10^9 spores/g) and *P. chlamydosporia* (with 1×10^6 spores/g) each at 20 g/ m^2 and field application of 5 tonnes of enriched FYM with the above bioagents each at 5 kg significantly reduced reniform nematodes in tomato by 72% over control. The yield increase was up to 21.7% with benefit: cost ratio of 4.9.

6.2.3.3. Root-knot Nematode, *Meloidogyne incognita* and Wilt, *Fusarium oxysporum* f. sp. *lycopersici* Disease Complex

(i) Symptoms: The effect of the nematode in combination with the fungus enhanced the suppression of plant growth than that of the fungus alone. Inoculation of the nematode and fungus exhibited a synergistic effect on growth retardation of plants. Maximum reduction in plant height (33.08 cm) was observed when nematode and fungus were inoculated simultaneously.

Jenkins and Coursen (1957) induced wilting in *Fusarium* wilt-resistant tomato variety 'Chesapeake' only when root-knot nematodes were present along with fungal inoculum. Furthermore, when *M. hapla* was combined with the fungus, only 60% of the plants wilted, whereas *M. incognita acrita* promoted wilt in 100% of the plants (Fig. 6.19).

Fig. 6.19. The simultaneous occurrence of both root-knot nematode (*Meloidogyne* spp.) and Fusarium wilt causing enhanced disease development and tomato yield loss.

(ii) Integrated Management

(a) Bioagents and Botanicals: Tomato roots that received *P. lilacinus* along with *T. harzianum* and neem cake were free from root-knot nematodes (*M. incognita*) and did not wilt due to *F. o.* f. sp. *lycopersici* till harvest. The roots were also free from *Fusarium* infection. The above treatment also reduced the percentage of wilt (10% compared to 90% in control), root-knot index (1.8 compared to 4.4 in control) and increased root colonization with bioagents (74% compared to 0% in control), parasitization of egg masses (56% compared to 0% in control) and eggs/egg mass (44% compared to 0% in control) (Nagesh *et al.*, 2006) (Table 6.19)).

Table 6.19. Effect of integration neem cake and bioagents on the management of *Meloidogyne incognita* infecting tomato

Treatment	Healthy plants (%)	Root-knot index	Colonization by bioagents		
			Root Colonization (%)	Colonization of egg masses/eggs (%)	
				egg masses	eggs
Control	10	4.4	---	---	---
P. lilacinus	30	3.6	48	48	46
T. harzianum	40	4.0	50	44	32
Neem cake	10	4.2	---	---	---
P. lilacinus + Neem cake	60	2.8	62	62	58
T. harzianum + Neem cake	70	3.8	68	36	32
P. lilacinus + *T. harzianum* + Neem cake	90	1.8	74	56	44
CD (P = 0.05)	**8.87**	**1.78**	**3.66**	**4.11**	**3.22**

(b) Bioagents, Cultural and Host Resistance: Deep ploughing and exposing soil to hot sun in summer, removal and burning of crop debris, soil application of *T. viride* and *P. lilacinus*, use of wilt resistant varieties like Utkal Pallavi, Utkal Deepti, Utkal Kumari, Utkal Urbasi, etc. help in controlling the disease complex.

(c) AMF and Botanicals: Bhagawati *et al.* (2000) demonstrated that although the mustard cake and AM fungus, *Glomus etunicatum* were effective in reducing the damage caused by *M. incognita* and *F. o.* f. sp. *lycopersici* on tomato, the performance of concomitant application of both (bioagent and botanical) was much better than the individual application.

6.2.3.4. Root-knot Nematode, *Meloidogyne* spp. and Root Rot, *Rhizoctonia solani* Disease Complex

(i) Symptoms: Abu-Elamayem *et al.* (1978) observed that damping-off of tomato was more severe in soil infested with both *M. javanica* and *R. solani* than with the fungus alone. *M. javanica* increased the extent of damage by pre- and post-emergence phases of damping off caused initially by *R. solani* in tomato.

M. incognita and *R. solani* are also frequently associated with tomato causing considerably greater damage to this crop (Haseeb, 2003). Highest reduction in fruit yield and plant height was observed in plants inoculated with nematode-fungus simultaneously followed by nematode 7 days prior to fungus. The highest root-knot index (3.5) was observed in plants inoculated with nematode alone followed by nematode inoculation 7 days prior to the fungus (Table 6.20). The highest root infection by the fungus was observed in simultaneous inoculation followed by nematode inoculation 7 days prior to the fungus (Kumar and Haseeb, 2009). The fact that prior inoculation of nematode caused more damage indicates that the roots are predisposed by *M. incognita* for subsequent damage by *R. solani*.

Table 6.20. Effect of *Meloidogyne incognita* and *Rhizoctonia solani* on plant growth and fruit yield of tomato cv. K-25.

Treatment	Fruit yield (kg/plant)	Plant height (cm)	Root-knot index	% root infection by *R. solani*
Untreated uninoculated	2.530	67.5	---	---
M. incognita (4000 J_2/4 kg soil)	1.280	50.8	3.50	---
R. solani (10g mycelial mat/4 kg soil)	1.440	53.7	---	30.0
M. incognita + *R. solani* simultaneously	0.487	26.7	2.00	63.5
M. incognita 7 days prior to *R. solani*	0.658	31.8	2.50	55.0
R. solani 7 days prior to *M. incognita*	0.821	40.0	1.35	48.5
CD (P = 0.05)	**---**	**2.41**	**0.13**	**2.65**

(ii) Integrated Management

(a) Two Bioagents: Combined application of *Pseudomonas aeruginosa* and *Paecilomyces lilacinus* significantly suppressed soil-borne root-infecting fungi such as *Macrophomina phaseolina, F. oxysporum, F. solani, R. solani* and *M. javanica*. *P. lilacinus* parasitized eggs and female of *M. javanica* and this parasitism was not significantly influenced in the presence of *P. aeruginosa* (Siddiqui *et al*., 2000).

P. aeruginosa-Bacillus subtilis treatment was the most effective in the suppression of root-rot disease complex with enhancement of plant growth (Siddiqui and Ehteshamul-Haque, 2001).

6.2.3.5. Root-knot Nematode, *Meloidogyne incognita* and Bacterial Wilt, *Ralstonia solancearum* Disease Complex

Pani and Das (1972) have reported the association of root-knot nematode with bacterial wilt of tomato.

(i) Symptoms: Haider *et al.* (1987) reported the significant reduction in the root-knot index and larval development of *M. incognita* in soil where *R. solanacearum* was present. *R. solanacearum* and *M. incognita* alone as well as in different combinations reduced plant growth and yield significantly with the nematode followed by the bacterium combination showing the maximum reduction in growth.

Napiere (1980) and Napiere and Quinio (1980) found that wilt disease development occurred earlier and with a higher mortality rate in both wilt-resistant and susceptible tomato cultivars grown in *R. solanacearum* and *M. incognita* – infested soil (Fig. 6.20).

(ii) Integrated Management

(a) Bioagents and Botanicals: Treatment of nursery bed with *Pseudomonas fluorescens* (10^9 cfu/g) and *Trichoderma harzianum* (10^6 cfu/g) each at the rate of 20g/m^2 and subsequent application of 5 MT of farm yard manure enriched with 5 kg each of *P. fluorescens* (10^9 cfu/g) and *P. lilacinus* (10^6 cfu/g) per hectare, significantly reduced *R. reniformis* and *M. incognita* in tomato roots by 74% and 70%, respectively; reduced the incidence of bacterial wilt; and increased the yield by 24.2%. Benefit: cost ratio (calculated for the additional cost of the bio-pesticides and additional returns accrued by the application of the bio-pesticide) was 4.4 (Rao *et al.*, 2009).

Fig. 6.20. Root-knot nematode and bacterial wilt disease complex

6.2.4. Validated IPM Technology for Tomato

6.2.4.1. Bangalore, Karnataka

Nursery

- Prepare raised seed bed of 15 cm height.
- Solarize soil for 3 weeks using transparent polythene sheet of 45 μm thickness.
- Grow leaf curl resistant hybrid/varieties like Avinash-2.
- Treat seed with *Trichoderma viride* at 4 g/kg of seed.
- Use nylon nets to prevent entry of white fly, etc.

Main field

- Apply neem cake at 250 kg/ha while planting or 20 DAP.
- Give wide spacing of 90 x 60 cm.
- Dip seedling roots in imidacloprid at 0.5 ml/l for 15 min before transplanting.
- Transplant 1 row of African marigold as trap crop for *Helicoverpa armigera* after every 14 rows of tomato.
- Spray imidacloprid/thiomethoxam at 0.5 ml/l at 15 DAP for white fly.
- Spray 5% NSKE at 15 DAP against leaf miner.

- Install pheromone traps at 5/ha 20 DAP for monitoring of *H. armigera*.
- Monitor top 3 leaves for *H. armigera* eggs at flowering.
- Release of *Trichogramma pretiosum* at 1 lakh/ha 6 times after appearance of adults.
- Spray *Ha*NPV at 250 LE/ha or neem soap 3 times at 28, 35 and 42 DAP.
- Regularly collect and destroy damaged fruits.
- Spray endosulfan at 650 g a.i./ha against *H. armigera*.
- Destroy leaf curl and wilt affected plants.
- Spray mancozeb/captan at 0.2% for the control of early and late blight.

6.2.4.2. Varanasi, Uttaranchal

Nursery

- Prepare raised seed bed of 15 cm height.
- Solarize soil for 3 weeks using transparent polythene sheet of 45 µm thickness.
- Treat seed with *Trichoderma viride* at 4 g/kg of seed.
- Spray 0.2% copper oxychloride.

Main field

- Spray imidacloprid at 0.5 ml/l at 15 DAT.
- Install pheromone traps at 5/ha, for monitoring of *H. armigera*.
- Spray *Ha*NPV at 250 LE/ha.
- Release of *Trichogramma bactrae* at 1 lakh/ha 6 times.
- Regularly collect and destroy damaged fruits.
- Apply pesticides like thiomethoxam and mancozeb at 0.2% based on needs.

6.2.4.3. Ranchi, Jarkhand

Nursery

- Prepare raised seed bed of 15 cm height.
- Solarize soil for 3 weeks using transparent polythene sheet of 45 µm thickness.
- Grow leaf curl resistant hybrid/varieties like Avinash-2.
- Treat soil with FYM enriched with *Trichoderma viride* at 1 kg/MT.

Main field

- Spray imidacloprid at 0.5 ml/l at 15 DAT.

- Install pheromone traps at 5/ha, for monitoring of *H. armigera*.
- Monitor top 3 leaves for *H. armigera* eggs at flowering.
- Release of *Trichogramma chilonis* at 1 lakh/ha 6 times after appearance of adults.
- Spray *Sl*NPV at 250 LE/ha twice at 15 days interval or *Ha*NPV for *H. armigera*.
- Regularly collect and destruction damaged fruits from time to time.
- Apply pesticides like endosulfan at 0.07% based on needs.

During the period 2001-04, IPM technology in tomato was validated and promoted in more than 40 ha area in 42 villages covering 88 families located 40 km from Bangalore. Similarly, near Varanasi IPM technology has been validated in 8 villages in about 40 ha area covering 100 families. Near Ranchi, IPM technology has been validated and promoted in 20 villages with the support of 100 farming families covering an area of 40 ha together. In IPM validation studies conducted at three locations (Bangalore, Varanasi and Ranchi), IPM fields recorded higher tomato fruit yields of 74.038, 14.250 and 22.293 MT/ha as compared to 45.056, 13.000 and 18.772 MT/ha in non-IPM fields, respectively (Sardana *et al.*, 2004) (Table 6.21). The cost of production of tomato by IPM was Rs.0.95/kg as against Rs.2.30 in non-IPM plots.

Table 6.21. Yield and economics of IPM in tomato at different locations

Centre	Yield (MT/ha)	Net returns (Rs.)	Benefit : Cost ratio
Bangalore - IPM	74.03	249721	4.82
- Non IPM	45.05	69704	0.61
Varanasi - IPM	14.25	39917	3.30
- Non IPM	13.00	38167	2.02
Ranchi - IPM	22.29	56705	1.87
- Non IPM	18.77	41776	1.32

Khaderkhan *et al.* (1998) observed that the farmers who adopted the IPM technology sprayed 8 times and spent Rs. 6,628/ha, where as the non-adopters applied 17 sprays and spent Rs. 11,362/ha. Thus, on an average non-adopters spent Rs. 4,734/ha extra money. Net returns in the farmers practice was Rs. 47,359/ha with a benefit: cost ratio of 1.08 as compared to Rs. 60,168/ha net profit and 1.53 benefit: cost ratio by IPM followers (Table 6.22)

Table 6.22. Economic analysis of IPM practice in tomato

Practice	Yield (t/ha)	Gross returns (Rs/ha)	Net returns (Rs/ha)	Benefit: cost ratio
Farmers practice (Non-IPM adopters)	49.400	91,375	47,359	1.08
IPM practice (IPM adopters)	62.280	99,450	60,168	1.53

6.3. BRINJAL, *SOLANUM MELONGENA*

6.3.1. Insect Pests

6.3.1.1. Shoot and Fruit Borer, *Leucinodes orbonalis*

This is the most important pest on brinjal not only in India, but also in all South Asian Countries. This pest has developed resistance against all groups of insecticides and management is very difficult. Often the extent of damage due to this pest reaches up to 70-80%. Damage is very severe during rainy season and early winter.

(i) Damage: Caterpillars feed inside the tender shoots before flowering and cause wilting of the affected shoots (Fig. 6.21). Later, the larvae bore into flowers, flower buds and the grown up larvae migrate and bore fruits contaminating them with excreta (Fig. 6.21). When the incidence is high, unopened flower buds swell and harbour the borer. Just before pupation, the grown up larvae come out of the fruits and flower buds to pupate in silky cocoons on plant parts or debris.

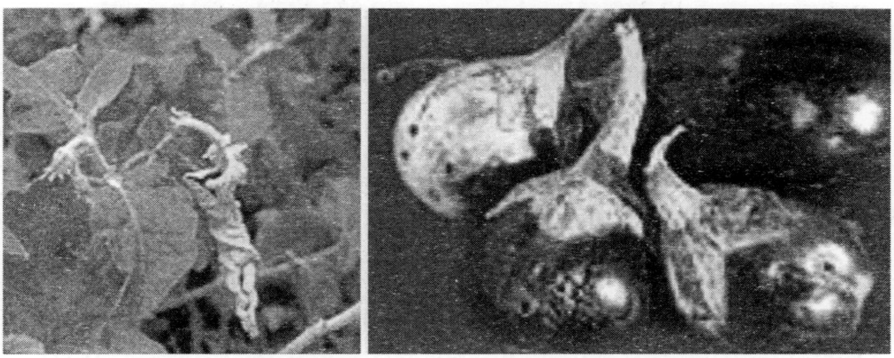

Fig. 6.21. Shoot and fruit borer damage on brinjal.

(ii) Integrated Management

(a) Botanicals and Cultural: Combination of clipping of shoots affected by the borer at weekly interval followed by neem cake application (250 kg/ha) at 30 days after planting and 4% pulverized NSPE/1% neem soap/1% pongamia soap sprays at 60, 75, 90 and 105 days after planting (DAP) was most effective with 22% borer damage as compared to 46% in control (Krishnamoorthy and Krishna Kumar, 2002) (Table 6. 23).

Table 6.23. Incidence of shoot and fruit borer of brinjal under different treatments

Treatments	Cumulative % of borer infestation	Yield (kg/plot)
Soil appln. of neem cake at 250 kg/ha at 30, 60 and 90 DAP.	30.05	5.96
Shoot clipping at weekly interval + neem cake at 30 DAP + 4% NSKE spray at 60, 75, 90 and 105 DAP.	22.04	6.72

Barrier crop of maize sown 10 days before brinjal planting (2 rows at 30 cm from brinjal and maize at 5 cm form plant to plant) + shoot clipping at weekly interval + 4% NSKE spray at 45, 60, 75 and 90 DAP.	23.51	6.53
Barrier crop as above + shoot clipping + endosulfan 700 g a.i./ha alternated with cypermethrin at 50 g a.i./ha at 45, 60, 75 and 90 DAP.	29.44	9.01
Cypermethrin at 50 g a.i./ha at 45, 60, 75 and 90 DAP.	36.62	9.59
4% NSKE spray at 45, 60, 75 and 90 DAP.	28.54	14.33
Control.	45.90	5.33
CD (P = 0.05).	**2.71**	**3.40**

(b) Bioagents and Chemicals: Minimum brinjal fruit infestation by *Leucinodes orbonalis* (3.74%) and maximum fruit yield (16.410 MT/ha) was recorded in treatment *Bacillus thuringiensis* var. *kurstaki* (*Btk*) + methomyl (1.0 ml + 0.8 g/litre of water) as against 12.34% fruit damage and 9.977 MT/ha fruit yield in untreated control. Treatment of *Btk* + endosulfan (1.0 ml + 0.75 ml/litre of water) was on par with the above treatment (Qureshi *et al.*, 1998) (Table 6.24).

Table 6.24. Bioefficacy of *Bacillus thuringiensis* var. *kurstaki* (*Btk*) and insecticides for the management of *Leucinodes orbonalis* on brinjal

Treatment	**Dose/litre of water**	**Fruit damage (%)**	**Fruit yield (MT/ha)**
Btk (Dipel 8 L)	1.5 ml	10.33	10.549
Btk (Dipel 8 L)	2.0 ml	8.78	12.073
Btk + Endosulfan 35 EC	1.0 ml + 0.75 ml	4.74	15.666
Btk + Methomyl 40 SP	1.0 ml + 0.8 g	3.74	16.410
Endosulfan 35 EC	1.5 ml	6.10	12.939
Methomyl 40 SP	1.6 g	5.60	14.983
Control (Untreated check)	---	12.34	9.977
CD (P = 0.05)	**---**	**3.07**	**2.000**

B. thuringiensis (Dipel) in combination with carbaryl (Baskaran and Kumar, 1980) or methomyl (Quereshi *et al.*, 1998) was found to be more effective in reducing borer damage in brinjal fields and giving maximum fruit yield. *B. thuringiensis* var. *kurstaki* at 1 ml + methomyl at 0.8 g/l of water was the best treatment in reducing the brinjal fruit damage to 3.74% with a higher fruit yield of 16.41 MT/ha. *Btk* at 1 ml + endosulfan at 0.75 ml/ l of water was also effective in reducing fruit damage to 4.74% (Quereshi *et al.*, 1998).

Setting of pheromone traps integrated with release of egg parasitoid *Trichogramma chilonis* at 5.0 lakh/ha from flower initiation, and spraying of *Bacillus thuringiensis* var. *krustaki* at 2.0ml/L once in ten days interval (a total of five sprays were given) recorded a mean infestation of 2.5 per cent, followed by 3.1 per cent fruit damages in treatment involving setting of pheromone trap integrated with weekly releases of egg parasitoid *T. chilonis*. A mean of 27.4% fruit damage was recorded when pheromone alone was erected (Ganga Visalakshy and Krishnamoorthy, 2009).

(c) Botanicals and Chemicals: Neem oil at 4% recorded the minimum infestation of brinjal shoot and fruit borer (9.07%) which was on par with endosulfan 0.07% + neem oil 2% (9.49%), endosulfan 0.07% (9.56%), endosulfan 0.07% + NSKE 5% (9.98%) and NSKE 5% (10.89%). The maximum fruit yield was obtained in neem oil 4% (24.48 MT/ha) which was on par with endosulfan 0.07% + neem oil 2% (23.53 MT/ha), carbaryl 0.05% (23.15 MT/ha) and endosulfan 0.07% (23.13 MT/ha) (Raja *et al*, 1998a) (Table 6.25).

Table 6.25. Effect of neem products and insecticides on the shoot and fruit borer infestation and yield of brinjal

Treatment	Fruit borer damage (%)	% decrease over control	Marketable yield (MT/ha)	% increase over control
Endosulfan 0.035%	16.6	61.38	20.15	54.28
Endosulfan 0.07%	9.56	77.76	23.13	77.11
Carbaryl 0.05%	10.89	74.66	23.15	77.26
Neem oil 2%	14.18	67.01	21.60	65.39
Neem oil 4%	9.07	78.90	24.48	87.44
NSKE 5%	10.89	74.66	21.85	67.30
Endosulfan 0.07% + Neem oil 2%	9.49	77.92	23.53	80.17
Endosulfan 0.07% + NSKE 5%	9.98	76.78	21.86	67.38
Trichogramma chilonis release at fortnightly interval	16.18	62.35	20.30	55.44
Control	42.98	---	13.06	---
CD (P = 0.05)	**2.83**	---	**1.61**	---

(d) Two Bioagents: Spraying *Bacillus thuringiensis* formulation (1%) at weekly interval followed by release of *Trichogramma chilonis* at 2,50,000/ha (50,000/ release-5 times at weekly intervals, starting from flowering) was found to reduce the borer incidence.

6.3.1.2. Epilachna Beetle, *Henosepilachna vignitioctopunctata*

(i) Damage: Both grubs and adult beetles scrape the leaves in semi-circular or half moon shaped fashion (Fig. 6.22). Heavy infestation results in leaf skeletonization. Pupation takes place on the plant itself.

(ii) Integrated Management

(a) Botanicals and Chemicals: Neem oil 2% + endosulfan 0.035% reduced the epilachna grub population by 57.3% and was comparable to endosulfan 0.07% alone (53.9%). The maximum fruit yield was obtained in endosulfan 0.07% alone (5.99

MT/ha) which was on par with endosulfan 0.07% + neem oil 2% (5.64 MT/ha) (Rajendran, 1998) (Table 6.26).

Fig. 6.22. Epilachna beetle and grub damage on brinjal leaf.

Table 6.26. Effect of neem oil and endosulfan on epilachna beetle on brinjal

Treatment	Epilachna grubs/5 plants	% reduction over control	Fruit yield (MT/ha)
Neem oil 1%	8.75	20.2	4.12
Neem oil 2%	7.90	27.3	4.56
Neem oil 3%	6.29	41.1	4.98
Neem oil 4%	5.30	49.2	5.49
Neem oil 2% + Endosulfan 0.035%	4.60	57.3	5.64
Endosulfan 0.035%	4.70	53.9	5.99
Teepol	10.30	13.8	4.41
Control	11.30	11.3	4.24
CD (P = 0.05)	**0.48**	---	**0.62**

6.3.2. Diseases

6.3.2.1. Damping-off, *Pythium aphanidermatum*

(i) Symptoms: The pathogen causes pre- and post-emergence damping off in the nursery beds. In Pre-emergence phase of disease, the young seedlings are killed before they emerge through the soil surface. In fact, the seeds may rot or the seedlings may be killed before the hypocotyl has broken the seed coat. The radical and the plumule, when they come out of the seed, undergo complete rotting. Since this happens under the soil surface, the disease is often not recognized at all by the farmer, who attributes the failure of emergence to poor quality of the seed.

The post-emergence mortality of seedlings is generally very conspicuous. This phase of the disease is characterized by the toppling over of infected seedlings any time after they emerge from the soil (Fig. 6.23) until the stem has hardened sufficiently to resist invasion. Infection usually occurs through the roots or at the ground level. The infected tissue appears soft, stained and water soaked. As the disease advances, the stem become constricted at the base and the plant collapse. Seedlings that are apparently healthy one day may have collapsed by the following morning. Generally, the cotyledons and leaves wilt slightly before the seedlings are prostrated, although sometimes they remain green and turgid until collapse occurs. In fields and nurseries, the disease usually radiates from initial infection points, causing large spots or areas in which nearly all the seedlings are killed.

Fig. 6.23. Damping-off symptoms on brinjal seedlings

(ii) Integrated Management

(a) Physical and Bioagents: Combination of the seed/root application of *T. harzianum* or *P. fluorescens* with soil solarization was very effective for the management of damping-off in brinjal nursery at farmers' field.

(b) Physical, Chemical and Bioagents: Soil solarization gave least damping-off followed by seed treatment with captan + soil drenching with captan which was on par with seed treatment with *Trichoderma viride* + soil application of FYM enriched with *T. viride* (Table 6.27) (Rahman *et al.*, 2002).

Table 6.27. Comparative efficacy of soil solarization, bioagents and fungicide against damping-off of brinjal

Treatment	% Disease index
Soil solarization with transparent polythene sheets for 30 days	1.50 (6.93)
Seed treatment with *Trichoderma viride* at 4g/kg + soil application of 50 kg FYM enriched with 500g of *T. viride*	4.74 (12.40)
Seed treatment with *Pseudomonas fluorescens* at 10g/kg + soil application of 50 kg FYM enriched with 2.5 kg of *P. fluorescens*	7.20 (15.37)

Seed treatment with *Trichoderma harzianum* at 4g/kg + soil application of 50 kg FYM enriched with 500g of *T. harzianum*	6.31 (14.40)
Seed treatment with *Azotobacter croococcum* at 16g/kg + soil application of 50 kg FYM enriched with 500g of *A. croococcum*	10.88 (19.17)
Seed treatment with captan at 2.5g/kg + soil drenching with captan at 0.25% at 6L/m^2	4.38 (11.55)
Control (Check)	42.82 (29.74)
CD at 5%	**(3.13)**

Figures in parentheses indicate the arc sin transformed values

6.3.2.2. Fusarium Wilt/Root Rot, *Fusarium oxysporum, F. solani*

(i) Symptoms: Affected plants show yellowing of leaves that progressively wilt and die from bottom upwards (Fig. 6.24). Woody stem and root tissue of diseased plants turn brown.

Fig. 6.24. Fusarium wilt on brinjal

(ii) Integrated Management

(a) Physical and Chemical/Botanical/Bioagent: The effect of solarization, and its combinations with dazomet (400 kg/ha), chicken manure (10 tonnes/ha) or *Trichoderma*, on soil-borne diseases (*Fusarium* spp.) were detected in eggplant. Solarization with dazomet and solarization and manure gave better disease control compared to other treatments (Table 6.28).

Table 6.28. Effect of integration of soil solarizaton, chemical, organic amendment and bioagent on disease incidence in eggplant

Treatment	% Disease
Soil solarization + Dazomet	20
Soil solarization + Chicken manure	20
Soil solarization + *Trichoderma*	31.2
Check	57.5

6.3.2.3. Collar Rot, *Sclerotinia sclerotiorum*

(i) Symptoms: Infection takes place on leaves, twigs, flowers and fruits (Fig. 6.25). Water-soaked lesions develop and the infected tissue is macerated by the pathogen. White mass of mycelium grow on the surface and later embedded sclerotia are formed (Fig. 6.25). These sclerotia become black after drying. The pathogen is soil-borne. The disease affects plant population, yield and quality of the fruits.

Fig. 6.25. Collar rot symptoms on brinjal

(ii) Integrated Management

(a) Physical and Chemical/Botanical/Bioagent: The effect of solarization, and its combinations with dazomet (400 kg/ha), chicken manure (10 tonnes/ha) or *Trichoderma*, on soil borne diseases (*Sclerotinia sclerotiorum*) was observed in the field. Solarization + *Trichoderma* controlled diseases better than solarization + dazomet or solarization + manure (Table 6.29).

Table 6.29. Effect of integration of soil solarizaton, chemical, organic amendment and bioagent on disease incidence in eggplant

Treatment	% Disease
Soil solarization + Dazomet	10.5
Soil solarization + Chicken manure	11.2
Soil solarization + *Trichoderma*	9.1

(b) Physical and Bioagents: Summer ploughing and seedling dip treatment with *T. viride* effectively controlled the disease incidence at all the growth stages up to the maturity of crop (185 to 205 DAT) (Jadon, 2009).

6.3.2.4. Root Rot, *Rhizoctonia solani*

(i) Symptoms: Generally early transplanted brinjal crop is much affected during the months of August-September. Lesions start on stem near collar region at or below the soil level and expand downward into the roots. Initially the bark become wet, soft with macerated tissue. Later on, drooping and wilting of the plant is observed. The disease is more common in poorly drained soil and the fields having prolonged excessive moisture.

(ii) Integrated Management

(a) Bioagents and Chemicals: Combined treatments of the fungicide PCNB and *T. harzianum* decreased the inoculum potential of *R. solani* and increased the disease control in comparison to separate treatments for the control of damping-off of eggplant (Hadar *et al.*, 1979).

6.3.3. Nematodes

6.3.3.1. Root-knot Nematode, *Meloidogyne* spp.

M. incognita was responsible for 27.30 to 48.55% loss in fruit yield of brinjal (Bhatti and Jain, 1977; Parvatha Reddy and Singh, 1981; Darekar and Mahse, 1988).

(i) Symptoms: Affected plants are normally stunted and eventually wilt and die. The most characteristic symptom is formation of root galls (knots) and these can be seen with the naked eye (Fig. 6.26). The infested roots eventually rot and affected plants die.

(ii) Integrated Management

(a) Bioagents and Botanicals: Integration of *P. chlamydosporia* (at 100 ml/ seed pan containing 1.2 x 10^4 spores/ml) in castor cake (at 40 g/seed pan) amended soil was effective in increasing the seedling weight and colonization of roots with the bioagent.

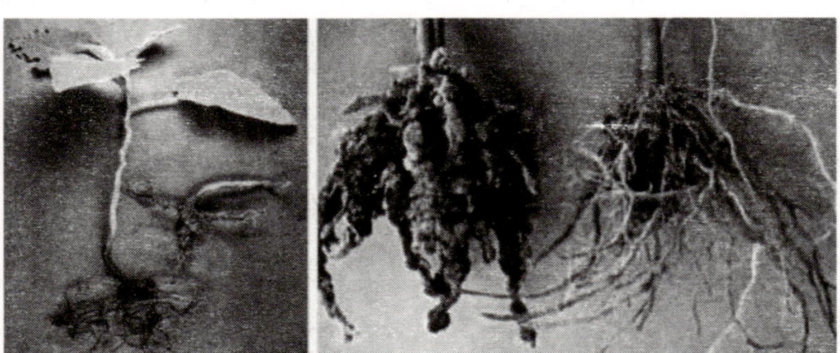

Fig. 6.26. Root-knot nematode on brinjal.

Brinjal seedlings raised in the above treatment transplanted in pots gave maximum increase in plant growth, root colonization and parasitization of eggs of *M. incognita* by *P. chlamydosporia*. The above treatment also gave least root galling and final nematode population both in soil and roots (Rao and Parvatha Reddy, 1993b) (Table 6.30).

Table 6.30. Effect of integration of castor cake and *Pochonia chlamydosporia* on root galling, final nematode population, root colonization and egg parasitization in eggplant infected with *Meloidogyne incognita*

Treatment	**Rot-knot index**	**Final nema popn. (soil + roots)**	***P. chlamydosporia* root colonization (cfu/g)**	**% parasitization of egg masses by *P. chlamydosporia***
Castor cake	3.2	3474	---	---
P. chlamydosporia	2.7	2863	3578	53
Castor cake + *P.chlamydosporia*	2.0	1746	5102	71
Control	4.3	6497	---	---
CD (P = 0.05)	**0.25**	**349.75**	**415.35**	**5.32**

Borkakaty (1993) observed that inoculation of *P. lilacinus* at 4 g/kg of soil in combination with mustard oil cake at 0.5 and 1.0 MT/ha increased plant growth with corresponding decrease in number of galls, egg masses and eggs/egg mass of *M. incognita* on brinjal.

Application of 10% neem cake extract (at 20 ml/pot) mixed with spores of *P. lilacinus* (at 1 x 10^6 spores/ml) was effective in increasing plant growth and reducing root galling and final nematode population both in soil and roots. The above treatment also gave maximum parasitization of egg masses of *M. incognita* and spore density of *P. lilacinus* in soil (Rao and Parvatha Reddy, 1994).

Table 6.31. Effect of root dip treatment with *Paecilomyces lilacinus* and neem leaf suspension on root galling, root colonization, spore density in soil and parasitization of eggs of *Meloidogyne incognita* infecting brinjal

Treatment	**Root-knot index**	**Final nema popn. (soil + roots)**	***P. lilacinus* root colonization (cfu/g)**	***P. lilacinus* spore density in soil (cfu/g)**	**Parasitization of eggs by *P. lilacinus* (%)**
P. lilacinus	2.8	3832	32540	20560	53
NLS - 5%*	3.1	5346	---	---	---
NLS - 10%*	2.7	4056	---	---	---
P. lilacinus + NLS - 5%	2.4	2964	36680	25640	59
P. lilacinus + NLS-10%	2.1	2356	39460	27690	64
Control	4.2	8258	---	---	---
CD (P = 0.05)	**0.46**	**785.32**	**3050.76**	**2850.78**	**4.59**

*Neem leaf suspension

Dipping of eggplant seedling (30 days old seedlings raised in sterilized soil) roots in 5 and 10% neem leaf suspensions mixed with *P. lilacinus* at 4 x 10^5 spores/ml for 30 min before transplanting gave significant reduction in root galling and final population of *M. incognita*. Further, significant increases were observed on the colonization of *P. lilacinus* on the roots of eggplant and parasitization of eggs of *M. incognita*, indicating the complementary interaction between these two components for the effective management of root-knot nematodes on eggplant (Rao *et al.*, 1997b) (Table 6.31).

Root-dip treatment of brinjal seedlings in neem cake extract based formulation of *P. lilacinus* (at 5 x 10^6 spores/ml) for 20 min and planted in pots gave significant increase in plant growth, root colonization, propagule density in soil and parasitization of eggs of *M. incognita* by *P. lilacinus* and drastic reduction in root galling, fecundity and final nematode population in soil and roots (Rao *et al.*, 1998d).

Application of castor cake extract based formulation of *T. harzianum* (at 500 ml/m^2 containing 9.9 x 10^3 spores/ml) to nursery beds of brinjal was effective in producing vigorous seedlings (with maximum seedling weight) with least root galling. The above treatment also increased root colonization and parasitization of *M. incognita* females by *T. harzianum* (Rao *et al.*, 1998c) (Table 6.32).

Table 6.32. Effect of oil cake based *Trichoderma harzianum* on root galling, root colonization, and egg parasitization of egg plant infected with *Meloidogyne incognita*

Treatment	No. of galls/100 seedlings	*T. harzianum* root colonization (cfu/g)	% Parasitization of eggs by *T. harzianum*
Neem cake extract (NCE) – 10%	62	---	---
Castor cake extract (CCE) – 10%	58	---	---
Pongamia cake extract (PCE) – 10%	64	---	---
10% NCE based *T. harzianum*	60	13746	43
10% CCE based *T. harzianum*	57	15625	51
10% PCE based *T. harzianum*	61	12242	42
T. harzianum (grown on paddy seeds)	69	9,648	30
Control	93	---	---
CD (P = 0.05)	**8.56**	**94.75**	**4.26**

Integration of neem seed powder/neem cake with *P. lilacinus* gave effective control of root-knot and reniform nematodes infecting brinjal (Table 6.33).

Soil application of poultry manure at 10 MT/ha a week prior to transplanting + *P. lilacinus* (10^9/conidia/g) at 25 kg spore dust with carrier/ha at the time of transplanting; or neem cake at 2 MT/ha a week prior to transplanting + *P. lilacinus* (10^9 /conidia/g) at 25 kg spore dust with carrier/ha at the time of transplanting; or carbofuran at 2 kg/ha in 2 equal splits (one at the time of transplanting and the other after 75 days) + *P. lilacinus* (10^9/conidia/g) at 25 kg spore dust with carrier/ha at the time of transplanting considerably reduced root galling and also gave higher brinjal fruit yield over control (Vyas *et al.*, 2009) (Table 6.34).

Table 6.33. Integrated management of *Meloidogyne incognita* and *Rotylenchulus reniformis* in brinjal using botanicals and biocontrol agents

Treatment	No. of galls/ plant	Nema popn. in soil (100 ml)	
		M. incognita	*R. reniformis*
Neem seed powder (NSP)	25	125	200
Neem cake (NC)	27	125	250
Paecilomyces lilacinus (*Pl*)	31	150	258
Pochonia chlamydosporia (*Pc*)	32	167	242
NSP + *Pl*	10	100	167
NSP + *Pc*	13	108	208
NC + *Pl*	14	108	192
NC + *Pc*	18	117	208
Control	41	200	375

Table 6.34. Effect of different organic amendments, bioagent and nematicide for the management of *Meloidogyne incognita* infecting brinjal

Treatment	Gall index (0-5 scale)	Yield MT/ha	ICBR
P. lilacinus at 25 kg/ha + poultry manure at 10 MT/ha	2.0	42.088	1: 18.5
P. lilacinus at 25 kg/ha + Mustard cake at 2 MT/ha	2.0	44.801	1: 7.7
P. lilacinus at 25 kg/ha + Neem cake at 2 MT/ha	1.8	41.745	1: 5.8
P. lilacinus at 25 kg/ha + carbofuran at 2 kg a.i./ha	1.9	41.498	1: 13.6
P. lilacinus at 25 kg/ha	2.5	39.941	1: 77.6
Poultry manure at 10 MT/ha	2.3	40.000	1: 18.5
Mustard cake at 2 MT/ha	1.9	41.810	1: 6.7
Neem cake at 2 MT/ha	1.9	39.954	1: 5.2
Carbofuran at 2 kg a.i./ha	2.0	40.530	1: 14.9
Control	**3.9**	**28.107**	---

(b) AMF and Botanicals: Brinjal seedlings raised in nursery treated with *G. fasciculatum* (at 250 g/m^2 containing 16 chlamydospores/g) planted in pots amended with castor cake (applied at 10 g/kg soil 15 days before transplanting) resulted in maximum plant growth, root colonization with *G. fasciculatum* and chlamydospore density in soil and least root galling, fecundity and final nematode population in soil and roots (Rao *et al.*, 1998) (Table 6.35).

Table 6.35. Effect of integration of castor cake and *Glomus fasciculatum* on root galling, final nematode population, root colonization and egg parasitization in eggplant infected with *Meloidogyne incognita*

Treatment	Root-knot index	Final nema popn. (soil + roots)	No. of eggs/egg mass	% root colonization by *G. fasciculatum*	*G. fasciculatum* chlamydospore density in 10 g soil
G. fasciculatum	6.2	5349	356	61	53
Castor cake – 5 g	5.7	4756	426	---	---
Castor cake – 10 g	5.1	3431	458	---	---
G. fasciculatum + Castor cake – 5 g	4.4	2934	295	73	72
G. fasciculatum + Castor cake – 10 g	4.1	2346	263	78	80
Control	8.3	9398	584	---	---
CD (P = 0.05)	**1.25**	**814.36**	**27.52**	**9.95**	**10.32**

Integration of *G. fasciculatum* and neem cake at 0.5 t/ha was found to be effective in increasing plant growth parameters and yield (30 MT/ha compared to 17 MT/ha in control) and in reducing the root-knot index (2.6 compared to 4.8 in control) and final nematode population (173.3/250 ml soil compared to 510.8/250 ml soil in control) (Borah and Phukan, 2000).

(c) Bioagents and AMF: Brinjal seedlings raised in seed pans treated with *G. mosseae* (at 100 g/seed pan containing 28-32 chlamydo spores/g) dipped in *P. lilacinus* spore suspension (containing 4 x 10^5 spores/ml) for 5 min. and transplanted in pots gave maximum increase in plant growth, root colonization, propagule density in soil of both *G. mosseae* and *P. lilacinus* and parasitization of eggs of *M. incognita.* The above treatment also gave least root galling, fecundity and final nematode population in soil and roots (Rao *et al.*, 1998) (Table 6.36).

Table 6.36. Effect of integration of *Glomus mosseae* and *Paecilomyces lilacinus* on root galling, final nematode population, root colonization and egg parasitization in eggplant infected with *Meloidogyne incognita*

Treatment	Root-knot index	Final nematode popn.	*P. lilacinus* root colonization (cfu/g)	% Root colonization by *G. mosseae*	Parasitization of eggs by *P. lilacinus* (%)
G. mosseae	3.0	4876	---	62	---
P. lilacinus	2.7	3252	35240	---	52
G. mosseae + *P. lilacinus*	2.3	2564	37270	64	65
Control	4.3	7746	---	---	---
CD (P = 0.05)	**0.32**	**541**	**875**	**NS**	**5**

(d) Bioagents, AMF and Botanicals: Inoculation of *G. fasciculatum* in the castor cake amended nursery beds followed by the root-dip treatment of mycorrhizal seedlings of brinjal in spore suspension of *P. lilacinus* was found effective for the management of *M. incognita* (Rao *et al.*, 1993b).

In nursery, soil application of neem cake at 400 g/m^2 along with *G. mosseae* (at 500 g/ m^2 containing 26-32 chlamydospores/g) and *P. lilacinus* (at 2 liters/m^2 containing 6 x 10^5 spores/ml) resulted in production of healthy and vigorous brinjal seedlings colonized with *G. mosseae* and *P. lilacinus* and with least root galling. In field, transplanting of brinjal seedlings raised in the above treatment gave maximum reduction in root galling, egg mass production, fecundity and final nematode population in soil and roots. The above treatment also gave highest fruit yield, root colonization with *G. mosseae* and *P. lilacinus* and egg parasitization with *P. lilacinus* (Rao and Parvatha Reddy, 2001) (Table 6.37).

Table 6.37. Effect of integration of *Glomus mosseae, Paecilomyces lilacinus* and neem cake on root galling, root colonization, egg parasitization and yield of egg plant infected with *Meloidogyne incognita*

Treatment	Rot-knot index	Yield (kg)/6 m^2	*P. lilacinus* root colonization (cfu/g)	% Root colonization by *G. mosseae*	Parasitization of eggs by *P. lilacinus* (%)
G. mosseae	7.2	6.2	---	47	---
Neem cake	6.6	5.6	---	---	---
P. lilacinus	7.0	5.3	41660	---	31
Neem cake + *G. mosseae*	5.9	7.3	---	58	---
Neem cake + *P. lilacinus*	5.7	6.9	59600	---	36
G. mosseae + *P. lilacinus*	5.4	6.5	44760	56	40
Neem cake + *G. mosseae* + *P. lilacinus*	4.8	7.8	51382	50	51
Control	8.7	5.3	---	---	---
CD (P = 0.05)	**0.49**	**0.89**	**1377.21**	**6.49**	**4.52**

(e) Two Bioagents: Integration of highly toxic fungus, *Aspergillus niger* (kills most of the infective second stage juveniles) and an egg parasite, *Cladosporium oxysporum* (invades and kills the eggs in egg sac) both at half the dosages significantly reduced *M. incognita* population and exhibited better plant growth than when either of the fungal bioagents in brinjal (Goswami and Singh, 2002).

In egg-plant, the nursery seedling stand index was good with seed treatment with *P. fluorescens* (50 g/kg) + soil application of *P. fluorescens* (10 g/m^2 seed bed) followed by soil application of neem based *T. harzianum* + *P. fluorescens* and neem based *P. fluorescens*, where the crop stand index was 4.6 and 4.0, respectively.

Combined soil application of *P. lilacinus* and *Aspergillus niger* at the time of transplanting brinjal is very effective in reducing root-knot nematodes.

(f) Bioagents and Chemicals: Split application of *T. harzianum* at 50 kg/ha (10^8 cfu/g) (before transplanting and 45 days after transplanting) + carbofuran 3G at 16.5 kg/ha was effective in increasing the brinjal fruit yield (Haseeb *et al.*, 2004).

(g) AMF/Bioagents and Chemicals: Integration of AM fungus *G. fasciculatum* at 25 g/m^2 (600 spores) with carbofuran at 0.5 kg a.i./ha gave significant increase in plant growth characters and yield with corresponding decrease in root galling and egg mass production, followed by *T. harzianum* at 2 g/kg soil + carbofuran at 0.5 kg a.i./ha (Saikia and Borah, 2008) (Table 6.38).

Table 6.38. Effect of integration of AMF/bioagents and carbofuran for the management of *Meloidogyne incognita* infecting brinjal

Treatment	% increase in yield	No. of galls/plant	No. of egg masses/ plant	Nema popn /200 ml soil	Nema popn in roots
G. fasciculatum- 25 g/m^2	70.12	51.73	29.30	265	2486
P. penetrans - 1 g/kg soil	56.49	77.86	61.93	324	3372
T. harzianum -2 g/kg soil	55.84	138.93	116.86	310	3078
Carbofuran – 2 kg a.i./ha	43.50	40.76	33.83	147	1608
G. fasciculatum- 25 g/m^2 + Carbofuran - 0.5 kg a.i./ha	108.44	20.63	12.30	170	1727
P. penetrans - 1 g/kg soil + Carbofuran - 0.5 kg a.i./ha	68.83	33.63	22.43	214	2320
T. harzianum -2 g/kg soil + Carbofuran - 0.5 kg a.i./ha	80.51	24.06	16.43	176	2187
Control	---	366.10	211.30	527	5558
CD (P = 0.05)	---	**4.944**	**17.33**	**16.6**	**18.55**

(h) Cultural and Chemicals: The planting of marigold combined with application of carbofuran at 1 kg a.i./ha controls *M. javanica* infestation on brinjal (Singh, 1991).

Nursery treatment with carbofuran at 0.3g a.i./m^2 along with main field treatment with ploughing + exposing the field and ploughing + covering with polythene sheets for 15 days improved the plant growth and yield of brinjal by 20-32% (Sheela *et al.*, 2002)..

(i) Botanicals and Chemicals: Application of aldicarb at 1.0 kg a.i./ha in the nursery beds along with neem cake at 400 kg/ha increased yield and reduced root galling (Singh and Gill, 1998) (Table 6.39).

Table 6.39. Integrated management of root-knot nematodes in brinjal with neem cake and nematicides.

Treatment	% Reduction in gall index	% Increase in yield
Aldicarb (1.0 kg a.i./ha) + Neem cake (400 kg/ha)	92.90	58.70
Carbofuran (1.0 kg a.i./ha) + Neem cake (400 kg/ha)	68.10	68.10

(j) Cultural and Physical: Integrating summer ploughing with soil solarization with polythene mulching effectively reduced *M. javanica* population in brinjal (Table 6.40) (Jain and Gupta, 1991).

Table 6.40. Integrated management of root-knot nematodes by summer ploughing/ fallowing and soil solarization in brinjal

Treatment	Reduction in nematode population (%)		Yield in kg/sq.m.	
	Hisar	Vellayani	Hisar	Vellayani
Ploughing + Fallowing for 15 days	87.00	66.20	28.98	---
Ploughing + Covering with polythene sheet for 15 days	94.90	77.00	18.84	78.00
No ploughing + No covering	77.60	---	---	---

(k) Physical and Botanicals Soil solarization of nursery beds (using 100 gauge LDPE clear film for 15 days) and application of neem cake at 200 g/sq.m. proved most effective in reducing root-knot nematode infestation and increasing yield in brinjal. Solarization of nursery beds for 15 days in summer and application of poultry manure at 200 kg/ha gave maximum yield (Jain and Gupta, 1991).

6.3.3.2. Reniform Nematode, *Rotylenchulus reniformis*

(i) Symptoms: General symptoms include reduced root systems, leaf chlorosis, overall stunting of host plants, and reduced yields and plant longevity. Female nematodes and their eggs are often visible when plant roots are viewed under a dissecting microscope.

(ii) Integrated Management

(a) Bioagents ands Chemicals: Integration of a bio-agent, *P. lilacinus* with carbofuran at 1 kg a.i./ha was found effective in the management of reniform nematode, *R. reniformis* infecting brinjal (Parvatha Reddy and Khan, 1989) (Table 6.41).

Table 6.41. Effect of *Paecilomyces lilacinus* and carbofuran on *Rotylenchulus reniformis* infecting brinjal

Treatment	Dose	Mature females in root/plant	Total nema population/ plant	Reproduction factor	% males
P. lilacinus	1 g/plant	12.00	1086	1.09	27.00
P. lilacinus	2 g/plant	7.33	834	0.83	48.79
P. lilacinus	4 g/plant	5.33	315	0.31	60.00
P. lilacinus + Carbofuran	2.0 g/plant + 0.5 kg a.i./ha	6.67	704	0.70	55.02
P. lilacinus + Carbofuran	2.0 g/plant + 1.0 kg a.i./ha	4.67	389	0.39	34.81

Carbofuran	0.5 kg a.i./ha	8.67	961	0.96	33.33
Carbofuran	1.0 kg a.i./ha	10.00	551	0.55	30.77
Control	---	25.50	2550	---	28.54
CD (P = 0.05)	---	**7.80**	**95.00**	---	---

(b) Botanicals, Bioagents and AMF: Inoculation of *G. fasciculatum* in the castor cake amended nursery beds followed by the root-dip treatment of mycorrhizal seedlings of brinjal in spore suspension of *P. lilacinus* is effective for the management of *R. reniformis* (Rao *et al.*, 1993b).

6.3.3.3. Root-knot Nematode, *Meloidogyne incognita* and Bacterial Wilt, *Ralstonia solanacearum* Disease Complex

(i) Symptoms: Eggplant is prone to many soil borne diseases among which the bacterial wilt in combination with root-knot nematode takes heavy toll every year all over the world (Naik, 2004) (Fig. 6.27). Association of pathogenic and above pathogenic levels of inoculum of both *M. incognita* and *R. solanacearum* increased the severity of wilt on brinjal crops.

The combined pathogenic effects of *M. incognita* and *R. solanacearum* on a resistant brinjal cultivar (Pusa purple cluster) provided synergistic effect towards the development of wilt symptoms and negatively influenced different plant growth parameters such as shoot length, shoot weight, root length and root weight.

Fig. 6.27. Root-knot nematode and bacterial wilt disease complex in brinjal.

(ii) Integrated Management

(a) Two Bioagents: Combined formulation of *T. harzianum* and *P. chlamydosporia* performed very well on disease complex. The yield was 2.024 kg/3m^2 followed by combination of *P. fluorescens* + *P. chlamydosporia* (1.782 kg) and *T. harzianum* + *P. fluorescens* (1.680 kg). Root gall index (RGI) and wilt disease incidence due to *R. solanacearum* (WDI) was minimum in combination treatment of *T. harzianum* + *P. chlamydosporia* followed by *P. fluorescens* + *P. chlamydosporia* and *T. harzianum* + *P. fluorescens* where the RGI was 1.8, 1.81 and 1.93, respectively and that of WDI was 22.18%, 25.13% and 27.28%, respectively (Naik, 2004).

(b) Cultural, Chemicals, Botanicals and Bioagents: Integration of two summer ploughing during May-June, half recommended dose each of carbofuran (0.75 kg a.i./ha), neem cake (7.5 g/spot), streptocycline (250 ppm at 30 ml/spot) and full dose of *T. harzianum* (150 g/spot) improved plant growth parameters and fruit yield with corresponding decrease in the nematode reproduction rate (*M. incognita*), bacterial population in the soil and bacterial wilt incidence (*R. solanacearum*) in brinjal (Hussain and Bora, 2008) (Table 6.42).

Table 6.42. Effect of integration of different practices for the management of disease complex in brinjal

Treatment	No. of galls/root system	No. of egg masses/root system	% wilt incidence	Yield (kg/4m^2)
2 Summer ploughings + full dose of carbofuran (1.5 kg a.i./ha)	65	59	27.77	3.21
2 Summer ploughings + full dose of neem cake (15 g/spot)	71	69	27.77	3.10
2 Summer ploughings + full dose of streptocycline (500 ppm at 30 ml/spot)	128	112	38.88	2.43
2 Summer ploughings + full dose of *T. harzianum* (150 g/spot)	67	63	27.77	3.10
2 Summer ploughings + ½ dose carbofuran + full dose *T. harzianum*	35	26	5.53	3.90
2 Summer ploughings + ½ dose neem cake + full dose *T. harzianum*	57	47	11.10	3.30
2 Summer ploughings + ½ dose streptocycline + full dose *T. harzianum*	59	55	24.99	3.20
2 Summer ploughings + ½ dose carbofuran + ½ dose neem cake + ½ dose streptocycline	28	20	5.53	4.20
2 Summer ploughings + ½ dose carbofuran + ½ dose neem cake + ½ dose streptocycline + full dose *T. harzianum*	24	17	5.53	4.90
Control	362	207	66.66	1.6
CD (P = 0.05)	**6.69**	**6.75**	**11.15**	**0.83**

6.3.3.4. Root-knot Nematode, *Meloidogyne incognita* and Wilt, *Fusarium* spp. Disease Complex

Brinjal was not susceptible to *F. oxysporum* unless *M. incognita* was also present (Smits and Noguera, 1982) (Fig. 6.28).

Fig.6.28. Root-knot *Fusarium* wilt complex in brinjal.

(i) Integrated Methods

(a) Bioagents and Chemicals: Treatment of nursery beds with carbofuran at 1 kg a.i./ha + *T. harzianum* at 50 kg/ha (having 10^8 cfu/g) was found highly effective in increasing the number and fresh weight of brinjal seedlings and maximum reduction in root-knot index due to *M. incognita*. However, minimum percent root infection by *F. solani* was observed in carbofuran at 1 kg a.i./ha + bavistin at 1 kg a.i./ha treated seedlings (Kumar *et al.,* 2009a).

6.3.4. Validated IPM Technology for Brinjal Pests and Diseases at Ghaziabad

Nursery

- Raised seed bed of 10-15 cm height to avoid flooding of bed.
- Soil solarization for three weeks during June using polythene sheets of 45 µm thickness.
- Soil treatment with *Trichoderma viride* at 100 g/kg FYM. Enrichment of *T. viride* for 3 weeks before mixing in soil.

Main field

- Erection of bird perches at 25/ha to facilitate predation of insects.
- Soil application of neem cake at 250 kg/ha.
- Installation of delta traps and yellow sticky traps at 5/ha for hopper and white fly.

- Pheromone traps installed at 12/ha for mass trapping as well as monitoring of *Leucinodes orbonalis*.
- Soil application of neem cake at 250 kg/ha along the plant rows 30 DAP.
- Three sprays of 5% NSKE against leaf hoppers, aphids, mites depending upon the appearance of the pests.
- Six releases of egg parasitoid, *Trichgramma chilonis* at 1 lakh/ha at weekly interval for shoot and fruit borer.
- Collection and destruction of egg masses, larvae and adults of hadda beetle.
- Clipping of borer damaged shoots and collection and destruction of damaged fruits.
- Rouging out of little leaf affected plants at monthly interval.
- One spray each of imidacloprid at 0.5 ml/litre and cypermethrin at 1 ml/litre in the season.

In brinjal crop, IPM technology has been validated in about 3 ha area in Raispur village near Ghaziabad during 2003-04. IPM fields gave higher yields of 61.3 MT/ha as compared to 50.9 MT/ha in non-IPM fields (Sardana *et al.*, 2004) (Table 6.43).

Table 6.43. Yield and economics of IPM in brinjal

Parameter	IPM	Non IPM	% increase
Yield (MT/ha)	61.30	50.90	20.43
Net returns (Rs/ha)	62700	28043	20.01
Benefit: cost ratio	2.27	1.02	122.55

6.4. CHILLI AND BELL PEPPER, *CAPSICUM* spp.

6.4.1. Insect Pests

6.4.1.1. Fruit Borer, *Helicoverpa armigera*

Tomato fruit borer, *H. armigera* is also a serious pest in chilli and capsicum. It affects the marketable value of the chilli crop to a great extent, if proper care is not taken.

(i) Damage: Young larvae feed on flower buds and young pod by making a circular hole. Later, the larvae feed on seeds usually with its head inside the pod and rest of the body outside. A circular hole is noticed at the base of the pedicel. Premature dropping of flower and pods is also noticed (Fig. 6.29).

(ii) Integrated Management

(a) Cultural and Bioagents: Planting of one row of African marigold after every 18 rows of chilli and spraying with *Ha*NPV was very efficient in controlling the larval population of *H. armigera* on chilli. There was a significant reduction in mean larval population of *H. armigera* (1.10 larvae/plant). The percentage fruit damage of 8.88% was lowest and the fruit yield was highest (2.822 MT/ha of dry chillies) in this treatment.

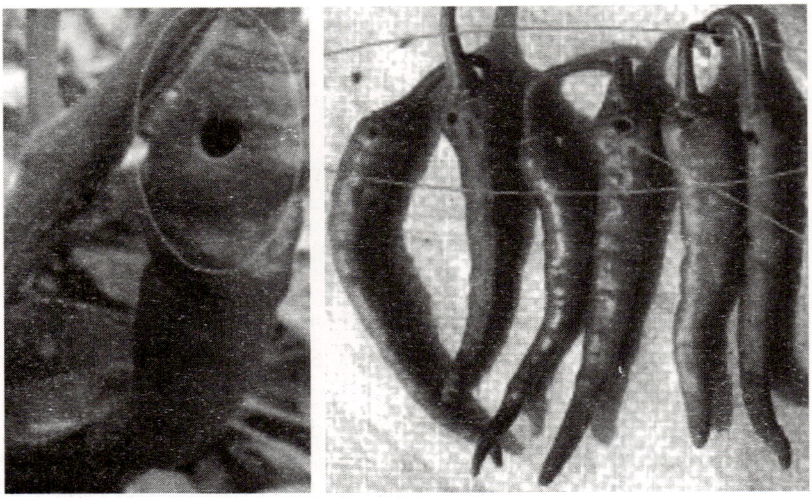

Fig. 6.29. Pod borer, *Helicoverpa armigera* damage on chilli fruits.

(b) Bioagents, Botanicals and Chemicals: IPM comprising of three releases of *Trichogramma japonicum* followed by sequential application of *Ha*NPV at 250 LE/ha, 3% neem oil and need based application of 0.07% endosulfan gave effective control of fruit borer and increased the fruit yield (22.4 MT/ha) as compared to non-IPM fields (18.2 MT/ha).

6.4.1.2. Thrips, *Scirtothrips dorsalis*

(i) Damage: Both adults and nymphs suck the sap from young leaves. Affected leaves curl upwards along the margin, and get crinkled and reduced in size (Fig. 6.30). When the incidence is severe, leaves drop and cause heavy reduction in yield. The pest is serious during dry monsoon periods and summer months.

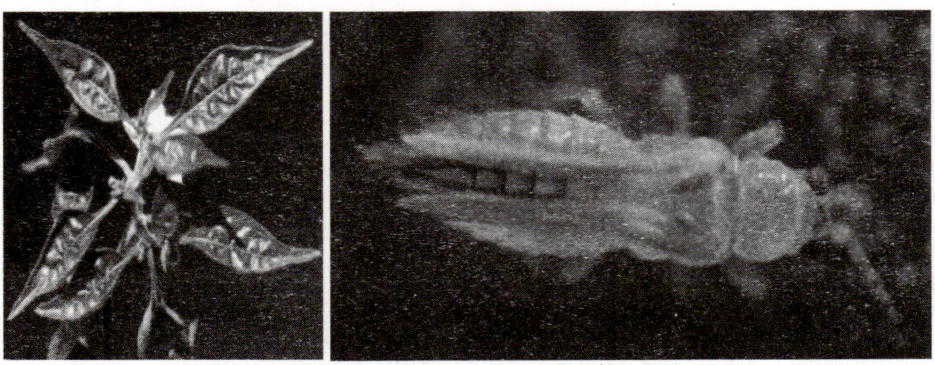

Fig. 6.30. Thrips on leaves and fruits of chilli.

(ii) Integrated Management

(a) Botanicals and Chemicals: Pongamia oil at 2 ml when mixed with 0.5 g acephate + 2 ml of sticker gave excellent control of thrips in chillies, even during summer (Krishnamoorthy and Krishna Kumar, 2002). The thrips damage rating was reduced to 20.67 in the above treatment as compared to 45.67 in control. Similarly, the yield was also increased in treated plots over control (Table 6.44).

Table 6.44. Thrips damage and yield under different management programmes in chilli (Summer 2002)

Treatments	Doses	Rating of thrips damage/5 leaves	Yield (kg/plot)
Acepahte + Pongamia oil	0.5 g+2 ml	20.67	1.69
Dimethoate + Neem oil	2 ml+2ml	30.00	1.26
Pongamia oil	20 ml	45.67	1.05
Neem oil	20 ml	40.67	3.50
Dimethoate	2 ml	25.00	0.90
Neem seed powder	4%	40.00	0.05
Acephate	0.75 g	10.67	1.35
Control	---	45.67	0.02
CD (P = 0.05)	---	**7.33**	**0.10**

6.4.2. Diseases

6.4.2.1. Damping-off, *Pythium aphanidermatum*

(i) Symptoms: The shrinking of the cortical tissue of the hypocotyl and toppling over of the infected seedlings takes place. It affects germination and stand of seedlings in nursery beds due pre-emergence damping-off. The pathogen is responsible for death of seedlings in nursery beds (Fig. 6.31).

Fig. 6.31. Damping-off in chilli nursery

(ii) Integrated Management

(a) Physical, Botanicals and Bioagents: Integration of soil solarization, application of neem cake and *T. viride* gave effective control of the disease.

(b) Physical, Chemical and Bioagents: Soil solarization gave least damping-off followed by seed treatment with *Trichoderma viride* + soil application of FYM enriched with *T. viride* (Table 6.45) (Rahman *et al.*, 2002).

Table 6.45. Comparative efficacy of soil solarization, bioagents and fungicide against damping-off of chilli

Treatment	% Disease index
Soil solarization with transparent polythene sheets for 30 days	1.28 (6.45)
Seed treatment with *Trichoderma viride* at 4g/kg + soil application of 50 kg FYM enriched with 500g of *T. viride*	4.70 (12.26)
Seed treatment with *Pseudomonas fluorescens* at 10g/kg + soil application of 50 kg FYM enriched with 2.5 kg of *P. fluorescens*	6.75 (14.82)
Seed treatment with *Trichoderma harzianum* at 4g/kg + soil application of 50 kg FYM enriched with 500g of *T. harzianum*	6.03 (13.97)
Seed treatment with *Azotobacter croococcum* at 16g/kg + soil application of 50 kg FYM enriched with 500g of *A. croococcum*	6.68 (14.80)
Seed treatment with captan at 2.5g/kg + soil drenching with captan at 0.25% at $6L/m^2$	4.75 (12.29)
Control (Check)	20.13 (26.22)
CD at 5%	**(3.09)**

Figures in parentheses indicate the arc sin transformed values

Fig. 6.32. Fusarium wilt of chilli

(c) Physical and Bioagents: Combination of the seed/root application of *T. harzianum* or *P. fluorescens* with soil solarization was very effective in management of damping-off of capsicum in nursery at farmers' field.

6.4.2.2. Wilt, *Fusarium oxysporum* f. sp. *capsici, F. solani*

(i) Symptoms: Symptoms include leaf chlorosis, vascular discolouration, and wilting of chilli pepper plants (Fig. 6.32). High temperature and high moisture are conducive to symptom development.

(ii) Integrated Management

(a) Two Bioagents: Combined seedling root-dip treatment and soil application of non-pathogenic *Fusarium* (Fo 52) and *P. fluorescens* gave maximum inhibition of wilt (Singh *et al.*, 2002).

6.4.2.3. Root Rot, Fruit Rot and Leaf Blight, *Phytophthora capsici*

(i) Symptoms: This is a disease of rainy season, characterized by small, water-soaked spots appearing on fruits leading to complete rot of fruits. On leaves, water-soaked bleached spots appear, resulting in blighting. It causes damage to plants and affect fruit yield. Infection of the root and lower portion of the stem leads to plant wilting (Fig. 6.33), which is the most conspicuous symptom of the disease. In furrow-irrigated fields, there is a row-delimited pattern of wilted plants (Fig. 6.33). Over time, wilting is displayed over the entire field. *Phytophthora capsici* produces specialized swimming cells, known as zoospores, which enable spread of the pathogen from plant to plant.

Fig. 6.33. Wilting of chilli plant and a field severely affected by *Phytophthora* root rot (Courtesy: D. Lindsey).

(ii) Integrated Management

(a) Bioagents and Chemicals: Seed treatment with carbendazim (0.2%) + seedling dip in *P. fluorescens* (10g/l) + 2 sprays of *P. fluorescens* at 45 and 60 days after transplanting + 2 sprays of hexaconazole at 75 and 90 days after transplanting recorded the least disease incidence of fruit rot (16.0 PDI) and powdery mildew

(26.7 PDI). This treatment also gave highest yield (8.5 q/ha) and highest net returns (Rs 25,940/ha) (Mesta *et al.*, 2009).

(b) Physical and Botanicals: Solarization and biofumigation with fresh sheep manure (if it is done from July to October) shown promise to be efficient disinfectants which increased plant growth, yield and decreased disease incidence (Table 6.46) (Bello *et al.*, 1997). The best results were obtained when biofumigation with solarization was applied at the end of August or the beginning of September. Sweet pepper yield using fresh sheep manure with soybean flour, or fresh sheep manure with chicken manure treatments, was similar to sweet pepper yield when MB was used. The reiteration of biofumigation with solarization over two or more years led to an improvement in pathogen and weed control, higher plants, an increase in yield and improvements in the soil physical properties, with a higher level of macro- and micronutrients as well as electrical conductivity.

Table 6.46. Effect of solarization and biofumigation on *Phytophthora* root rot in plastic greenhouses (Season 1998/99).

Treatments	Plants height (cm)	Commercial yield (kg m^{-2})	% plants with *Phytophthora*
MB 60 g/m^2	163.8 a	9.7 a	0.4 a
Biofumigation and solarization (4 weeks)	135.3 b	8.7 b	9.4 b
Untreated	122.4 c	4.3 c	37.9 c

No significant differences between values with the same letter (P> 0.05)

6.4.2.4. Southern Blight, *Sclerotium rolfsii*

(i) Symptoms: Early symptoms consisted of water-soaked lesions on crown and lower stem tissue in contact with the soil. Plant foliage became pale green and wilted, followed by a complete collapse of the plant. A dense white mycelial mat formed on the lower stem and crown with 1- to 2-mm-diameter, spherical, tan-to-dark brown sclerotia (Fig. 6.34).

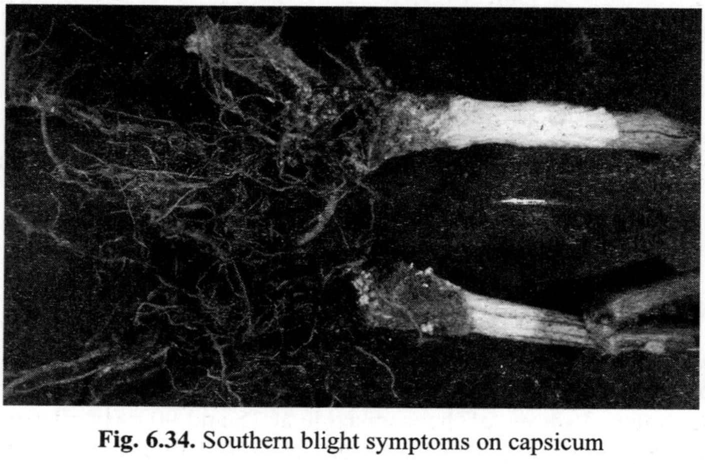

Fig. 6.34. Southern blight symptoms on capsicum

(ii) Integrated Management

(a) Physical and Bioagents: Solarization of fallow soil in raised beds for 6 weeks and application of a bran prill formulation of *T. virens* significantly reduced the southern blight disease incidence in bell pepper and survival of sclerotia of *S. rolfsii* to depths of 30 cm (Ristaino *et al.*, 1996).

(b) Bioagents and AMF: Sreenivasa (1997) reported significant reduction in sclerotial bodies produced by *S. rolfsii* when chili crop was co-inoculated with AM fungi and *T. harzianum*.

(c) Bioagents and Chemicals: Effective control of *S. rolfsii* on bell pepper with a combination of *T. harzianum* and ridomil has been reported.

6.4.2.5. Powdery Mildew, *Levillula taurica*

(i) Symptoms: Pepper powdery mildew grows unseen on the undersurface of leaves (Fig. 6.35), within the leaf tissue for a latency period of up to 21 days. Disease monitoring, early detection and prevention of pepper powdery mildew is critical. By the time pepper powdery mildew is detected in a greenhouse many more leaves are already infected but do not show any disease symptoms. In addition, pepper plants can become defoliated and do not recover as quickly. Pepper powdery mildew does not infect the fruit or stems but can quickly destroy unprotected leaves and eventually the entire pepper crop.

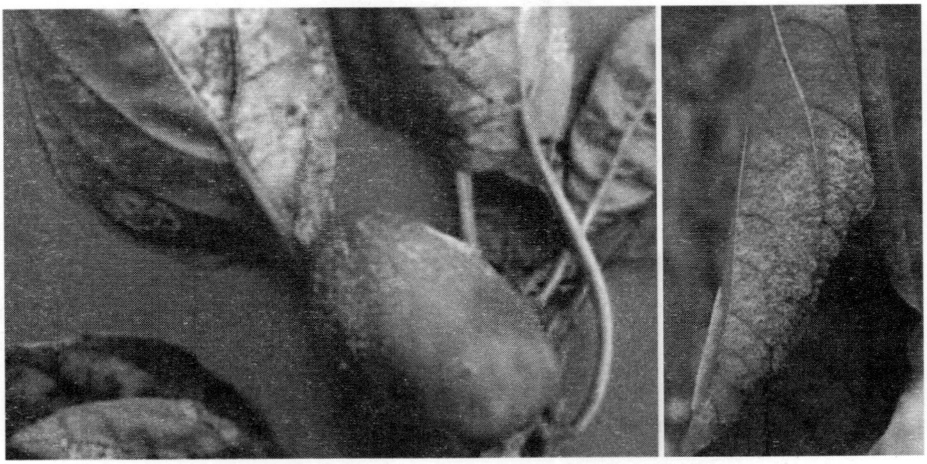

Fig. 6.35. Powdery mildew on chilli leaf.

(ii) Integrated Management

(a) Bioagents and Chemicals: Seed treatment with carbendazim (0.2%) + seedling dip in *P. fluorescens* (10g/l) + 2 sprays of *P. fluorescens* at 45 and 60 days after transplanting + 2 sprays of hexaconazole at 75 and 90 days after transplanting recorded the least disease incidence of fruit rot (16.0 PDI) and powdery mildew (26.7 PDI). This treatment also gave highest dry chilli yield (0.85 t/ha) and highest net returns (Rs 25, 940/ ha) (Mesta *et al.*, 2009).

6.4.2.6. Alternaria Rot, *Alternaria alternata*

(i) Symptoms

- The fungus is reported to enter wounds (sunscald or punctures).
- Dusty black spores on fruit spots are characteristic.
- In most instances this disease follows blossom-end rot, but it also follows injuries, chilling, and other decays
- On the fruit, large greenish-brown to brown lesions covered with grayish-brown mold are produced (Fig. 6.36).

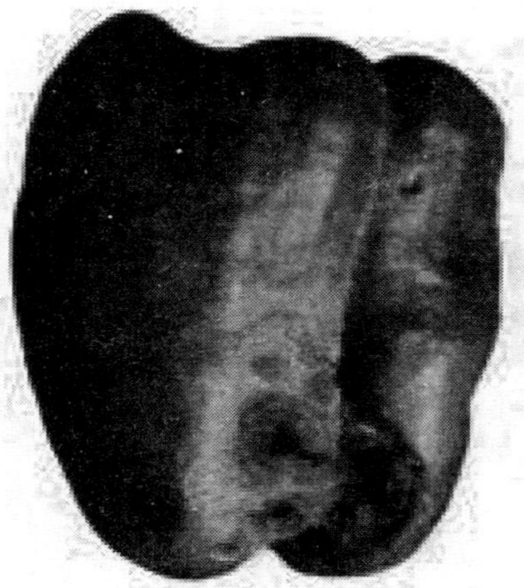

Fig. 6.36. Alternaria fruit rot on capsicum

- Similar lesions on the lower-part of the fruit are characteristic of *Alternaria* rot following blossom-end rot.
- The larger lesions may show alternating light and dark-brown concentric zones.
- Shipping peppers under standard refrigeration will check the development of this rot, but when the fruit is removed from refrigeration the decay will advance rapidly at moderate to warm temperatures.

(ii) Integrated Management

(a) Physical and Chemical: Methyl-isothiocyante (MITC) at 0.56 mg/ml in combination with low density polyethylene film (LDPF) solarization reduced the fungal infection (*A. alternata*) in the bell pepper fruit better than a commercial fungicide (Captan) without any detrimental effects on fruit quality (Troncoso-Rojas *et al.*, 2008).

6.4.2.7. Leaf Curl Virus

(i) Symptoms: The Gemini virus causes enation leaf curl symptom. Reduction in size of leaves, shortening of veins, puckering, sometimes mottling and stunting of plants are common symptoms of the disease (Fig. 6.37). Veins are prominently visible on the lower side of the leaves. The infected plants remain stunted and bear no fruits.

Fig. 6.37. Chilli leaf curl Gemini virus

(ii) Integrated Management (of insect-borne viruses)

(a) Cultural, Chemicals and Botanicals: At Coimbatore, soil application of carbofuran at 1 kg a.i./ha + covering the nursery bed with nylon net of 400 mesh followed by 2 sprays of 2% neem oil at 15 and 35 DAP recorded lesser virus incidence and highest yield with benefit: cost ratio of 3.3.

(b) Cultural and Chemicals: At Sabour, Use of nylon net and soil application of carbofuran at 1 kg a.i./ha in seed bed and application of carbofuran at 1.5 kg a.i./ha in the main field a week after transplanting + 3 sprays of nuvacron at 1 ml/litre was most effective in managing the disease with benefit: cost ratio of 5.64.

Use of nylon net cover in nursery beds, soil application of carbofuran at 1 kg a.i. /ha in combination with foliar application of 0.1% imidacloprid at 10 days interval effectively reduced the vector population as well as the disease incidence.

6.4.3. Nematodes

6.4.3.1. Root-knot nematodes, *Meloidogyne* spp.

(i) **Symptoms:** Nematode infestations damage the plant roots, and therefore symptoms reflect poorly functioning root systems. Aboveground symptoms of severe root-knot infestations include patches of chlorotic, stunted, necrotic, or wilted plants. Nematode-infested plants are more susceptible to moisture or temperature stress and exhibit stress symptoms earlier than other plants. Furthermore, root systems that have been damaged by nematodes are often more susceptible to infection by soil-inhabiting fungi such as *Fusarium* and *Verticillium* species. Feeding by root-knot nematodes results in characteristic galls on roots. Severely galled roots may appear malformed and the root system shortened and thickened (Fig. 6.38). Root galls caused by root-knot nematodes on sweet pepper are frequently small.

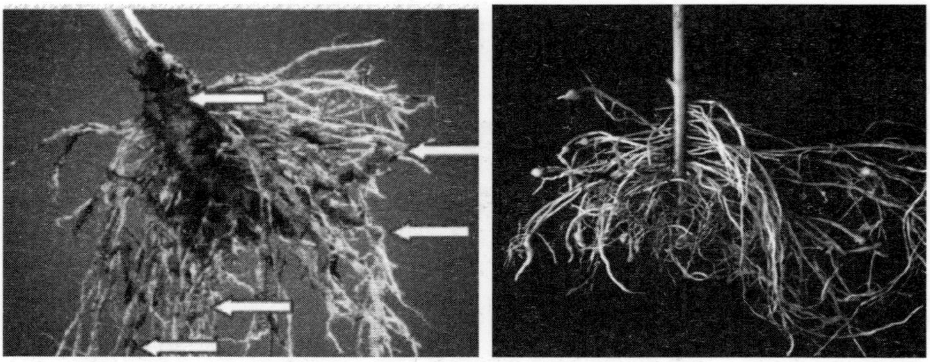

Fig. 6.38. Root-knot nematode on bell pepper.

(ii) Integrated Management

(a) Two Bioagents: Integration of *P. lilacinus* with *P. chlamydosporia* in nursery was found most effective in increasing the plant growth of capsicum seedlings and reducing the nematode population both in soil and roots by 59 and 72%, respectively. When capsicum seedlings raised in nursery treated with *P. lilacinus* + *P. chlamydosporia* were transplanted in the main field, there was reduction in root gall index by 51% and increase in fruit yield by 44% compared to control (Naik, 2004) (Table 6.47).

Table 6.47. Effect of integration of bioagents for the management of *Meloidogyne incognita* and fruit yield of capsicum.

Treatment	Dose/m^2	Root-knot index (1-10 scale)	Yield in kg/2 m^2
P. chlamydosporia	50 g (10^6 cfu/g)	6.20	2.58
T. harzianum	50 g (10^6 cfu/g)	7.20	2.36
P. lilacinus	50 g (10^6 cfu/g)	6.70	2.61
P. chlamydosporia + *P. lilacinus*	25 g + 25 g	4.35	3.28
T. harzianum + *P. lilacinus*	25 g + 25 g	4.75	2.86
P. chlamydosporia + *T. harzianum*	25 g + 25 g	5.35	2.07
P. chlamydosporia + *T. harzianum* + *P. lilacinus*	17 g + 17 g + 17 g	5.10	2.05
Carbofuran	25 g	7.80	2.30
Control	---	8.75	2.04
CD (P = 0.05)	---	**0.44**	**0.77**

Integrated management of *M. incognita* infecting capsicum was achieved by seed treatment with *P. fluorescens* at 50g/kg combined with nursery bed treatment with *P. chlamydosporia* at 50g/m^2. The above treatment was significantly effective in increasing the seedling growth and root colonization with bioagents and in reducing root galling (Table 6.48). These seedlings when transplanted in field significantly reduced nematode population both in soil and roots, root galling and increased root colonization by bioagents, propagule density in soil, parasitization of eggs and fruit yield. The main purpose of these studies was to raise capsicum seedlings that are colonized by *P. chlamydosporia* before transplanting so that they could carry the bio-agent to the field. During this process the field soil would be enriched with the propagules of the bio-agent for two or three seasons (Rao *et al.*, 2004) (Table 6.49).

Table 6.48. Effect of integration of bioagents on plant growth and management of *Meloidogyne incognita* infecting capsicum in nursery

Treatment/Dose	Seedling weight (g)	No. of galls/10 seedlings	Root colonization of *P. chlamydosporia* (cfu/g)	Root colonization of *P. fluorescens* (cfu/g)
P. chlamydosporia - nursery treat (25 g/m^2)	3.6	62	15,896	---
P. chlamydosporia - nursery treat (50 g/m^2)	3.8	59	17,459	---
P. fluorescens - seed treat (50g/kg)	4.0	55	---	12,563
P. fluorescens - seed treat (50g/kg) + *P. chlamydosporia* - nursery treat (25 g/m^2)	4.5	51	16,256	12,249
P. fluorescens - seed treat (50g/kg) + *P. chlamydosporia* - nursery treat (50 g/m^2)	4.4	53	16,789	11,897
Control	3.2	82	---	---
CD (P = 0.05)	**0.34**	**7.96**	**959.72**	**729.65**

Table 6.49. Effect of integration of bioagents on plant growth and management of *Meloidogyne incognita* infecting capsicum under field conditions

Treat.*	Root-knot index (1-10 scale)	Yield (kg/ 6 m^2)	% eggs parasitized by *P. chlamydosporia*	Root colonization (cfu/g)		Propagule density (cfu/g soil)	
				Pc	*Pf*	*Pc*	*Pf*
T1	6.3	4.4	41.00	22,456	---	19,569	---
T2	5.6	4.7	45.00	25,789	---	22,845	---
T3	6.0	5.1	---	---	22,568	---	15,987
T4	4.8	5.3	40.00	21,679	21,789	19,234	16,843
T5	4.4	5.5	42.67	24,587	21,567	23,221	15,359
Control	8.1	4.0	---	---	---	---	---
CD 5%	**0.45**	**0.28**	**3.78**	**1246.76**	**1089.86**	**1178.32**	**873.65**

* T1 - *P. chlamydosporia* - nursery treat (25 g/m^2), T2 - *P. chlamydosporia* - nursery treat (50 g/m^2), T3 - *P. fluorescens* - seed treat (50g/kg), T4 - *P. fluorescens* - seed treat (50g/kg) + *P. chlamydosporia* - nursery treat (25 g/m^2), T5 - *P. fluorescens* - seed treat (50g/kg) + *P. chlamydosporia* - nursery treat (50 g/m^2).

(b) Biofumigation and Solarization: The best results were obtained when biofumigation with solarization was applied at the end of August or the beginning of September. Sweet pepper yield using fresh sheep manure with soybean flour, or fresh sheep manure with chicken manure treatments, was similar to sweet pepper yield when MB was used. The reiteration of biofumigation with solarization over two or more years led to an improvement in pathogen and weed control, higher plants, an increase in yield and improvements to the soil physical properties, with a higher level of macro- and micronutrients as well as electrical conductivity (Bello *et al.*, 2004).

(c) Physical and Chemicals: Experiments in Sicily (Cartia and Greco, 1987; Cartia *et al.*, 1988, 1989), aimed at the control of soil-borne pathogens of pepper in greenhouse culture, yield and fruit size from solarized soil averaged 12.2 and 2.3 times, respectively, over those in controls. Higher yield was obtained with soil solarization, alone or in combination with reduced dosages of methyl bromide, and with methyl bromide alone (6 times over control) than with DD (5.4 times the control).

(d) Botanicals and Bioagents/Chemicals: Soil application of poultry manure at 10 MT/ha a week prior to transplanting + *P. lilacinus* (10^9/conidia/g) at 25 kg/ha spore dust with carrier at the time of transplanting; or neem cake at 2 MT/ha a week prior to transplanting + *P. lilacinus* (10^9 conidia/g) at 25 kg/ha spore dust with carrier at the time of transplanting; or carbofuran at 2 kg/ha in 2 equal splits (one at the time of transplanting and the other after 75 days) + *P. lilacinus* (10^9 conidia/g) at 25 kg/ha spore dust with carrier at the time of transplanting considerably reduced root galling and also gave higher capsicum fruit yield over control (Vyas *et al.*, 2009) (Table 6.50).

Table 6.50. Effect of different organic amendments, bioagent and nematicide for the management of *Meloidogyne incognita* infecting capsicum

Treatment	Gall index (0-5 scale)	Yield (MT/ha)	ICBR
P. lilacinus at 25 kg/ha + Poultry manure at 10 MT/ha	2.0	42.088	1: 18.5
P. lilacinus at 25 kg/ha + Mustard cake at 2 MT/ha	2.0	44.801	1: 7.7
P. lilacinus at 25 kg/ha + Neem cake at 2 MT/ha	1.8	41.745	1: 5.8
P. lilacinus at 25 kg/ha + Carbofuran at 2 kg a.i./ha	1.9	41.498	1: 13.6
P. lilacinus at 25 kg/ha	2.5	39.941	1: 77.6
Poultry manure at 10 MT/ha	2.3	40.000	1: 18.5
Mustard cake at 2 MT/ha	1.9	41.810	1: 6.7
Neem cake at 2 MT/ha	1.9	39.954	1: 5.2
Carbofuran at 2 kg a.i./ha	2.0	40.530	1: 14.9
Control	3.9	28.107	---

(e) Physical and Botanicals/Bioagents/Chemicals: The effect of solarization, and its combinations with dazomet (400 kg/ha), chicken manure (10 tonnes/ha) or straw (500 kg/ha), on soil-borne nematodes and crop were demonstrated at various sites. Combinations of solarization with basamid (400 kg/ha), chicken manure (10 tonnes/ha) or *Trichoderma* were the most effective applications for the management of root-knot nematodes infecting bell pepper. MB and solarization + straw (500 kg/ha) were partially effective. However, it can be said that all treatments effectively controlled nematodes (Table 6.51). Galling index overall averages of six pepper plastic houses were 0.10, 0.89, 0.98, 1.92, 2.04, and 5.95 for solarization + *Trichoderma*, solarization + manure, solarization + dazomet, MB, solarization + straw, and check, respectively.

Table 6.51. Effects of treatments on the nematode population as percentage of non-treated check during the growing season of pepper.

Date	Treatments and control rates (%)				
	Solarization + Basamid	Solarization + *Trichoderma*	Solarization + Manure	Solarization + Straw	Methyl bromide
03-10-2000	100.0	100.0	100.0	100.0	100.0
19-10-2000	98.8	100.0	100.0	100.0	100.0
07-11-2000	100.0	100.0	100.0	100.0	100.0
28-11-2000	89.9	100.0	80.0	100.0	97.8
19-12-2000	100.0	100.0	100.0	100.0	100.0
09-01-2001	96.3	99.2	79.9	99.3	100.0
30-01-2001	99.7	100.0	100.0	100.0	99.3
20-02-2001	100.0	100.0	100.0	100.0	97.0
13-03-2001	100.0	99.3	99.9	98.1	97.9
03-04-2001	79.7	98.7	96.2	94.3	91.6
25-04-2001	99.5	95.5	99.8	70.9	90.4
16-05-2001	72.9	88.8	94.4	93.2	87.4
06-06-2001	74.6	91.4	95.8	99.6	59.4
23-07-2001	40.8	59.3	84.9	0.0	0.0

100 is complete control and 0 shows no control

6.4.3.2. Root-knot Nematode, *Meloidogyne incognita* and Bacterial Wilt, *Ralstonia solanacearum* Disease Complex

(i) Symptoms: Capsicum is prone to many soil-borne diseases among which the bacterial wilt (*R. solanacearum*) in combination with root-knot nematode (*M. incognita*) takes heavy toll every year all over the world (Fig. 6.39) (Naik *et al.*, 2003). The disease ratings of bacterial wilt in treated plants (inoculated with *M. incognita* one month later with *R. solanacearum*; inoculated with *R. solanacearum* and 2 weeks later with the nematodes; or inoculated with nematodes and *R. solanacearum* simultaneously) were higher than ratings in plants inoculated with bacteria only. It is concluded that a complex infection of root-knot nematodes and *R. solanacearum* increased disease severity and reduced the resistance of bacterial wilt-resistant chilli plants.

Fig. 6.39. Meloidogyne incognita and Ralstonia solanacearum disease complex in capsicum

(ii) Integrated Management

(a) Two Bioagents: The seedling stand was good when the nursery beds were combinedly treated with neem based formulations of *T. harzianum* + *P. fluorescens* followed by seed treatment with talc based *P. fluorescens* + nursery bed treatment with neem based *P. fluorescens* (Naik *et al.*, 2003) (Table 6.52).

Table 6.52. Effect of integration of bioagents on nursery stand of capsicum seedlings

Treatment	Nursery stand of seedlings*
Seed treatment with talc based *P. fluorescens*	3.4
Nursery bed treatment with neem based *P. fluorescens*	3.2
Seed treatment with talc based *P. fluorescens* + Nursery bed treatment with neem based *P. fluorescens*	4.2
Nursery bed treatment with *T. harzianum* + *P. fluorescens*	4.4
Control	3.0

* 3 – Poor, 4 – Good, 5 – Very good

Combined application of neem based formulations of *P. fluorescens* and *P. chlamydosporia/T. harzianum* at 40 g/m^2 in nursery beds and transplanting these seedlings in the main field resulted in significant reduction in disease index and root-knot index in capsicum to the tune of 70% and increased the crop yield by 37% (Rao *et al.*, 2002a) (Table 6.53). Combinations of the bioagents did not affect the colonization of the individual bioagents on the roots and hence the transplants carried the bioagents to the main field. This has resulted in the effective management of the pathogens involved in the disease complex.

Table 6.53. Effect of integration of neem-based bioagents on plant growth and management of root-knot and bacterial wilt disease complex and yield of capsicum.

Treatment	Seedling weight (g)	Root-knot index (1-10)	Disease index (1-9)	Yield in kg/4 m^2
Seed treat. with *P. fluorescens*	421	5.6	6.4	4.3
Seed treat. with neem-based *P. fluorescens*	428	5.2	6.7	4.7
Nursery treat. with *P. fluorescens*	435	4.6	5.4	4.8
Nursery treat. with *P. chlamydosporia*	364	4.4	7.3	3.2
Nursery treat. with *T. harzianum*	374	4.8	7.0	3.4
Nursery treat. with *P. fluorescens* + *P. chlamydosporia*	463	4.1	5.2	4.0
Nursery treat. with *P. fluorescens* + *T. harzianum*	493	3.8	3.5	5.1
Control	340	8.7	8.2	2.6
CD (P = 0.05)	**27.20**	**0.49**	**0.38**	**0.25**

Integration of seed treatment with *P. fluorescens* and nursery bed treatment with *P. chlamydosporia* at 25 or 50 g/m^2 increased plant height, weight of seedlings and root colonization with bioagents; and reduced root galling in capsicum (Table 6.54) (Naik *et al.*, 2003). The seedlings were highly vigorous which were colonized by the bioagents and reached the main field when transplanted.

Table 6.54. Effect of integration of bioagents on plant growth, root galling and root colonization by bioagents in capsicum under nursery conditions

Treatment*	Plant height (cm)	Seedling weight (g)	No. of galls/10 seedlings	*P. chlamydosporia* root colonization (cfu/g)	*P. fluorescens* root colonization (cfu/g)
T_1	12.58	3.6	62	15,896	---
T_2	14.63	3.8	59	17,459	---
T_3	15.87	4.0	55	---	12,563
T_4	18.54	4.5	51	16,256	12,349
T_5	17.42	4.4	53	16,789	11,897
T_6	11.23	3.2	82	---	---
CD (P = 0.05)	**1.67**	**0.34**	**7.96**	**959.72**	**729.65**

* T_1 - Nursery bed treatment *P. chlamydosporia* at 25 g/m^2, T_2 - Nursery bed treatment *P. chlamydosporia* at 50 g/m^2, T_3 - Seed treatment with *P. fluorescens*, T_4 - T_1 + T_3, T_5 - T_2 + T_3, T_6 - Control.

The seedlings raised in nursery beds treated with *P. chlamydosporia* at 25 or 50 g/m^2 + seed treatment with *P. fluorescens* when transplanted in the main field gave least root galling and increased fruit yield, colonization of roots with the bioagents, and parasitization of eggs by *P. chlamydosporia* (Table 6.55) (Naik *et al.*, 2003).

Table 6.55. Effect of integration of bioagents on fruit yield, root galling and root colonization by bioagents in capsicum under field conditions

Treatment*	Root-knot index	Fruit yield ($kg/6m^2$)	*P. chlamydosporia* root colonization (cfu/g)	*P. fluorescens* root colonization (cfu/g)	% eggs parasitized by *P. chlamydosporia*
T_1	6.3	4.4	22,456	---	41.00
T_2	5.6	4.7	25,789	---	45.00
T_3	6.0	5.1	---	22,568	---
T_4	4.8	5.3	21,679	21,789	40.00
T_5	4.4	5.5	24,587	21,567	42.67
T_6	8.1	4.0	---	---	---
CD (P = 0.05)	**0.45**	**0.28**	**1246.76**	**1089.86**	**3.78**

* T_1 - Nursery bed treatment *P. chlamydosporia* at 25 g/m^2, T_2 - Nursery bed treatment *P. chlamydosporia* at 50 g/m^2, T_3 - Seed treatment with *P. fluorescens*, T_4 - $T_1 + T_3$, T_5 - $T_2 + T_3$, T_6 - Control.

Naik (2004) reported that combination of *T. harzianum* along with *P. fluorescens* increased the yield (3.320 $kg/3m^2$ plot) followed by combination of *T. harzianum* and *P. chlamydosporia* (2.960 $kg/3m^2$). The least gall index was present where the combination of *T. harzianum* and *P. fluorescens* was used (1.5) followed by *T. harzianum* + *P. chlamydosporia* (1.7) and *P. fluorescens* + *P. chlamydosporia* (1.7). But among combination treatment of bio-agents, all the treatments were on par (*P. fluorescens* + *T. harzianum*, *P. fluorescens* + *P. chlamydosporia*, *T. harzianum* + *P. chlamydosporia*). Results related to the percent bacterial wilt disease (*R. solanacearum)* incidence also showed similar trend in which *T. harzianum* + *P. fluorescens* performed very well and the percent incidence was 16.90 followed by *T. harzianum* + *P. chlamydosporia* (17.20) (Table 6.56).

Table 6.56. Effect of integration of bioagents on root galling, wilt disease incidence and fruit yield of capsicum

Treatment	No. of fruits /plant	Fruit yield ($kg/3m^2$)	Root-knot index	Wilt disease incidence (%)
Neem + wheat bran formulation (8:2) of *T. harzianum*	31.2	2.30	1.90	23.8 (29.20)
Liquid broth formulation of *P. fluorescens*	34.2	2.40	1.75	21.4 (27.56)
Neem + pongamia + wheat bran formulation (4:4:2) of *P. chlamydosporia*	32.8	2.04	1.92	26.7 (31.11)
T. harzianum + *P. fluorescens*	46.2	3.32	1.50	16.9 (24.20)
T. harzianum + *P. chlamydosporia*	40.6	2.96	1.70	17.2 (24.50)

P. fluorescens + *P. chlamydosporia*	40.0	2.48	1.70	17.8 (24.95)
Neem + wheat bran	27.2	1.93	2.25	30.5 (33.52)
Neem + pongamia + wheat bran	29.8	2.00	2.00	36.4 (37.11)
Control	26.2	1.78	3.80	96.0 (78.46)
CD (P = 0.05)	**7.38**	**0.41**	**0.32**	**3.28**

Naik *et al.* (2003) reported that soil application of organically developed *P. lilacinus* (50 g/m^2 of formulated product containing 10^6 spores/g) along with pure culture of *Bacillus pumilis* (10^8 cfu/ml) to the nursery beds of bell pepper increased plant growth, root galling and root colonization by the bioagents. The above seedlings when transplanted in field significantly reduced root galling and wilt incidence, and increased root colonization, propagule density in soil, egg parasitization and fruit yield in the main field (Tables 6.57 and 6.58).

Table 6.57. Effect of integration of bioagents on plant growth and management of disease complex in bell pepper in nursery beds

Treatment*	Seedling length (cm)	Seedling weight (g)	No. of galls /10 plants	Root colonization (cfu/g)	
				P. lilacinus	*B. pumilis*
T1	12.63	3.8	52	13,342	---
T2	15.53	4.0	50	16,539	---
T3	15.17	4.2	55	---	14,431
T4	18.43	4.4	44	12,984	13,964
T5	19.25	4.6	40	16,839	14,291
T6	10.53	3.3	76	---	---
CD (P = 0.05)	**1.29**	**0.37**	**6.28**	**1087.95**	**976.75**

* T1 – Nursery bed treatment with *P. lilacinus* (25 g/ m^2), T2 - Nursery bed treatment with *P. lilacinus* (50 g/ m^2), T3 - Nursery bed treatment with *B. pumilis*, T4 – T1+ T3, T5 – T2 + T3, T6 – Control.

Table 6.58. Effect of integration of bioagents on root galling, wilt incidence, root colonization, propagule density and yield of bell pepper under field conditions

Treat-ment*	Root-knot index	Mort-ality %	Yield kg/6m^2	Root colonization (cfu/g)		Propagule density (cfu/g soil)	
				P. lilacinus	*B. pumilis*	*P. lilacinus*	*B. pumilis*
T1	6.5	68.90	4.8	30,653	---	21,346	---
T2	5.3	52.45	5.4	34,289	---	23,467	---
T3	7.0	34.62	7.5	---	25,875	---	17,321
T4	4.3	25.86	8.2	31,764	21,459	20,457	15,347
T5	4.1	17.69	8.5	34,762	20,764	22,689	14,569

T6	8.4	91.45	0.5	---	---	---	---
CD (P=0.05)	**0.58**	**8.94**	**0.35**	**1176.52**	**1087.54**	**1234.87**	**897.34**

* T1 – Nursery bed treatment with *P. lilacinus* (25 g/ m^2), T2 - Nursery bed treatment with *P. lilacinus* (50 g/ m^2), T3 - Nursery bed treatment with *B. pumilis*, T4 – T1+ T3, T5 – T2 + T3, T6 – Control.

6.4.3.3. Root-knot Nematode, *Meloidogyne javanica* and Wilt, *Fusarium oxysporum* Disease Complex

(i) Symptoms: The disease complex involving *Meloidogyne javanica* and *Fusarium oxysporum/F. solani* inflict severe losses to chilli crop (Haseeb, 2003). The simultaneous occurrence of both root-knot nematode and Fusarium wilt disease caused enhanced disease development and chilli yield loss (Fig. 6.40).

Fig. 6.40. Root-knot and Fusarium wilt complex in chilli. Front-Infected, Back-Healthy

(i) Integrated Management

(a) Bioagents and Botanicals: Integration of *T. harzianum* at 50 kg/ha with neem seed powder at 250 kg/ha in nursery beds was found to be highly effective in increasing the number of germinated seedlings (87/0.5 m^2 compared to 39/0.5 m^2 in control) and fresh weight of seedlings (146.9g compared to 61g in control). The above treatment was also effective in reducing root galling due to *M. incognita* (0.7 RKI compared to 3.5 RKI in control) and percent root infection by *F. solani* (5% compared to 41.5% in control) in nursery beds (Table 6.59) (Kumar *et al.*, 2009b).

Table 6.59. Effect of biocontrol agents and organic amendments against *M. incognita-F. solani* disease complex in chilli

Treatment	No. of seedlings emerged/bed	Fresh total wt. of seedlings/ bed (g)	Root-knot index	% root infection
Untreated control	39	61.0	3.5	41.5
Trichoderma harzianum	67	110.3	1.5	15.0
Aspergillus niger	62	100.0	2.3	21.5
Paecilomyces lilacinus	60	97.5	1.5	15.0
Pseudomonas fluorescens	65	105.7	1.9	17.3
Neem seed powder	64	105.0	1.3	15.5
Farm yard manure	45	71.9	3.3	35.5
T. harzianum + Neem seed powder	87	146.9	0.7	5.0
A. niger + Neem seed powder	77	128.7	1.0	13.0
P. lilacinus + Neem seed powder	74	123.0	0.9	13.5
P. fluorescens + Neem seed powder	81	136.7	0.7	5.5
T. harzianum + Farmyard manure	80	134.5	1.2	7.7
A. niger + Farmyard manure	71	115.3	2.0	15.5
P. lilacinus + Farmyard manure	70	113.0	1.3	23.5
P. fluorescens + Farmyard manure	75	124.9	1.5	13.0
Carbofuran	72	117.0	0.6	25.5
CD (P = 0.05)	**3.53**	**7.91**	**0.06**	**0.65**

(b) Two Bioagents: *Pseudomonas aeruginosa* and *Paecilomyces lilacinus* when used together significantly reduced infection of the disease complex on chilli (Perveen *et al.,* 1998). Use of *P. aeruginosa* and *P. lilacinus* significantly ($P<0.05$) increased plant height of chilli. *P. aeruginosa* and *P. lilacinus* used alone or together significantly ($P<0.05$) reduced infection of *M. javanica* and root infecting fungi viz., *Macrophomina phaseolina, Rhizoctonia solani, F. solani* and *F. oxysporum* on chilli. *P. aeruginosa* was more effective than *P. lilacinus* in reducing the *M. javanica* infection (Table 6.60).

Table 6.60. Effect of *Pseudomonas aeruginosa* and *Paecilomyces lilacinus* on plant height and control of root rot disease complex in chilli

Treatment	Plant height (cm)	Root-knot index	Infection %			
			M. phaseolina	***R. solani***	***F. solani***	***F. oxysporum***
Control	10.5	3.3	31	19	75	37
P. lilacinus (PL)	14.5	2.9	19	0	56	44
P. aeruginosa (PA)	16.5	2.5	12	6	75	21

PL + PA	14.7	2.1	12	25	69	12
CD (P = 0.05)	**2.2**	**0.34**	**6.1**	**6.1**	**6.1**	**6.1**

Bare root-dip treatment with *Pseudomonas aeruginosa* along with or without *T. harzianum, T. koningii* and *T. hamatum* significantly controlled infection of roots by *F. solani* and *M. javanica on chilli.* Combined use of *T. harzianum* along with *P. aeruginosa* caused the greatest reduction in root galling by M. javanica *(Siddiqui* et al., *1999).*

6.4.3.4. Root-knot Nematode, *Meloidogyne incognita* and Damping-off, *Pythium aphanidermatum* Disease Complex

(i) Symptoms: *P. aphanidermatum* and *R. solani* were both found to interact with *M. incognita* on chilli, causing some loss of nematode resistance in two cultivars tested (Hasan, 1985a,b). The interaction of *P. debaryanum* with this nematode appeared to be due to physiological response of the plants to nematode infection, making the roots more susceptible to invasion by the fungus.

(ii) Integrated Management

(a) Bioagents and Botanicals: Biological control of *P. aphanidermatum-M. incognita* disease complex in chilli with organic amendments (FYM and neem cake), antagonistic organisms *T. viride* and *T. harzianum* (against *P. aphanidermatum*) and *Paecilomyces lilacinus* (against *M. incognita*) was reported by Karthikeyan *et al.* (1999).

Chapter -7

BULBOUS VEGETABLE CROPS

7.1. ONION, *ALLIUM CEPA* AND GARLIC, *ALLIUM SATIVUM*

7.1.1. Diseases

7.1.1.1. Damping-off, *Pythium* sp., *Rhizoctonia solani, Fusarium* sp.

(i) **Symptoms:** Cold, wet soils often encourage the development of damping-off symptoms very early in the seedling's growth. Seedlings may fall over and die as a result of breakdown of plant tissues at the soil line (Fig. 7.1). Sometimes damping-off occurs before the seedling even emerges. The disease is usually caused by Pythium, Rhizoctonia or Fusarium fungi, either alone or in combination. Damping-off can occur in the field, or in the nursery if conditions are too wet.

Fig. 7.1. Damping-off of onion seedlings

(ii) Integrated Management

(a) Physical and Bioagents: Seed treatment with *T. harzianum* at 5 g/kg and sowing in solarized soil helps in reducing the disease incidence.

7.1.1.2. Basal Rot, *Fusarium oxysporum* f. sp. *cepae*

(i) Symptoms: The early symptoms in the field are yellowing of leaves and tip dieback. As the disease progresses, the whole plant may collapse and, if the plant is pulled, it often comes out without any roots attached since they have decayed. The basal plate of the onion becomes pinkish-brown and secondary bacterial rots may develop in the affected area (Fig. 7.2). If infection occurs late in the season, the symptoms may not show up until the onions are in storage. Basal rot generally occurs when soil temperatures are very warm (optimum 29°C).

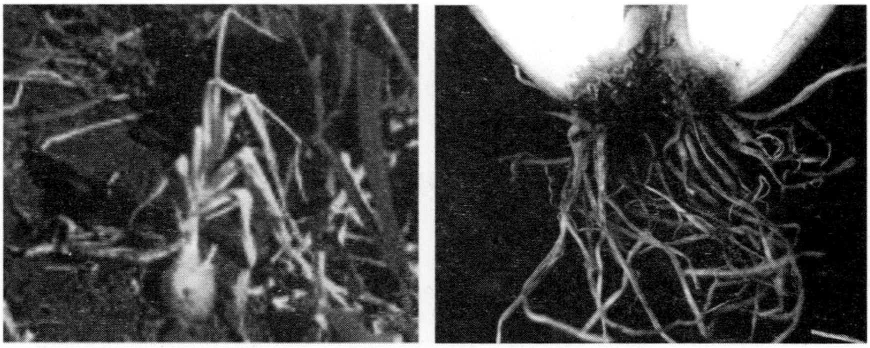

Fig. 7.2. Basal rot on onion bulbs.

(ii) Integrated Management

(a) Bioagents and Botanicals: *T. viride* at 1250 g + 50 kg FYM if applied to soil before planting gives good control of basal rot in seed crop.

7.1.1.3. White Rot, *Sclerotium cepivorum*

White rot is a very destructive disease that begins in the field and can carry over into storage.

Fig. 7.3. White rot of onion.

(i) Symptoms: The first above-ground symptoms are yellowing and dieback of the leaf tips, followed by a collapse of the affected leaves. When the bulbs and roots are examined, a white, fluffy mold and soft rot will be observed. Masses of tiny black sclerotia can also be seen within this mold (Fig. 7.3). These sclerotia remain in the soil for many years. Infected bulbs can rot in storage boxes and stain other bulbs. White rot typically develops in patches in the field and is less of a problem when soils are warm (higher than 24°C) and dry.

(ii) Integrated Management

(a) Bioagents and Chemicals: Seed treatment with captan (for control of damping-off) along with *T. atroviride* and application of procymidone (for control of onion white rot) during later stages is very effective (McLean *et al.*, 2001).

(b) Bioagents and Botanicals: Integration of *T. atroviride* (pellet formulation) with organic amendments including poultry manure, spent mushroom compost and certified green compost gave effective control of onion white rot disease.

(c) Physical and Bioagents: Application of *Bacillus subtilis, T. harzianum, T. viride,* and *T. virens* to the solarized soil effectively controlled the disease. Incorporation of *T. harzianum* in soil after soil solarization effectively controlled S. *cepivorum* in soil and increased the control from 79 to 98%. A significant population growth of antagonist occurred attaining 10^8 cfu/g of soil (Pereira *et al.,* 1996).

7.1.2. Nematodes

7.1.2.1. Root-knot Nematodes, *Meloidogyne* spp.

(i) Symptoms: Above-ground symptoms on onions heavily infected with M. *hapla* are those of general stunting, uneven growth; thicker necks and smaller bulbs; and also delayed maturity. The diagnostic symptoms are found on roots as galls or root thickenings of various sizes and shapes with protruding egg masses (Fig. 7.4). On onions, galls are usually small and barely noticeable, often no more than slight swellings. Bulb weight of onions was reduced by as much as 70% in heavily infested sections of commercial fields. In field micro plots, bulb weight of the onion cultivars Norstar and Paragon were reduced by about 50% at an infestation level of 20 eggs/cc soil.

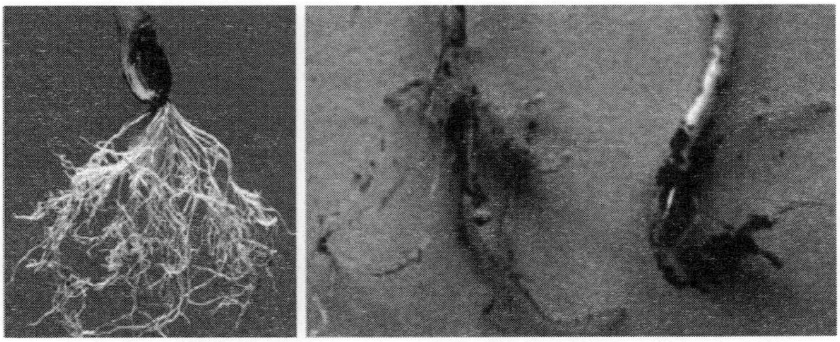

Fig. 7.4. Root-knot nematode on onion.

(ii) Integrated Management

(a) Bioagents and Botanicals: Root dip treatment of onion seedlings in neem cake suspension + *P. lilacinus* for 1 hr before transplanting proved effective.

Seed treatment with *P. fluorescens* (10^9 cfu/g) at 10 g/kg and subsequent soil application of 5 tons of FYM enriched with 5 kg each of *P. fluorescens* (10^9 cfu/g) and *Pochonia chlamydosporia* (10^6 cfu/g) per ha significantly reduced *M. incognita* population in roots by 69% and increased bulb yield by 21% (Anon, 2012).

Integration of neem cake with *P. lilacinus* was found effective and increased bulb yield (Table 7.1).

Table 7.1. Effect of neem cake and bioagents on the bulb yield of onion

Treatments	Yield (t/Ac)
Neem cake (NC)	12.972
Paecilomyces lilacinus (PL)	14.022
Pseudomonas fluorescens (PF)	13.746
NC+PL	17.347
NC+PF	16.842
PL+PF	14.749
Control	12.476

Chapter - 8

CRUCIFEROUS VEGETABLE CROPS

8.1. CABBAGE, *BRASSICA OLERACEA* VAR. *CAPITATA* AND CAULIFLOWER, *BRASSICA OLERACEA* VAR. *BOTRYTIS*

8.1.1. Insect Pests

8.1.1.1. Diamondback Moth (DBM), *Plutella xylostella*

This is a major pest of cruciferous crops, particularly cabbage and cauliflower during January-June months and also during drought periods in monsoon.

(i) Damage: The first instar larvae mine the epidermal surface of the leaves. Second instar onwards the larvae feed externally by making holes in the leaves (Fig. 8.1).

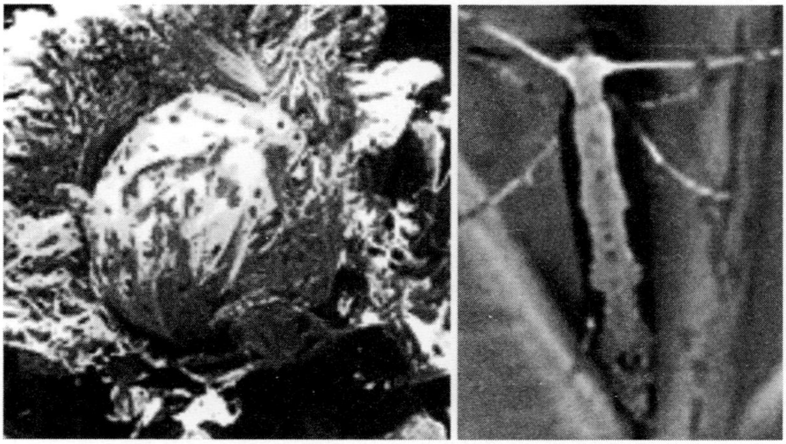

Fig. 8.1. Diamondback moth damage on cabbage.

(ii) Integrated Management

(a) Cultural and Bioagents: IPM using Indian mustard as a trap crop involves planting of paired rows of mustard after every 25 rows of cabbage/cauliflower (Fig. 8.2) and spraying of 4% NSKE at primordial formation. Two more sprays of 4% NSKE may be given at 10-15 day interval after the first spray. The IPM gave 60% more yield and 152% more returns than pure cabbage crop (Table 8.1) (Khaderkhan *et al.*, 1998b). IPM controls diamondback moth (*Plutella xylostella*); leaf webber

(*Crocidolomia binotalis*); stem borer (*Hellula undalis*); aphids (*Brevicorne brassicae, Hyadaphis erysimi*) and bug (*Bagrada cruciferarum*) (Srinivasan and Krishna Moorthy, 1991).

Fig. 8.2. IPM in cabbage and cauliflower using Indian mustard as a trap crop.

Table 8.1. Economics of cabbage IPM

Practice	Yield (MT/ha)	% increase	Net returns (Rs.)	% increase
Farmers' practice	20	---	19,817	---
IPM practice	32	60	49,251	152

In another study, the benefit-cost ratio in IPM and non-IPM plots was 2.42 and 0.83, respectively (Table 8.2) (Krishna Moorthy *et al.*, 2003).

Table 8.2. Economics of cabbage IPM

Treatment	Yield (MT/ha)	Net returns (Rs/ha)	Benefit: cost ratio
IPM plots	55	30085	2.42
Non-IPM plots (farmers' practice)	35	5090	0.83

(b) Bioagents and Chemicals: Maximum morality of diamondback moth larvae was achieved (98.03%) in treatment with emamectin benzoate + *Beauveria bassiana* at 3.0 + 0.75g/L of water along with 15.536 t/ha yield potential followed by emamectin benzoate + *B. bassiana* at 2.5 + 0.50 g/lit. of water with 96.54% morality over the control (Vishwakarma *et al.*, 2009).

Integration of gibberellic acid at 1000 ppm + *Pseudomonas fluorescens* at 5 kg/ha + *Bacillus thuringiensis* var. *kurstaki* at 1 kg/ha in alternation with *P. xylostella* granulosis virus at 1.5 x 10^{13} OB/ha recorded significantly lower incidence of *P. xylostella* (4.33 and 5.67 larvae/10 plants as compared to 43.00 in control) and increased curd yield of cauliflower (32.33 t/ha and 35.33 t/ha as compared to 21.33 t/ha in control) (Mohanasundaram and Dhandapani, 2009).

(c) Bioagents and Botanicals: The number of DBM larvae was lowest in cabbage treated with 4 alternate sprays of 3% NSKE and 0.1% *Bt* alternately at 25 days interval.

8.1.1.2. Tobacco Caterpillar, *Spodoptera litura*

(i) Damage: It is a polyphagous pest. The young larvae up to third instar feed gregariously and skeletonize the leaves (Fig. 8.3). Grown up larvae completely devour the leaves and other plant parts in case of severe incidence. When the incidence is high they attain cut worm status, hide during daytime and come out during night and devastate the crop. During severe infestation, the entire crop may be defoliated overnight.

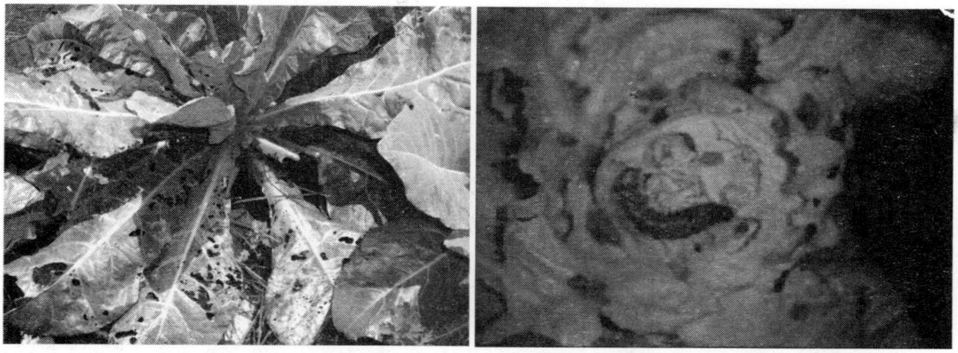

Fig. 8.3. Cabbage head damaged by *Spodoptera litura*.

(ii) Integrated Management

(a) Bioagents and Chemicals: Under field conditions, half the dosage of *Sl*NPV (2.6 x 10^6 POB/ml) + 50 ppm endosulfan was very effective in reducing the leaf damage caused by *S. litura* on cauliflower (Chaudhari and Ramakrishnan, 1980).

8.1.2. Diseases

8.1.2.1. Damping-off, *Pythium aphanidermatum*

(i) Symptoms: The fungus attacks in the seedling stage causing damping-off disease (Fig. 8.4).

Fig. 8.4. Damping-off of crucifer seedlings

(ii) Integrated Management

(a) Bioagents and Chemicals: Mukherjee *et al.* (1989) reported effective control of *P. aphanidermatum* on cauliflower with a combination of *T. harzianum* and ridomil.

8.1.2.2. Brown Rot, *Rhizoctonia solani*

(i) Symptoms: A dark, water soaked lesion initially appears on the stem. Later stems become wiry and slender at the point of the lesion. Diseased crucifer plants transplanted to the field grow poorly, are stunted, and may eventually die, especially if there is inadequate moisture shortly after transplanting. If infected plants remain alive, the stem becomes tough and woody (Fig. 8.5). Plants that survive usually mature late and fail to produce a marketable head.

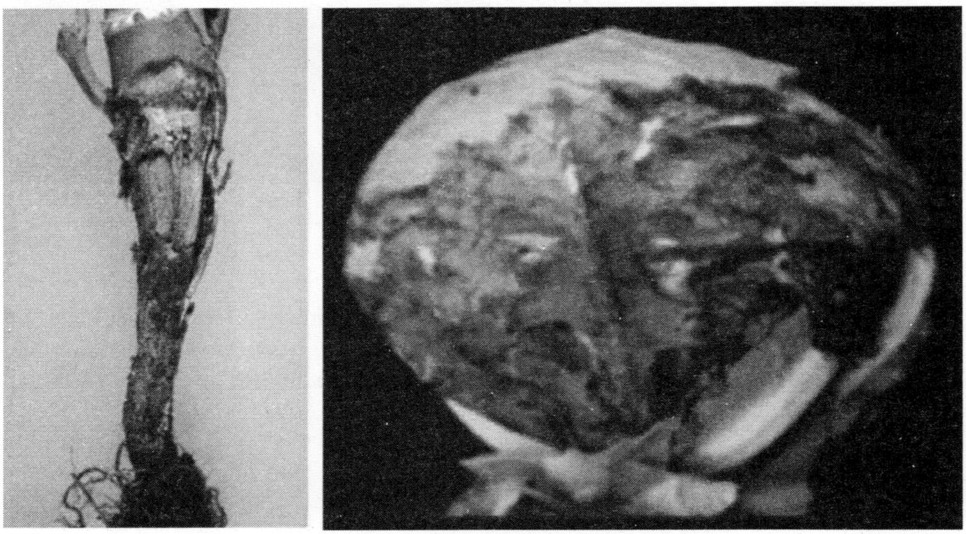

Fig. 8.5. Wire stem symptoms on lower stem and head rot.

(ii) Integrated Management

(a) Physical, Botanicals, Bioagents and Chemicals: The seed treatment with carbendazim at 2g/kg seed, raising seedling in solarized beds, crop raising in green manure field + neem cake 25 kg/ha with soil treatment by *Trichoderma viride* at 2 kg/ha gave the lowest disease intensity of 11.50 with maximum curd yield 29.958 t/ha with highest percent disease control (51.00) over control treatment (Dabbas *et al.*, 2009).

8.1.2.3. Alternaria Black Spot, *Alternaria brassicola*

(i) Symptoms: On leaves several small, dark brown zonate spots are produced expanding rapidly to form circular lesions up to 1 cm in diameter. The enlargement of lesions may lead to formation of concentric circles which coalesce in the centre (Fig. 8.6). In humid weather, the fungus may cause bluish growth in the centre of the spots which hampers the photosynthetic activity of the plant thereby hampering the overall productive potential of the plants. Sometimes cauliflower heads are also infected by the pathogen. The heads show browning, starting at the margin of the individual flowers or flower clusters.

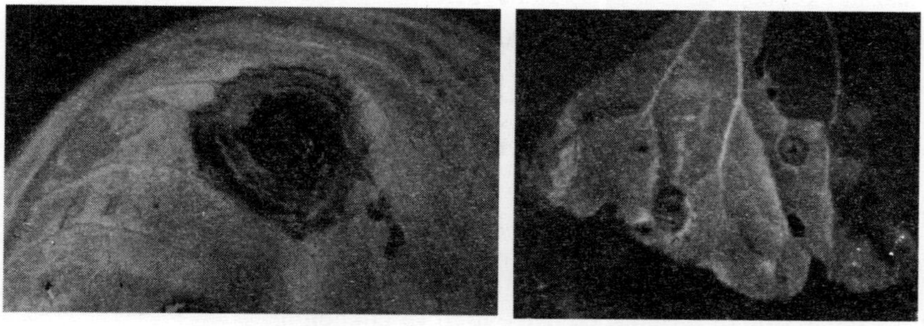

Fig. 8.6. *Alternaria* leaf spot on cabbage.

(ii) Integrated Management

(a) **Bioagents and Botanicals:** Seed treatment with *T. viride* and *P. fluorescens* and soil treatment of nursery beds with neem cake enhanced seed germination and seedling stand.

8.1.3. Nematodes

8.1.3.1. Root-knot Nematode and Club Root Disease Complex

(i) Integrated Management

(a) Two Bioagents: PGPR strains (*Pseudomonas fluorescens, Bacillus subtilis*) combined with fungal biocontrol agents (*Trichoderma viride, T. harzianum*) were found to be effective in reducing the nematode-fungal disease complex in cabbage (Loganathan *et al.*, 2001). The bioformulation mixture of *P. fluorescens, T. viride* and chitin effectively reduced the disease complex in cabbage and cauliflower both under greenhouse and field conditions (Samiyappan, 2003) (Table 8.3).

Table 8.3. Efficacy of bioformulation mixtures against root-knot nematode - club root disease complex in cabbage under greenhouse conditions.

Treatment	Club root index	Nematode incidence	
		Population	Root-knot index
Trichoderma viride	25.99 (30.65)	129	2.66
Pseudomonas fluorescens	28.20 (32.07)	112	2.33
T. viride + *P. fluorescens*	25.33 (30.22)	114	2.33
T. viride + Chitin	25.44 (30.29)	108	2.33
P. fluorescens + Chitin	25.66 (30.43)	111	2.00
T. viride + *P. fluorescens* + Chitin	22.22 (28.12)	108	2.00

Chitin alone	31.70 (34.26)	139	3.00
Carbendazim	19.90 (26.49)	264	4.66
Carbofuran	39.90 (39.17)	106	1.66
Carbendazim + Carbofuran	15.00 (22.79)	103	1.66
Plasmodiophora brassicae alone	48.90 (44.37)	0.033	0.133
Meloidogyne incognita alone	0.03 (0.60)	280	5.00

8.1.4. Validated IPM Technology for Cabbage

8.1.4.1. Bangalore, Karnataka

Nursery

- Preparation of raised seed beds up to 15 cm height.
- Soil drenching with 0.2% copper oxychloride.
- Seed treatment with *Trichoderma harzianum* at 4 g/kg.
- Use of nylon net to avoid entry of sucking pests.
- Spray *Bt* at 0.5 ml/l one day before planting.
- Spray Dithane M- 45/Ridomil at 0.2% against downy mildew.

Main Field

- Growing of 2 rows of mustard after every 25 rows of cabbage as trap crop at the time of planting.
- Wider spacing of 60 x 45 cm.
- Installation of light traps at 5/ha for trapping adult DBM.
- Spray *Bacillus thurungiensis* at 1 g/l if DBM is noticed early or spray 5% NSKE/1% neem soap/pongamia soap at primordial stage. Repeat 3-4 sprays.
- Removal of disease affected basal leaves from time to time.
- Need based sprays of chlorothalonil/mancozeb at 0.2% for *Alternaria* and blitox + streptocycline for black rot.

8.1.4.2. Varanasi, Uttaranchal

Nursery

- Preparation of raised seed beds up to 15 cm height.
- Soil solarization for 3 weeks with 45 µm polythene sheet.
- Seed treatment with *Trichoderma harzianum* at 4 g/kg.
- Use of nylon net to avoid entry of sucking pests.

Main Field

- Spray *Bt* at 1 g/l if one DBM is noticed per plant.

- Spray 5% NSKE at primordial stage i.e. 15-20 DAP. Repeat 3-4 sprays.
- Release of *Trichogramma bactrae* at 0.75 lakh/ha at weekly interval.
- Removal of disease affected heads and bottom leaves from time to time.
- Need based application of pesticides like mancozeb at 0.2%.

8.1.4.3. Ranchi, Jarkhand

Nursery

- Preparation of raised seed beds up to 15 cm height.
- Soil solarization for 3 weeks with 45 μm polythene sheet.
- Soil treatment with FYM enriched with *T. harzianum* at 1 kg/MT.

Main Field

- Spray Neemarin at 5% when DBM appeared early.
- Spray *Bt* at 1 g/l if one DBM is noticed per plant.
- Spray 1% pongamia soap at 15-20 DAP.
- Installation of pheromone traps at 5/ha for *Spodoptera litura* monitoring.
- Need based application of fungicides like chlorothalonil at 0.2% for *Alternaria*.
- Spray of insecticides like endosulfan at 0.07% or triazophos at 0.05%.

During the period 2001-04, IPM technology in cabbage was validated and promoted in more than 40 ha area in 42 villages covering 88 families located 40 km from Bangalore. Similarly, near Varanasi also IPM technology has been validated in 8 villages in about 40 ha area covering 100 families. Near Ranchi, IPM technology in cabbage has been validated and promoted in 20 villages with the support of 100 farming families covering an area of 40 ha. In IPM validation studies conducted at three locations (Bangalore, Varanasi and Ranchi), IPM fields recorded higher yields of 62.013, 24.080 and 34.750 MT/ha as compared to 58.013, 21.170 and 28.760 MT/ha, respectively in non-IPM fields (Sardana *et al*., 2004) (Table 8.4).

Table 8.4. Yield and economics of IPM in cabbage at different locations

Centre	Yield (MT/ha)	Net returns (Rs.)	Benefit: Cost ratio
Bangalore - IPM	62.01	80934	3.04
- Non IPM	58.01	40661	0.61
Varanasi - IPM	24.08	37500	3.70
- Non IPM	21.17	33333	1.89
Ranchi - IPM	34.75	231386	2.80
- Non IPM	28.76	156900	1.84

Chapter - 9

MALVACEOUS VEGETABLE CROPS

9.1. OKRA, *ABELMOSCHUS ESCULENTUS*

9.1.1. Insect Pests

9.1.1.1. Fruit Borer, *Earias vitella*

Earias vittella, commonly called as spotted bollworm, is very destructive to okra. It is active through out the year reaching peaks during March-May and August-October. The pest has about 8 to 12 overlapping generations in a year. *E. insulana* is found in drier regions.

(i) Damage: Female lays green coloured eggs with longitudinal ridges on buds, flowers and fruits. When the crop is young, larvae bore into tender shoots and tunnel downwards, which wither, drop down and growing points are killed (Fig. 9.1). With the formation of fruits, the caterpillars bore inside these and feed on inner tissues which become deformed in shape with no market value (Fig. 9.1). Like brinjal shoot and fruit borer, the infestation is seen on shoots before flowering and after flowering they feed exclusively on fruits.

Fig. 9.1. Shoot and fruit borer damage on okra.

(ii) Integrated Management

(a) Botanicals and Chemicals: The fruit damage was reduced to a minimum of 7.22% in endosulfan 0.07% + NSKE 5% which was on par with neem oil 4% (7.53%) and endosulfan 0.07% + neem oil 2% (9.29%). The marketable fruit yield was maximum in neem oil 4% (13.77MT/ha) which was on par with endosulfan 0.07% (13.40 MT/ha), endosulfan 0.07% + NSKE 5% (13.17 MT/ha), NSKE 5% (13.08 MT/ha) and neem oil 2% (12.98 MT/ha) (Raja *et al.*, 1998b) (Table 9.1).

Table 9.1. Effect of neem products and endosulfan on fruit borer and yield of okra

Treatment	Fruit borer infestation (%)	% reduction over control	Marketable fruit yield (MT/ha)
Endosulfan 0.07%	10.55	61.48	13.40
Endosulfan 0.07% + neem oil 2%	9.29	66.08	13.28
Neem oil 2%	10.95	60.02	12.98
Neem oil 4%	7.53	72.51	13.77
NSKE 5%	10.64	61.15	13.08
Endosulfan 0.07% + NSKE 5%	7.22	73.64	13.17
Trichogramma chilonis release at fortnightly interval	13.91	49.22	12.20
Control	28.39	---	10.67
CD (P = 0.05)	**2.68**	---	**1.12**

(b) Bioagents, Botanicals and Chemicals: Module II (First spray of monocrotophos at flowering followed by two subsequent sprays of a combination of *Bacillus thuringiensis* var. *kurstaki* and methomyl at fortnightly intervals) recorded minimum fruit borer infestation (4.21%) and maximum fruit yield (4.067 MT/ha) as against 41.59% fruit borer damage and 1.658 MT/ha fruit yield in the untreated control. However, statistically comparable results (fruit borer damage – 5.30% and yield – 4.009 MT/ha) were obtained when the second spray of *Btk* – methomyl (module II) was replaced by nimbecidine (module III) (Mathur *et al.*, 1998) (Table 9.2).

Table 9.2. Bioefficacy of different IPM modules for management of *Earias vitella* and yield of okra

Module	Treatment	Dose/ha	% fruit damage	Fruit yield (MT/ha)
I	(i) Monocrotophos 36 SL	1.0 Litre	8.80	3.410
	(ii) *Btk* (Dipel-8L) + Endosulfan 35 EC	1.0 Litre + 0.625 Litre		
	(iii) *Btk* (Dipel-8L) + Endosulfan 35 EC	1.0 Litre + 0.625 Litre		
II	(i) Monocrotophos 36 SL	1.0 Litre	4.21	4.067
	(ii) *Btk* (Dipel-8L) + Methomyl 40 SP	1.0 Litre + 0.625 Litre		
	(iii) *Btk* (Dipel-8L) + Methomyl 40 SP	1.0 Litre + 0.625 Litre		
III	(i) Monocrotophos 36 SL	1.0 Litre	5.30	4.009
	(ii) Nimbecidine (Azadirachtin-300 pm)	1.5 Litre		
	(iii) *Btk* (Dipel-8L) + Methomyl 40 SP	1.0 Litre + 0.625 Litre		

IV	(i) Monocrotophos 36 SL	1.0 Litre	12.42	2.583
	(ii) Endosulfan 35 EC	1.25 Litre		
	(iii) Endosulfan 35 EC	1.25 Litre		
V	(i) Nimbecidine (Azadirachtin-300 pm)	1.5 Litre	16.73	2.100
	(ii) *Beauveria bassiana* (Dispel)	1.5 Litre		
	(iii) *Btk* (Dipel-8L)	1.5 Litre		
VI	Control (Untreated check)			
	CD (P = 0.05)	---	**2.24**	**0.424**

Soil application of neem cake (with 8% oil) at 250 kg/ha when combined with sprays of 1% neem soap (NS), 4% neem seed powder extract (NSPE), 1% *Bacillus thuringiensis* (*Bt*) reduced the okra fruit borer (*Earias vitella*) and leaf hopper (*Amrasca biguttula biguttula*) significantly and increased marketable okra fruit yield in all the three seasons as compared to spray treatments of NS, NSPE, *Bt* and indoxacarb alone (Table 9.3).

Table 9.3. Effect of neem products and *Bacillus thuringiensis* on fruit borer, leaf hopper and yield of okra

Treatment	**Fruit borer incidence**	**Plant hopper incidence**	**Marketable yield (MT/ha)**
Neem cake – 250 kg/ha	15.09	26.92	9.37
Neem cake – 250 kg/ha + Neem seed powder extract 4%	9.19	19.83	9.81
Neem cake – 250 kg/ha + Neem soap 1%	9.77	19.50	9.89
Neem cake–250 kg/ha+*Bacillus thuringiensis* 1%	8.74	17.50	11.06
Neem cake – 250 kg/ha + Indoxacarb	9.23	26.33	10.54
Neem seed powder extract 4%	16.97	35.75	9.58
Neem soap 1%	16.49	38.92	9.84
Bacillus thuringiensis 1%	22.85	42.58	10.97
Indoxacarb	15.00	39.17	10.73
Control	26.61	58.92	7.20
CD (P = 0.05)	**4.28**	**1.36**	**2.66**

9.1.1.2. Aphids, *Aphis gossypii*

(i) Damage: This is a polyphagous pest, feeding in colonies and completely covers the shoot tips, buds and lower surface of leaves (Fig. 9.2). Aphids multiply parthenogenetically in large numbers in a very short period. Both nymphs and adults suck the sap due to which plants loose their vitality. Leaves curl downwards and there is retardation of growing shoots. They also excrete honeydew on which sooty mould develops which hampers the photosynthetic activity of the plant.

Fig. 9.2. Aphid infestation on okra leaves.

(ii) Integrated Management

(a) Bioagents and Chemicals: Spraying of *V. lecanii* spores at 10 x 10^6/ml + 0.005% quinalphos + 0.05% Teepol before the onset of rainy season is effective against aphids.

9.1.1.3. Pod Borer, *Helicoverpa armigera*

(i) Damage: This pest is occasionally serious during monsoon season on okra (Fig. 9.3). Damage is restricted to the apical end only. Since okra fruits are harvested once in 2-3 days, damage by *H. armigera* is not very high.

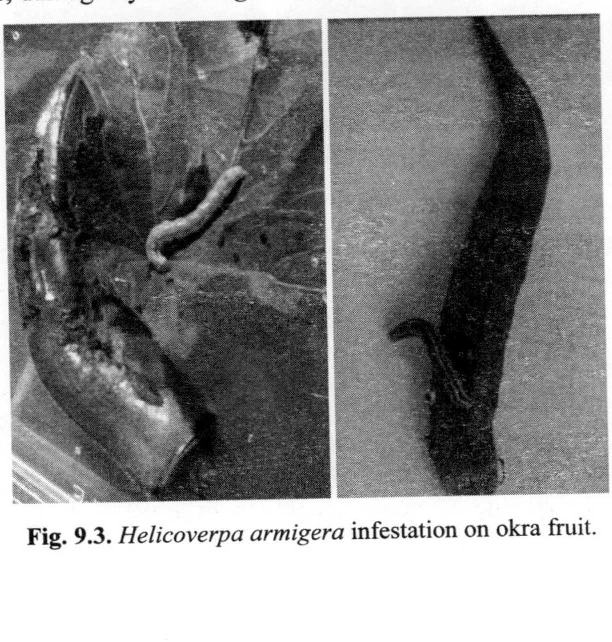

Fig. 9.3. *Helicoverpa armigera* infestation on okra fruit.

(ii) Integrated Management

(a) Two Bioagents: The release/application of biocontrol agents viz., *Trichogramma chilonis* (45 and 60 DAS), *Chrysoperla carnea* (45 and 60 DAS), *Bacillus thuringiensis* var. *kurstaki* (45 and 75 DAS), and *Ha*NPV (60 DAS) were superior in reducing the larval population and fruit damage by *H. armigera* (Praveen and Dhandapani, 2001).

9.1.2. Diseases

9.1.2.1. Yellow Vein Mosaic Virus (YVMV)

This is the most important and devastating virus disease of okra. Sastry and Singh (1974) estimated that if the plants are infected within 20 days after germination, the loss in yield was recorded up to 98%. The plants that infected at 35 and 50 days after germination, the loss in yield was estimated to be 83 and 49%, respectively.

(i) Symptoms: YVMV is caused by Gemini virus and transmitted by whitefly, *Bemisia tabaci*. Characteristic symptoms appear as yellow vein and veinlets leaving green tissue in interveinal area (Fig. 9.4). Severely infected leaves sometimes become completely yellow. Fruits also change in colour to yellow and become hard in early stage of development. The infected plants are stunted and bear very few yellow coloured fruits.

Fig. 9.4. YVMV infection on okra leaves.

(ii) Epidemiology: Vector population and virus incidence were more during March to June when the atmospheric temperature remained high and humidity is less which is favourable for whitefly multiplication and spread of the disease.

(iii) Integrated Management

(a) Cultural and Chemical: Sowing of 4 to 5 rows of sorghum or pearl millet or maize all round the okra field at least 60 days before sowing okra has been found beneficial for the management of YVMV. Border cropping along with 3 to 4 foliar sprays of either dimethoate or monocrotophos both at 0.1% at 10 days interval has been found more effective (Singh, 1990).

9.1.3. Validation of Okra IPM at Ghaziabad, Uttar Pradesh

- Planting of yellow vein mosaic virus resistant (YVMV) hybrids *viz.* Sun-40 and Makhmali.
- Sowing of sorghum/maize as border crop.
- Installation of yellow sticky polythene traps smeared with castor oil, and delta traps set up for whitefly and other small sucking pests.
- Erection of bird perches at 25/ha for facilitating predation of borer larvae.
- Installation of pheromone traps at 5/ha for monitoring *Earias vitella*.
- Three sprays of 5% NSKE for hopper, whitefly and mites starting at 28 DAS.
- Five releases of *Trichogramma chilonis* at 1 lakh/ha starting from 42 DAS at weekly interval.
- Rouging out YVMV affected plants from time to time.

In okra crop, IPM technology has been validated in about 3 ha area in Raispur village near Ghaziabad during 2003-04. The incidence of various insect pests and diseases recorded was invariably high in non-IPM fields during 2003-04 (Table 9.4).

Table 9.4. Incidence various insect pests and diseases in okra at Raispur village during 2003-04

Pest/disease incidence	IPM	Non-IPM
Leaf hopper/3 top leaves	3.20	16.22
Fruit borer (%)	1.15	8.02
Blister beetle/plot	7.77	7.00
Yellow vein mosaic virus (%)	3.11	32.06

IPM fields gave higher yields of 10.305 MT/ha as compared to 7.246 MT/ha in non-IPM fields (Sardana *et al.*, 2004) (Table 9.5).

Table 9.5. Yield and economics of IPM in okra

Parameter	IPM	Non IPM	% increase
Yield (MT/ha)	10.30	7.24	42
Net returns (Rs/ha)	64797	34678	86
Benefit: cost ratio	1.28	0.72	77

9.1.4. Nematodes

9.1.4.1. Root-knot Nematodes, *Meloidogyne* spp.

M. incognita was responsible for 28.08 to 90.90% loss in fruit yield of okra (Bhatti and Jain, 1977; Parvatha Reddy and Singh, 1981), while *M. javanica* caused 20.20 to 41.20% loss in yield (Jain *et al.*, 1986).

(i) Symptoms: Diseased plants are stunted, yellow and have a tendency to wilt in hot weather. Diseased plants appear in patches in the field. The root system of diseased plants is heavily galled and devoid of lateral roots in the final stage (Fig. 9.5). Plant growth and yield are negatively affected.

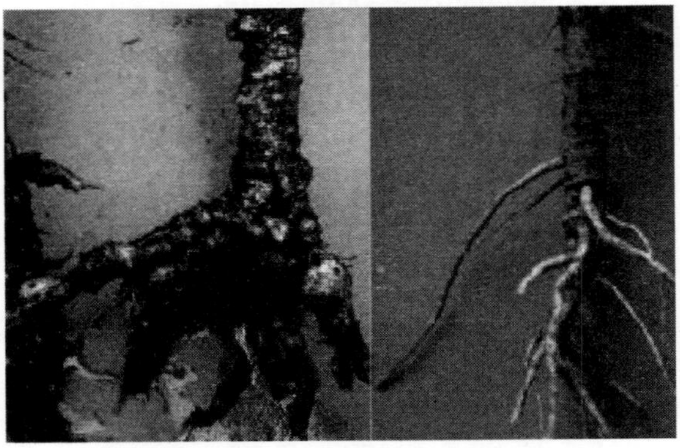

Fig. 9.5. Okra roots damaged by root-knot nematodes. Note gall or root-knots (left) and healthy roots (right).

(ii) Integrated Management

(a) Bioagents and Botanicals: Soaking of okra seeds in 10% castor cake suspension mixed with spores of *P. lilacinus* (1.5×10^6 spores/ml) for 30 min. and sowing in soil drenched with 10% castor cake suspension at 20 litres/6 m^2 was effective in reducing root galling, final nematode population of *M. incognita* and increasing the fruit yield, root colonization, propagule density in soil and parasitization of eggs by *P. lilacinus* (Rao *et al.*, 1997a) (Table 9.6).

Table 9.6. Effect of integration of bioagents and botanicals for the management of *Meloidogyne incognita* infecting okra

Treatment	Root-knot index	Yield (kg/6m^2)	% egg parasitization
Seed treat. with *P. lilacinus*	7.6	5.2	43
Seed treat. with castor cake suspension	7.2	5.0	---
Soil drenching with castor cake suspension	7.5	5.4	---
Seed treat. with *P. lilacinus* + castor cake suspension	5.5	5.7	46
Seed treat. with *P. lilacinus* & castor cake suspension + soil drenching with castor cake suspension	4.7	6.8	53
Seed treat. with castor cake suspension + soil drenching with castor cake suspension	5.9	6.3	---
Castor cake – 1 MT/ha	5.6	6.5	---
Castor cake – 2 MT/ha	4.6	7.0	---
Carbofuran – 1 kg a.i./ha	5.2	6.4	---

Carbofuran – 2 kg a.i./ha	4.5	6.7	---
Control	9.2	4.7	---
CD (P = 0.05)	**0.87**	**0.72**	**2.24**

Integration of *Paecilomyces lilacinus* with neem cake gave effective control of *M. incognita* on okra.

Application of 5 per cent inoculum of *Arthrobotrys conoides* to the pot soil amended with FYM, effectively reduced the larval penetration of *M. incognita* and root galling was reduced by 30-40 per cent in okra (Srivastava and Swarup, 1986).

(b) Botanicals and Chemicals: An integrated management of *M. incognita* infecting okra using neem or karanj oil cake at 0.5 t/ha along with carbofuran at 1 kg a.i./ha gave maximum reduction in root galling with consequent increase in okra fruit yield (Parvatha Reddy and Khan, 1991).

Application of subabool (*Leucaena leucophila*) leaves at 40 g/kg soil which were allowed to decompose for four weeks before sowing + carbofuran at 1 kg a.i./ha while sowing of okra seeds resulted in minimum galling (35.4 galls/plant) in comparison to control (64 galls/plant) and better plant growth parameters in okra (Paruthi *et al.*, 1987).

(c) Bioagents, Chemicals and Botanicals: The combined treatment with *P. lilacinus* at 4 g/kg soil + carbosulfan 25 EC at 0.2% + poultry manure at 2.5 t/ha + FYM at 2.5 t/ha gave maximum increase in plant growth parameters and yield (9.2 t/ha compared to 2.0 t/ha in control), and decrease in number of galls (83.5 compared to 183.6 in control), egg masses per root system (23.6 compared to 70.5 in control) and final nematode population in soil (200 compared to 585 in control) (Das and Sinha, 2005).

(d) Physical, Cultural, Botanical and Chemical : Summer ploughing + seed treatment with carbofuran at 3% a.i. w/w + main field treatment with aldicarb at 1 kg a.i./ha led to 76-79% decrease in nematode population and 35.1% higher yield over untreated check. Further, Summer ploughing + mulching transparent polythene sheet + seed treatment with carbosulfon at 3% a.i. w/w led to 32.5% higher yield. Summer solarization + treated seeds + use of neem cake at 400 kg/ha was most effective treatment and gave 50% higher okra yield.

(e) Physical and Botanicals: Integration of soil solarization for 15 days in summer and application of neem cake at 200 kg/ha is effective in the management of root-knot nematodes and in getting higher yields.

(f) Physical and Chemicals: Soil solarization with single layer of polyethylene mulch for 20 days during June and application of carbofuran at 0.5 kg a.i./ha gave least root galling (42/plant compared to 245/plant in control) (Sharma *et al.*, 2005).

9.1.4.2. Reniform Nematode, *Rotylenchulus reniformis*

(i) Integrated Management

(a) Botanicals and Chemicals: Application of aldicarb at 1 kg a.i./ha + neem cake at 0.5 t/ha followed by carbofuran at 1 kg a.i./ha + neem cake at 0.5 t/ha proved most effective in reducing the *R. reniformis* population and in increasing the growth of okra plants (Krishna Rao *et al.*, 1987).

(b) Cultural and Chemicals: Deep ploughing (20 cm) followed by fallowing for one month, fallowing for one month after weeding, integration of aldicarb application at 0.8 kg a.i. per ha at sowing after either of the cultural practices or deep ploughing (20 cm) together with carbofuran or aldicarb seed treatment resulted in the control of the reniform nematode and better yield of okra (Lakshmanan and Sivakumar, 1981).

9.1.4.3. Root-knot Nematode, *Meloidogyne incognita* and Wilt, *Fusarium oxysporum* f. sp. *vasinfectum* Disease Complex

Meloidogyne-Fusarium disease complex has been considered important on many crops including okra leading to reduction in its productivity.

(i) Integrated Management

(a) Bioagents and Botanicals: Soil application of 25 g/m^2 of *Pseudomonas fluorescens* (2 x 10^6 cfu/g) or *Pochonia chlamydosporia* (2 x 10^6 cfu/g) enriched deoiled neem cake has proved to be an effective treatment in combating the damage caused by *M. incognita* and *F. oxysporum* f. sp. *vasinfectum* to the tune of 68 and 57%, respectively. This treatment also increased the yield of okra fruits by 24% under field conditions (Chaaya *et al.*, 2010).

9.1.4.4. Root-knot Nematode, *Meloidogyne incognita* and Root Rot, *Rhizoctonia solani* Disease Complex

(i) Symptoms: Plants in untreated field soil or in sterilized soil inoculated with both organisms, developed a root rot in about 42 days. If the nematode preceded the fungus by three weeks, the root rot was more severe and appeared within 14-21 days. The fungus penetrated either directly or through ruptures in the root created by the mature female nematode. *R. solani* colonized nematode giant cells and root xylem cells. Vascular discolouration occurred both in roots and stem, however no fungus was isolated from stems.

M. incognita predisposed roots to *R. solani* which resulted in severe root rot and subsequent plant death. Okra plants inoculated with either *R. solani* or *M. incognita* alone were free of root decay for the entire period of six-week study. Three weeks after nematode and fungus inoculation, black sclerotia of *R. solani* were visible on nematode- induced galls, while on non-galled portions on the same root system were free of sclerotia. Prior to root rot development, *R. solani* demonstrated marked preference for root galls on nematode infected roots. It is hypothesized that the leakage of nutrients from the root was responsible for attracting the fungus to the galls and for initiating sclerotial formation.

Five weeks after inoculation, distinct brown lesions were observed only on the galls of plants inoculated with both *M. incognita* and *R. solani*. Lower leaves of plants were chlorotic and suffered premature leaf drop.

M. incognita and *R. bataticola* when inoculated simultaneously in soil, reduce the germination of seeds in okra (Chhabra and Sharma, 1981). Combined attack of both these pathogens cause significantly greater damage to the crop than that of the damage caused by either pathogen alone (Bhagawati *et al*., 2007).

(ii) Integrated Management

(a) Bioagents and Botanicals: Chaitali *et al.* (2003) observed that *T. viride* combined with neem cake controlled the disease complex better than *T. viride* combined with groundnut cake in okra.

(b) Two Bioagents: The lowest pre-emergence (5.01%) and post-emergence (7.73%) damping-off were observed in the treatment where both carbosulfan 25 SD and carbendazim 50 WP were applied together as seed treatment (T_{10}) which was at par with the treatment with dual application of *T. harzianum* (6.26%) and *P. fluorescens* (7.73%) as seed treatment (T_4). As evidenced from the results, the seed treatment was found to be significantly superior to soil application of bioagents (Bhagawati *et al*., 2009) (Table 9.7).

Table 9.7. Effect of bioagents and chemicals for the management of disease complex caused by *Meloidogyne incognita* and *Rhizoctonia solani*

Treatment*	Pre-emergence damping-off (%)	Post-emergence damping-off (%)	Plant height (cm)	No. of galls/root system	No. of egg masses/root system
T_1	32.98	34.34	27.30	236.00	138.25
T_2	11.05	13.99	36.40	70.50	44.68
T_3	10.86	12.52	36.82	74.75	46.25
T_4	6.26	7.73	44.26	54.25	26.25
T_5	20.45	24.95	31.85	102.25	58.75
T_6	18.69	25.14	32.76	107.50	60.25
T_7	18.88	23.38	36.10	72.50	42.50
T_8	20.26	24.94	32.58	76.00	40.26
T_9	18.78	24.76	31.40	146.00	62.30
T_{10}	5.01	7.73	43.60	51.75	30.75
T_{11}	0.00	0.00	45.94	0.00	0.00
CD (P = 0.05)	**3.94**	**4.25**	**3.26**	**0.71**	**0.67**

* T_1- *M. incognita* + *R. solani* (Inoculated & untreated control), T_2 - *M. incognita* + *R. solani* + *T. harzianum* (seed treatment), T_3 - *M. incognita* + *R. solani* + *P. fluorescens* (seed treatment), T_4 - *M. incognita* + *R. solani* + *T. harzianum* + *P. fluorescens* (seed treatment), T_5 - *M. incognita* + *R. solani* + *T. harzianum* (soil application), T_6 - *M. incognita* + *R. solani* + P. fluorescens (soil application), T_7 - *M. incognita* + *R. solani* + *T. harzianum* + *P. fluorescens* (soil application), T_8 - *M. incognita* + *R. solani* + carbosulfan (seed treatment), T_9 - *M. incognita* + *R. solani* + carbendazim (seed treatment), T_{10} - *M. incognita* + *R. solani* + carbosulfan + carbendazim 50 (seed treatment), T_{11} – Uninoculated & untreated control.

The maximum plant height was recorded in the uninoculated and untreated control (T_{11}) followed by the treatment receiving *T. harzianum* and *P. fluorescens* as seed treatment (T_4) which were at par with the treatment receiving carbosulfan 25 SD and carbendazim 50 WP as seed treatment (T_{10}) (Bhagawati *et al.*, 2009) (Table 9.6).

The minimum number of galls and egg masses in roots were recorded in the treatment receiving carbosulfan 25 SD and carbendazim 50 WP as seed treatment (T_{10}) which was on par with the treatment receiving *T. harzianum* and *P. fluorescens* as seed treatment (T_4). Further, the treatment with *T. harzianum* and *P. fluorescens* as seed treatment was found to be significantly better in reducing the host infection and nematode multiplication than soil application of both these bioagents (Bhagawati *et al.*, 2009) (Table 9.6).

Chapter - 10

ROOT VEGETABLE CROPS

10.1. CARROT, *DACUS CAROTA*

10.1.1. Diseases

10.1.1.1. Soft Rot, *Erwinia carotovora* s. sp. *carotovora*

(i) **Symptoms:** The disease causes a soft, watery, slimy rot. Water soaked irregular lesions appear on tubers (Fig. 10.1). The rotted tissues are grey to brown and may have a foul odour. It decays the core of the root. Also prolonged wet weather favours disease development. It is a serious transit and storage problem if affected carrots are not discarded. In the field, tops of rotted carrots turn yellow and wilt as roots breakdown. Mostly the disease occurs after the harvest.

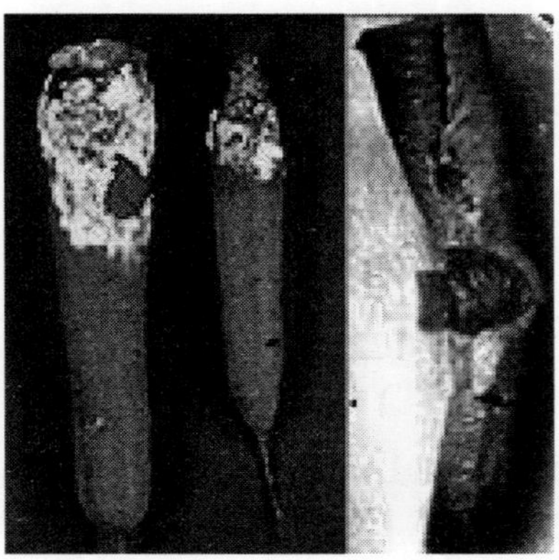

Fig. 10.1. Soft rot on carrot.

(ii) Integrated Management

(a) Bioagents and Botanicals: Seed treatment with *Pseudomonas putida* (10^9 cfu/g) at 10 g/kg and subsequent application of 5 tons of FYM enriched with 5 kg each of *P. putida* (10^9 cfu/g) and *T. harzianum* (10^6 cfu/g) per ha significantly reduced soft rot to 4.4% compared to 24% in control. Significant increase in yield (24%) was also observed. Benefit: cost ratio calculated for marginal cost of biopesticides and returns accrued by application of biopesticides was 5.3 (Anon, 2012).

10.1.2. Nematodes

10.1.2.1. Root-knot Nematode, *Meloidogyne incognita*

(i) Symptoms: Devi (1993) reported that *M. incognita* was responsible for 56.64% loss in yield of carrots. The nematode larvae feed on roots causing the swellings or knots that are characteristic of root-knot infection. Roots are often stunted and deformed. Root-knot nematodes develop characteristic forking of the roots in carrots (Fig. 10.2).

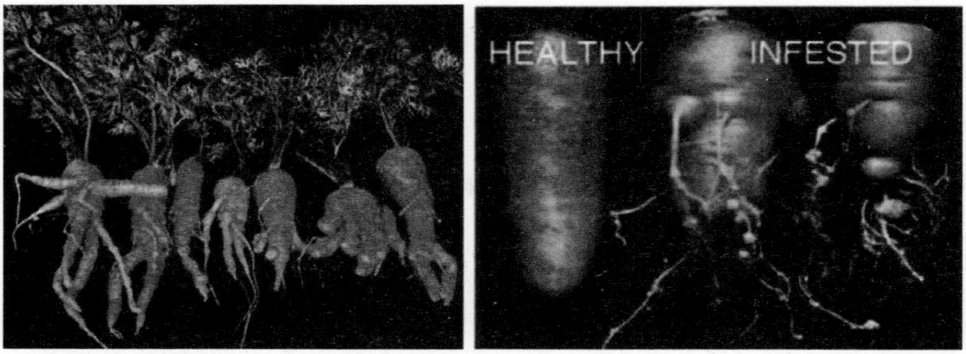

Fig. 10.2. Forking of carrots incited by root-knot nematodes

(ii) Integrated Management

(a) Bioagents and Botanicals: Soil application of *Paecilomyces lilacinus* (2 x 10^6 cfu/g) at 2.5 kg + FYM at 2.5 MT/ha gave minimum number of galls and final nematode population (120 and 12.5, respectively) compared to control (440 and 235, respectively). The highest yield was also recorded in the above treatment (9.34 MT/ha) over control (6.54 MT/ha).

Integration of neem cake with *P. lilacinus* and *Trichoderma harzianum* was found effective and increased carrot yield (Table 10.1).

Table 10.1. Effect of neem cake and bioagents on yield of carrot

Treatment	Yield (t/acre)
Neem cake (NC)	20.120
NC + *Trichoderma harzianum* (TH)	22.830
NC + *Paeciliomyces lilacinus* (PL)	23.347
NC + PL + TH	26.742
Control	18.245

Seed treatment with *Pseudomonas putida* (10^9 cfu/g) at 10 g/kg and subsequent application of 5 tons of FYM enriched with 5 kg each of *P. putida* (10^9 cfu/g) and *T. harzianum* (10^6 cfu/g) per ha significantly reduced reniform and root-knot nematode population in roots by 70 and 77%, respectively. Significant increase in yield (24%)

was also observed. Benefit: cost ratio calculated for marginal cost of biopesticides and returns accrued by application of biopesticides was 5.3 (Anon, 2012).

Seed treatment with *T. harzianum* at 10 g (with 1 x 10^6 cfu/g)/kg and *P. fluorescens* at 10 g (with 1 x 10^9 cfu/g)/kg and subsequent field application of 5 tons of enriched FYM with *T. harzianum* (with 1 x 10^6 cfu/g) and *P. fluorescens* (with 1 x 10^9 cfu/g) per ha significantly reduced root-knot and reniform nematodes in carrot roots by 79 and 75%, respectively. The yield increase was to the tune of 29.8% with a benefit: cost ratio of 13.6.

Significantly lowest number of second stage juveniles (J2's) was recovered from the soil incorporated with broccoli left-over materials and *Trichoderma* inoculant. Galls and egg masses in secondary roots were highest in unamended-inoculated soil which was significantly different from broccoli-amended soil with polarization and *Trichoderma* inoculant. The yield was significantly highest in broccoli-amended soil with solarization and *Trichoderma* inoculant In general, the treatments with broccoli residues and *Trichoderma* inoculant were able to decrease root-knot nematode population and significantly increase the yield relative to untreated soil (Pedroche *et al.*, 2009).

(b) Cultural and Chemical: A soil fumigant at the beginning of two consecutive carrot crops, followed by two year onion production and at last a cover crop sudax in the course of 5 year rotation can control *M. hapla* problem (Bird, 1981).

10.1.2.2. Root-knot Nematode, *Meloidogyne incognita* and Wilt, *Erwinia carotovora s.* sp. *carotovora* Disease Complex

(i) Integrated Management

(a) Bioagents and Botanicals: The neem cake enriched with *Pseudomonas fluorescens* and *Paecilomyces lilacinus* applied at 10 g/m^2 increased the root colonization of both the bioagents and reduced the incidence of *M. incognita* and *E. carotovora* s. sp. *carotovora* by 68 and 56%, respectively. There was also a significant increase in the yield of carrot to the tune of 23% (Sowmya *et al.*, 2010).

10.2. RADISH, *RAPHANUS SATIVUS*

10.2.1. Diseases

10.2.1.1. Damping-off, *Rhizoctonia solani*

(i) Integrated Management

(a) Bioagents and Botanicals: When applied to soil at rates of 0.04-0.15 g/kg (dry weight basis), wheat-bran cultures of *Trichoderma harzianum* protected radish seedlings from damping-off induced by *R. solani* and also increased radish germination in non-infested soils. Protection lasted for five successive weekly plantings (Henis *et al.*, 1978).

(b) Bioagents and Chemicals: Combined treatments of the fungicide PCNB at 4 μg/g and *T. harzianum* decreased the inoculum potential of *R. solani* and increased the disease control in comparison to separate treatments for control of damping-off of radish (Henis *et al*., 1978).

Integration of banodanil and *T. harzianum* were found effective for the control of *Rhizoctonia* pre-emergence damping-off of radish (Lifshitz *et al*., 1985).

(c) Two Bioagents: BINAB a mixture of *T. harzianum* and *T. polysporum* has been registered for the control of *Rhizoctonia* damping-off complex (Utkhede and Gupta, 1996).

10.3. BEET ROOT, *BETA VULGARIS*

10.3.1. Diseases

10.3.1.1. Storage Rot, Sclerotium rolfsii

(i) Symptoms: Sclerotium root rot or southern root rot can be a very destructive disease of beet root in some areas. Symptoms appear as poor top growth with **wilting** occurring as the tap root is decayed by the fungus. Under high temperatures, plants will eventually wilt permanently. The pathogen is characterized by **cottony mycelial growth** on the surface of the tap root with small (1-3 mm) spherical **sclerotia** that are tan to dark tan when mature (Fig. 10.3).

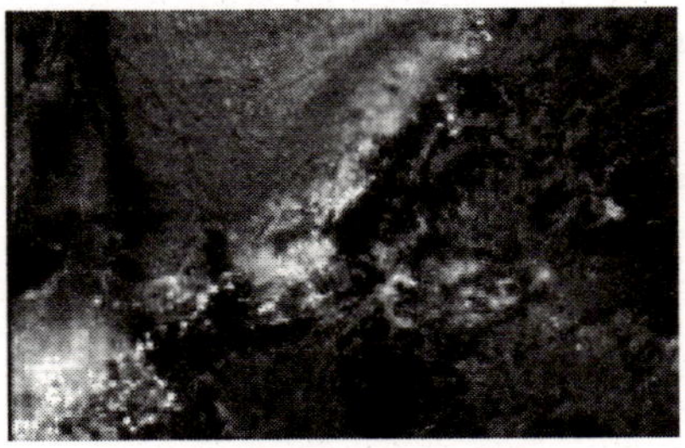

Fig. 10.3. Storage rot symptoms on beet root

(ii) Integrated Management

(a) Two Bioagents: The conidial mixture of 2 bioagents *T. pseudokoningii* (effective in reducing the mycelial growth) and *T. virens* (good colonizer of sclerotia) was effective in reducing the incidence of storage rot.

Chapter - 11

LEGUMINOUS VEGETABLE CROPS

11.1. PEA, *PISUM SATIVUM*

11.1.1. Diseases

11.1.1.1. Damping-off, *Pythium ultimum*

(i) Symptoms: *Pythium ultimum* commonly cause seed rot as well as pre- and post-emergence damping-off of pea. Root rot of older plants also occurs, and often results in root-pruning that significantly reduce root length. Infected roots are typically brown in colour and soft and watery to the touch. Infected plants are frequently stunted and pale green to yellow in colour.

***(ii)* Integrated Management**

(a) Bioagents and Chemicals: Integration of soil application of wheat bran based formulation of *T. harzianum* and giving 2 sprays of 0.05% carbendazim starting from flowering stage at 15 days interval was found effective (Kapoor and Sharma, 2000).

(b) Bioagents, Botanicals and Chemicals: Soil amendment with Lantana camera (10 MT/ha), application of Trichoguard (*T. viride*) at 2.5 kg in 10 kg FYM/ha (Kapil and Kapoor, 2002) followed by 1-2 sprays of 0.05% carbendazim was found effective in reducing damping-off disease of pea.

11.1.1.2. Powdery Mildew, *Erysiphe polygonum*

(i) Symptoms: Mealy white patches appear on both sides of leaves, stems, branches, pods and tendrils (Fig. 11.1). These patches originate as minute discoloured specks from which a powdery mass radiates on all sides and covers a large area of aerial parts. It causes considerable damage and may result in 20-30% losses in pod number and quality. If it attacks early, plants fail to bear fruit or the pods get chaffy.

(ii) Integrated Management

(a) Bioagents and Botanicals: In organically grown pea, biopriming of pea seeds with *Trichoderma harzianum* + *Pseudomonas fluorescens* was most effective in improving the seedling stand. Foliar application of bioagents significantly reduced the incidence of powdery mildew disease and increased the yield of pea (Table 11.1).

Fig. 11.1. Powdery mildew on pea leaves and pod.

Table 11.1. Effect of different biocontrol agents applied through seed, FYM and/ or foliar application on powdery mildew disease and yield of pea

Treatment	Seedling stand/m^2	Powdery mildew incidence (%)	Yield (kg/ha)
Seed biopriming with *Trichoderma harzianum* + sprays with *Pseudomonas fluorescens*	56	13	355
FYM colonized with *Trichoderma harzianum* + sprays with *Pseudomonas fluorescens*	63	16	352
Seed biopriming with *Trichoderma harzianum* + FYM colonized with *T. harzianum* + sprays with *T. harzianum*	69	12	388
Control	52	36	301
CD (P = 0.05)	**6**	**10**	**53**

11.1.1.3. White Rot, *Sclerotium rolfsii*

(i) Symptoms: It is a ubiquitous fungus with a broad host range. Initially wet rotting of bark is observed. Entire bark of the plant near collar region rots. The characteristic symptoms include white, cottony fungus growth is observed on affected portion as well as on parts in contact with soil. Gradually this hyphal mat is converted into small, mustard-like sclerotia that survive in the soil.

(ii) Integrated Management

(a) Bioagents and Botanicals: Soil application of *Trichoderma* spp. supplemented with organic matter and green manuring is beneficial.

11.1.1.4. White Mold, *Sclerotinia sclerotiorum*

(i) Symptoms: The disease is commonly found on pea which is responsible for complete killing of plants. Symptoms observed include wet and soft rotting of the tissues. White fungus growth is observed on the rotted portion. Embedded sclerotia covered with white mycelium are formed on the infected portion as well as on inner portion of the pith and fruits.

(ii) Integrated Management

(a) Botanicals and Bioagents: Addition of *Eupatorium/Ageratum conyzoides* plant material with wheat bran based formulation of *T. harzianum* at the time of sowing was found effective in managing root rot/wilt complex of pea (Kaur, 1999).

11.1.1.5. Wilt, *Fusarium oxysporum* f. sp. *pisi*

(i) Symptoms: Infection leads to drooping and wilting of plants at early stage of plant growth (Fig. 11.2). Vascular discolouration of stem and reddish appearance in the pith extends towards roots. The roots turn black and rot (Fig. 11.2). Plant growth is checked, foliage turns yellow and downward curling of stipules and leaflets takes place. The entire plant wilts and the stem shrivels. Epinasty of affected plants is a characteristic symptom.

Fig. 11.2. Fusarium wilt of pea.

(ii) Integrated Management

(a) Bioagents and Botanicals: Soil amendment with green manure and application of *Trichoderma harzianum* is most beneficial.

Fig. 11.3. L- Rhizoctonia patches in green pea, R- pea roots rotted by Rhizoctonia solani

11.1.1.6. Rhizoctonia Root Rot

(i) Symptoms: It primarily rots seeds and causes pre- and post-emergence damping-off. Roots of older plants can be rotted and stems can be infected to cause death of plants (Fig. 11.3). Mild lesions and root rot can result in stunted and stressed plants. A firm, dry, brown to reddish-brown decay or sunken lesion appears on the root and stem below or near the soil line. Plants may be stunted and may wilt or break off.

(ii) Integrated Management

(a) Bioagents and Chemicals: By itself, B. *subtilis* provided more disease protection than the fungicide Anchor, but Anchor combined with B. *subtilis* was even more effective.

11.2. FRENCH BEAN, *PHASEOLUS VULGARIS*

11.2.1. Insect Pests

11.2.1.1. Bean Fly, *Ophiomyia phaseoli;* Leaf Hopper, *Empoasca kerri;* Leaf Miner, *Liriomyza trifolii*

(i) Damage: The bean fly larvae mine the leaf lamina, veins, midrib, petiole and enter the stem. Larval feeding in stem results in mortality or reduction of the plant growth (Fig. 11.4).

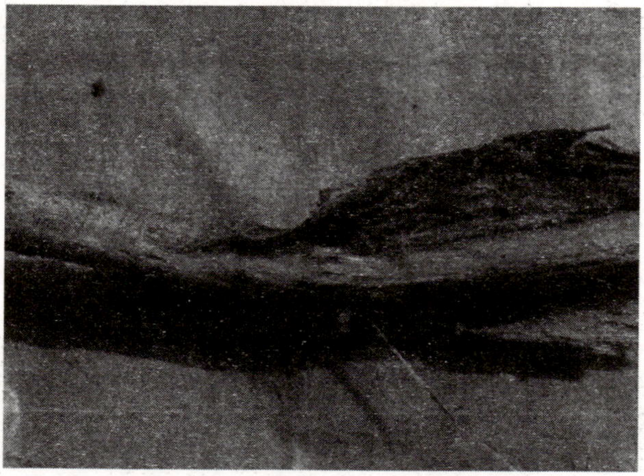

Fig. 11.4. Stem fly damage on stem of French bean.

The greenish yellow nymphs and adults of leaf hopper suck the sap. during September-October months after the onset of North East monsoon, the pest becomes very serious causing withering of the foliage and plants.

Leaf miner mines the leaves below epidermis in zig zag manner generally in basal leaves and feed on chlorophyll (Fig. 11.5).

(ii) Integrated Management

(a) Botanicals and Chemicals: Spraying 4% NSKE/1% neem or pongamia soap 10 DAS combined with spraying of endosulfon at 15 DAS during rainy season is effective against bean pests.

Fig. 11.5. Serpentine leaf miner damage on bean

11.2.2. Diseases

11.2.2.1. Web Blight, *Rhizoctonia solani*

(i) Symptoms: At emergence, the infection on hypocotyl and stem results in damping-off symptoms, but afterwards, elongated, sunken, reddish brown lesions are produced on the stem at ground level. On the foliage, circular to irregular brown spots having distinct borders appear. Leaf scald symptoms are common. In rainy season, the disease assumes serious proportions. Extensive damage results under continuous wet weather along with high temperature.

(ii) Integrated Management

(a) Bioagents and AMF: Integration of *Glomus mosseae* with *T. viride* gave total protection against *R. solani* infection in French bean (Girija Ganesan, 1999).

11.2.2.2. Collar Rot, *Sclerotinia sclerotiorum*

(i) Symptoms: The pathogen causes *Sclerotinia* wilt or white mold and also stem rot under certain conditions. The disease frequently occurs after a period of warm, humid weather. It can be recognized by the white fungus growth and large (2 to 5 mm) black bodies (sclerotia) in the pith of the stem.

(ii) Integrated Management

Bioagents and Chemicals: *Coniothyrium minitans* in combination with other control measures such as fungicide applications late in the growing season has been suggested (Trutmann *et al.*, 1982).

11.3. COWPEA, *VIGNA UNGUICULATA*

11.3.1. Nematodes

11.3.1.1. Root-knot Nematode, *Meloidogyne incognita*

Parvatha Reddy and Singh (1981) reported that *M. incognita* was responsible for 28.60% loss in pod yield of cowpea.

(i) Symptoms: Symptoms of damage induced by root-knot nematode include patches of stunted and yellowed plants. Severe damage can lead to reduced number of leaves and buds. At high densities severe root galling occurs (Fig. 11.6). Visual symptoms of damage first occurred at 1,000 and 10,000 juveniles/500g of soil.

Fig. 11.6. Heavy galling of cowpea roots infected with *Meloidogyne incognita* (Courtesy: F.E. Caveness).

(ii) Integrated Management

(a) Botanicals and Bioagents: Hasan and Jain (1992) reported that soil application of *P. lilacinus* cultured on sorghum seeds together with certain organic matter effectively reduced the incidence of *M. incognita* and increased the crop yield of cowpea.

(b) Cultural and Chemical: Summer ploughing along with seed treatment with carbosulfon 3% w/w or seed soaking in monocrotophos at 0.1% for 6 hr. gave effective control of root-knot nematodes and increased the cowpea yield.

(c) AMF and Botanicals: Combined application of *G. fasciculatum* and Achook, a neem product was very effective in reducing root-knot nematode population in cowpea (Jain and Hasan, 1995).

Integration of chopped leaves of *Prosopis fuliflora* with *G. fasciculatum* to cowpea increased the spore production and root colonization of *G. fasciculatum* which resulted in reduced *M. incognita* population.

11.3.1.2. Reniform Nematode, *Rotylenchulus reniformis*

(i) Symptoms: The reniform nematode *R. reniformis* is known to attack and cause growth reduction of cowpea. This nematode is most likely to cause or contribute to yield losses on cultivated cowpea.

(ii) Integrated Management

(a) Physical and Chemical: Summer ploughing along with seed treatment with carbosulfan 3% w/w or seed soaking in monocrotophos at 0.1% for 6 hr. gave effective control of reniform nematodes and increased the cowpea yield.

11.3.1.3. Root-knot Nematode, *Meloidogyne incognita* and Root Rot, *Macrophomina phaseolina* Disease Complex

(i) Integrated Management

(a) Botanicals and AMF: Devi and Goswami (1992) demonstrated that *G. fasciculatum* together with mustard cake helped in reducing the disease severity caused by *M. incognita* and *M. phaseolina* in cowpea. They observed that the pre-establishment of AMF checked the entry of *M. incognita* larvae as also colonization of pathogenic fungus.

11.3.1.4. Root-knot Nematode, *Meloidogyne incognita* and Wilt, *Fusarium oxysporum* Disease Complex

The wilt fungus, *F. oxysporum* and root-knot nematode, *M. incognita* co-infect cowpea. Histopathological studies revealed that in nematode + fungus inoculated cowpea roots, conidia of *F. oxysporum* could be observed in the cortex as well as in xylem vessels adjacent to the giant cells but not inside the giant cells induced by *M. incognita* (Singh *et al.*, 2007).

(i) Integrated Methods

(a) Chemicals and Botanicals: The minimum gall diameter index was reported in reduced dose of both neem cake and carbofuran (Singh *et al.*, 2007).

11.4. PIGEON PEA, *CAJANUS CAJAN*

11.4.1. Insect Pests

11.4.1.1. Pod Borer, *Helicoverpa armigera*

(i) Damage: The larvae feed for short time on the tender leaflets, flower buds and tender roots. Slowly it enters and feed on the seeds inside the pods. The half portion of larva remains inside while feeding on the developing seeds (Fig. 11.7).

(ii) Integrated Methods

(a) Two Bioagents: The least population of *H. armigera* was observed in virus (*Ha* NPV at 2 x 10^6 PIB's/ml) + nematode [DD-136 (*Steinernema feltiae*) at 3 x 10^3 infective juveniles/ml] treated plots (0.2) when compared to control (12.6). The percentage of pod damage was significantly lower in virus + nematode (1.07) when compared to control (40.31). Significant increase in pod yield was obtained in virus + nematode treated plots (10.7 kg) when compared to control (3.2 kg) (Narayanan and Gopalakrishnan, 1988).

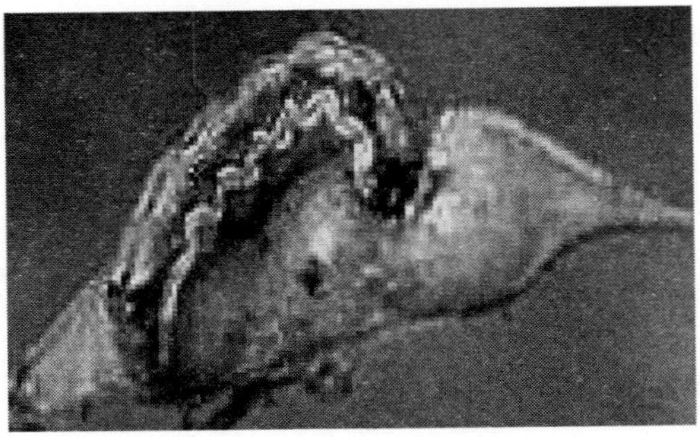

Fig. 11.7. Pod borer damage on pigeon pea.

11.4.2. Diseases

11.4.2.1. Wilt, *Fusarium udum*

(i) Symptoms: The disease is characterized by slow wilting of the plant. The symptoms can be observed after a month of sowing or at the flowering or pod formation stage. The affected plants become yellow in colour followed by drooping and finally the whole plant dries up (Fig. 11.8). The symptoms resemble as if the plant is suffering from the drought. The disease can be diagnised whenever the affected stem is cut and opened where the browning of the xylem vessels could be clearly seen (Fig. 11.8). It has been recorded that wilting can be partial or total. However, it has been observed that partial wilting is mainly associated with lateral root infection, whereas tap root infection may cause complete wilting.

Fig. 11.8. L - Dead plant due to Fusarium wilt, R - **Blackening of the xylem**

(ii) Integrated Management

(a) Bioagents and Chemicals: Wilt of pigeon pea was successfully managed by integration of *T. harzianum* and *T. virens* with carboxin (Mukhopadhyay, 1994).

Upadhyay and Roy (1997) reported effective management of Fusarium wilt by integration of *T. virens* with fytolan.

(b) Bioagents and AMF: Application of *B. subtilis, Bradyrhizobium japonicum* and *Glomus fasciculatum* used either alone or in combination increased shoot dry weight, number of nodules, phosphorus content and reduced nematode multiplication and wilting index in pigeon pea (Siddiqui and Mahmood, 1995).

(c) Bioagents and Botanicals: Pigeon pea seed treatment with bacterial antagonists (*Pseudomonas putida, Bacillus subtilis*) at 10 g/kg seed + soil application at 2.5 kg/ha of above bioagents mixed with FYM gave 22 to 28% Fusarium wilt control and increased the yield by 12 to 14% (Ramanujam *et al.*, 2003) (Table 11.2).

Table 11.2. Field performance of bacterial antagonists against Fusarium wilt of pigeon pea

Treatment	Formulation used and cfu/g	Disease control (%)	Increase in yield (%)
Seed treat. with *Pseudomonas fluorescens* at 10 g/kg + Soil appln. at 2.5 kg/ha mixed with FYM	Talc $10^7 - 10^8$	22	6
Seed treat. with *Pseudomonas putida* at 10 g/kg + Soil appln. at 2.5 kg/ha mixed with FYM	Talc $10^6 - 10^7$	28	12
Seed treat. with *Bacillus subtilis* at 10 g/kg + Soil appln. at 2.5 kg/ha mixed with FYM	Talc $10^6 - 10^7$	26	14

11.4.3. Nematodes

11.4.3.1. Cyst Nematode, *Heterodera cajani*

The cyst nematode has been found associated with pigeon pea in a number of regions of India.

Fig. 11.9. Root damage symptoms of pearly root caused by *Heterodera cajani* in pigeon pea

(i) Symptoms: The most important characteristic symptom is the presence of cysts on the root surface. Identification of "pearly root" caused by the presence of white females is a useful symptom of *H. cajani* infestation in pigeon pea at the vegetative stage (Fig. 11.9). The symptoms of nematode injury include stunting, reduced leaf lamina size and yellowing on cotyledonary leaves. Flowers and pods are reduced in size and number and the root system may also be poorly developed. The cyst nematode retarded emergence of leaves and reduced the number of flowering buds, flowers, growing pods and yield.

(ii) Integrated Management

(a) Two Bioagents: Integration of *T. harzianum* at 5 kg/ha along with *P. chlamydosporia* at 2 kg/ha significantly increased the plant height and pigeon pea yield (0.56 kg/plant as compared to 0.40 kg/plant in control) and reduced the seedling mortality, eggs per cyst, parasitization of cysts by the bioagents and cyst nematode population (Table 11.3).

Table 11.3. Effect of bioagents on plant growth and management of *Heterodera cajani* infecting pigeon pea

Treatment	Seedling mortality	Yield (kg/plant)	eggs/cyst
T. harzianum – 5 kg/ha	13.21	0.529	36.8
P. chlamydosporia – 2 kg/ha	13.67	0.545	30.1
T. harzianum – 5 kg/ha + *P. Chlamydosporia* – 2 kg/ha	9.20	0.560	24.0
Carbofuran – 2 kg a.i./ha	2.16	0.634	18.4
Control	18.40	0.404	45.1
CD (P = 0.05)	**1.14**	**0.0053**	**3.45**

(b) Bioagents and Botanicals: Combined treatment of 10% neem seed kernel powder + *Trichoderma viride* at 10 g/kg seeds was found to reduce *H. cajani* in pigeon pea by 58% and increased yield by 32%. Application of neem cake at 100 kg/ha + *Trichoderma viride* at 2.5 kg/ha was also found to reduce *H. cajani* by 62% and increased pigeon pea yield by 34%.

Seed treatment with neem seed kernel powder at 10% w/w + *Pseudomonas fluorescens* at 10 g/kg seed was highly effective which recorded 30.11% reduction in nematode population and 29.10% increase in yield over control. Soil application of neem cake at 10 g/m^2 + *Pseudomonas fluorescens* at 2.5 kg/ha was also highly effective which recorded 31.63% % reduction in nematode population and 17.88% increase in yield over control.

(c) Physical, Botanicals and AMF: Soil solarization (transparent polythene sheet of 400 gauge thickness for a period of 4 weeks) and application of AM fungi at 100 kg/ha provided good plant growth and yield of pigeon pea by effectively controlling the detrimental effects of *H. cajani.* Maximum reduction in nematode population was observed in soil solarization + soil application of neem seed powder at 50 kg/ha + AMF, and soil solarization + seed treatment with neem seed powder at 10% w/w + AMF (Nageswari and Mishra, 2005).

11.4.3.2. Root-knot Nematode, *Meloidogyne* spp. and Wilt, *Fusarium udum* Disease Complex

(i) Symptoms: Among plant-parasitic nematodes, *M. incognita* is considered most serious threat to its cultivation. Similarly, among fungi, *Fusarium* wilt is one of the most destructive diseases of pigeon pea. The wilt disease complex caused by *F. udum* in association with *M. incognita* has been reported as the most severe constraint in the cultivation of pigeon pea. When both these pathogens attack the crop, intensity of damage is increased by several folds.

Simultaneous or sequential inoculation of *M. incognita* and *F. udum* increased the severity of the disease. Highest reduction in plant height, fresh/dry weight was observed in plants inoculated with *M. incognita* and *F. udum* simultaneously followed by *M. incognita* prior and *F. udum* 7 days later and *F. udum* prior and *M. incognita* 7 days later. Reproduction of *M. incognita* was enhanced in the presence of *F. udum* but per cent root colonization by *F. udum* was suppressed in the presence of *M. incognita*. Highest final nematode population and gall index of *M. incognita* were observed in simultaneous inoculation of *M. incognita* and *F. udum* and lowest in *F. udum* prior and *M. incognita* 7 days later, while highest per cent root colonization was found in *F. udum* prior and *M. incognita* 7 days later followed by *M. incognita* and *F. udum* simultaneously and *M. incognita* prior and *F. udum* 7 days later (Table 11.4) (Perveen *et al.,* 1998).

Table 11.4. Effect of *Meloidogyne incognita* and *Fusarium udum* on plant growth and root galling and disease index in pigeon pea

Treatment	Plant height (cm)	Dry plant wt. (g)	Gall index	Disease index
Control	33.9	1.03	---	---
M. incognita	32.6	0.73	1.75	0.00
F. udum	32.6	0.81	---	45.32
M. incognita + *F. udum* simultaneouly	23.6	0.38	2.15	38.21
M. incognita (pre) + *F. udum* (post)	27.3	0.57	2.75	32.14
F. udum (pre) + *M. incognita* (post)	30.6	0.65	1.25	42.33
CD (P = 0.05)	**1.3**	**0.031**	**0.143**	**0.273**

(ii) Effect of Disease Complex on Wilt Resistant Accessions

Among 10 pigeon pea accessions (identified as resistant to Fusarium wilt) (Vishwadhar and Chaudhary, 2000) evaluated against combined infection of *M. javanica* and *F. udum* under pot culture conditions, increase in wilting was observed in five accessions namely, ICP 8859, AWR 74/15, KPL 44, ICPL 89049 and ICPL 12745. In these accessions wilting started 30 days after inoculation of fungus. Maximum wilting was observed in AWR 74/15 (60%) followed by ICP 8859 (50%) and ICPL 89049 (50%). Wilting increased from 8 to 33% in KPL 44, 15 to 60% in AWR 74/15, 25 to 50% in ICP 8859 and ICPL 89049, and 15 to 50% in ICP 12745 when *M. javanica* was present with *F. udum*. Whereas in other 5 accessions (KPL 43, PI 397430, BWR 370, GPS 33 and ICPL 89048), wilting was not influenced much in the presence of *M. javanica* (Singh *et al*., 2004).

The plant height was significantly lower in combined inoculation compared to either nematode or fungus alone. The root-knot index varied from 3.0 to 4.5 in these accessions in both treatments having nematode alone or nematode and fungus together. The lowest root-knot index was observed in KPL 43 (1.5) and GPS 33 (1.75) (Table 11.5) (Singh *et al*., 2004).

Table 11.5. Effect of combined inoculation of *M. javanica* and *F. udum* on plant growth and disease complex in different accessions of pigeon pea

Pigeon pea accessions	Plant height (cm)	Gall index	% Wilt incidence
KPL 43	25.9	2.45	0
KPL 44	17.4	1.97	33
AWR 74/15	18.3	1.56	60
ICP 8859	14.7	3.39	50
ICPL 89049	24.3	2.17	50
PI 397430	24.8	3.06	0
BWR 370	28.4	3.17	0
GPS 33	32.0	2.25	0
ICPL 89048	28.3	3.25	0
ICP 12745	19.4	2.50	50
CD (P = 0.05)	**2.40**	**1.60**	---

(iii) Integrated Management

Botanicals and AMF: The treatment constituting FYM, karanj oilseed cake and arbuscular mycorrhizal fungus, *Glomus fasciculatum* reduced the disease incidence caused by root knot nematode, *M. incognita* and root wilt fungus, *F. udum* on pigeon pea to a great extent with the most promising improvement in plant growth parameters (Goswami *et al*., 2007).

11.4.3.3. Cyst Nematode, *Heterodera cajani* and Wilt, *Fusarium udum* Disease Complex

(i) Symptoms: The wilt disease complex caused by *F. udum* in association with *Heterodera cajani* has been reported as the most severe constraint in the cultivation of pigeon pea. Inoculation with *F. udum* and *H. cajani* together significantly increased wilt severity in pigeon pea seedlings compared with inoculation of the fungus alone.

(ii) Integrated Methods

(a) Bioagents, Botanicals and Chemicals: In a trial conducted in the field during 2002, no wilting was observed in plots treated with neem seed powder + dimethoate/*T. harzianum*/*P. lilacinus*/latex, and neem seed powder and *T. harzianum* alone. Neem seed powder + *T. harzianum* was found to be the most effective treatment in increasing the yield and suppressing the pathogens, followed by carbofuran, neem seed powder + *P. lilacinus*, neem seed powder + dimethoate and neem seed powder + latex. Data regarding *F. udum* infection in roots gathered at the pre-flowering stage (90 DAS) indicated that all the treatments maintained significant protection of the roots as compared to control (Haseeb and Shukla, 2005) (Table 11.6).

Table 11.6. Effect of seed treatment with bioagents, chemicals and botanicals on wilt and cyst nematodes infecting pigeon pea under field conditions during 2002.

Treatment	**Grain yield (kg)/100 m^2**	**% wilting of plants**	**% root infection by *F. udum***	**Nematodes / g of root**	**Cysts & larvae/ 100 ml soil**
Carbofuran	23.47	5.0	35	9	27
Neem seed powder (soil application)	19.55	0.0	20	23	45
Neem seed powder (seed treatment)	18.33	0.0	20	30	43
T. harzianum	15.46	0.0	25	31	44
P. lilacinus	12.24	5.0	35	36	55
Dimethoate	11.75	10.0	50	39	58
Latex of Calotropis	11.34	10.0	50	39	59
NSP (soil) + Dimethoate	21.05	0.0	15	16	36
NSP (soil) + *T. harzianum*	23.35	0.0	15	10	28
NSP (soil) + *P. lilacinus*	21.85	0.0	20	14	34
NSP (soil) + latex	20.75	0.0	15	18	43
Control	5.35	45.0	75	51	67
CD (P = 0.05)	**1.22**	2.8	**3.9**	**0.68**	**3.1**

(b) Bioagents and AMF: The detrimental effects of a disease complex on pigeon pea involving the sedentary endoparasite *H. cajani* and the fungus *F. udum* were reduced following application of the fungi *P. lilacinus* and *P. chlamydosporia* together with the arbuscular mycorrhizal fungus *Gigaspora margarita* (Siddiqui and Mahmood, 1995).

Combined application of *T. harzianum, P. chlamydosporia* and *G. mosseae* enhanced the activity against *H. cajani* – *F. udum* wilt disease complex in pigeon pea.

11.4.4. Validated IPM technology for Pigeon Pea

11.4.4.1. Gulburga, Karnataka

- Fall ploughing to expose pupae to hot sun and natural enemies.

- Mixing of sorghum or mesta seeds at 250 g/ha with pigeon pea seeds, which act as live bird perches.
- Installation of pheromone traps at 5/ha to monitor the pests.
- Erecting the branched twigs at 20/ha which act as bird perches.
- First spraying with ovicide indoxacarb or methomyl at 300 g/ha or profenophos at 2 litres/ha.
- Second spraying with 5% NSKE or commercial neem formulation (1500 ppm Azadirachtin) at 2 litres/ha.
- Third spraying with *Ha*NPV at 250 LE/ha.
- Fourth spraying with indoxacarb at 300 ml/ha or chlorpyriphos at 2.5 litres/ha or quinalphos or endosulfan at 2 litres/ha.
- If necessary, spraying of synthetic pyrethroids at 500 ml/ha.
- In case of scarcity of water, dust 4% endosulfan followed by 1.5% quinalphos or 0.4% fenvalerate at 25 kg/ha.

The IPM module demonstrated on the large scale during 2001-02 (2 villages) and 2002-03 (4 villages) indicated the higher benefit: cost ratio (Tables 11.7 and 11.8) (Sharma *et al*., 2004).

Table 11.7. Economics of IPM in Gulburga (Karnataka) during 2001-02

Particulars	**Sannur**		**Farthabad**	
	IPM	**Non-IPM**	**IPM**	**Non-IPM**
Cost of plant protection (Rs./ha)	1915	2873	1915	2607
Total cost of cultivation (Rs./ha)	6685	7498	6685	7267
Seed yield (q/ha)	7.690	7.300	9.700	8.050
Gross income (Rs./ha)	12689	12045	16005	13282
Profit over non-IPM (Rs./ha)	1457	---	3305	---

Table 11.8. Economics of IPM on pigeon pea at different locations in Gulburga district of Karnataka during 2002-03

Centre	**Yield (q/ha)**	**Profit over non-IPM (Rs/ha)**
Sannur - IPM	9.64	1600
- Non-IPM	9.25	
Farthabad - IPM	10.81	2687
- Non-IPM	9.75	
Tadtegnoor - IPM	10.50	1400
- Non-IPM	10.25	
Kodla - IPM	15.62	9590
- Non-IPM	9.50	

11.4.4.2. Nanded, Maharashtra

Pre-sowing

- Deep ploughing and exposure of soil to hot summer to kill pupating larvae and fungal propagules.
- Use of FYM enriched with *Trichoderma viride* at 50 kg/ha mixed along with neem cake to reduce disease as well as nematode population.
- In areas where termite and cut worm is a problem, seed treatment with chlorpyriphos at 8 ml/kg or pre-sowing mixing of soil with chlorpyriphos dust is recommended as they are relatively low cost and initially protect the seedlings from cut worms.
- Synchronized sowing of multiple pest resistant varieties such as Sharad, Asha, Maruti, Bahar, Abhaya, BSMR-736.
- Ridge planting to prevent incidence of blight caused by *Phytophthora dreschleri*.

Sowing Time

- Intercrop sorghum to reduce wilt, conserve beneficial insects and serve as bird perches.
- Early planting around mid–June in North West Plain Zone to avoid *Helicoverpa armigera*.
- Use marigold as a trap crop on border or interspersed with crop for pod borer control.

Post-sowing

- Monitoring through regular field scouting along with pheromone traps (5-10/ha) to assess population build up of borers and their management. A threshold level of 5-6 moths/trap/day is indicative of its peak activity period and warrants initiation of management practices.
- Rogue out and destroy sterility mosaic virus affected plants.
- Mechanical shaking of plants and collection of larvae for preparation of *Ha*NPV locally to cut down the cost.
- Spray 5% NSKE or neem based formulation against *H. armigera* and other Lepidopterous pests.
- Spray with *Ha*NPV at 450 LE/ha along with UV retardant.
- Spray eco-friendly pesticide like endosulfan if pest complex exceeds ETL.

The data on % bud and pod damage due to *H. armigera* and leaf roller and plume moth damage indicated that IPM had a significant edge over non-IPM despite use of chemical pesticides. The seed yield obtained for IPM and non-IPM fields was 1.096 and 0.724 tons/ha, respectively (Table 11.9) (Sharma *et al.*, 2004).

Table 11.9. Comparative level of damage in IPM and non-IPM fields during 2002-03

Type of damage	IPM	Non-IPM
Bud damage (%)	18	24
Pod damage (%)	22	27
Pod borer damage (early) (%)	55	61
Pod borer damage (late) (%)	58	83
Leaf roller (No./plant)	1.01	36
Plume moth (No./plant)	0.13	0.30
Yield (t/ha)	1.096	0.724

11.4.4.3. Varanasi, Uttar Pradesh

- Deep summer ploughing to destroy immature stages and pathogen propagules.
- Prior to sowing, soil application of *Trichoderma harzianum* at 10 g/kg of FYM for controlling the pigeon pea wilt.
- Sowing on ridges to control *Phytophthora.*
- Sowing of high yielding varieties with pest/disease tolerance - Asha and Maruthi against wilt and PSM for Andhra Pradesh and Karnataka, and Bahar for Uttar Pradesh (Varanasi).
- Installation of pheromone traps at 10/ha in the month of September.
- Erection of bird perches at 25/ha for facilitating predation of *Helicoverpa* larvae.
- One spray with 2% neem oil.
- Two applications of 5% NSKE in September and October.
- Spray *Ha* NPV at 500 LE/ha (1.5 x 10^{12} POB)/ha in September and October – when larvae are small.
- Shaking of plants 5 times a day starting from October for short duration pigeon pea.
- Need based spray with endosulfan at 2 litres/ha.

As a result of implementation of IPM, the grain yield was found to be twice (0.765 t/ha) in comparison to non-IPM fields (0.375 t/ha). Besides, the quality of environment improvement, favourable benefit: cost ratio (2.01) was recorded in IPM fields. The incidence of leaf roller and *Phytophthora* blight incidence in IPM and non-IPM fields is presented in Table 11.10 (Sharma *et al.*, 2004).

Table 11.10. Comparative incidence of pest/disease in IPM and non-IPM fields

Pest/disease	Short duration cv. UPAS-120		Long duration cv. Bahar	
	IPM	Non-IPM	IPM	Non-IPM
Leaf roller (No./plant)	5	15	15	30
Phytophthora blight (%)	2-3	25	2-3	30

11.5. CLUSTER BEAN, *CYAMOPSIS TETRAGONOLOBA*

11.5.1. Nematodes

11.5.1.1. Root-knot Nematode, *Meloidogyne javanica* and Wilt, *Fusarium solani* Disease Complex

(i) Integrated Management

(a) Two Bioagents: *Pseudomonas aeruginosa* and *Paecilomyces lilacinus* when used together significantly reduced infection of the disease complex on cluster bean (Perveen *et al*., 1998).

11.5.1.2. Root-knot Nematode, *Meloidogyne javanica* and Root Rot/Wilt, *Macrophomina phaseolina, Rhizoctonia solani, Fusarium solani, F. oxysporum* Disease Complex

(i) Integrated Methods

Two Bioagents: *Pseudomonas aeruginosa* and *Paecilomyces lilacinus* used alone or together significantly ($P<0.05$) reduced infection of root knot nematode *M. javanica* and root infecting fungi viz., *M. phaseolina, R. solani, F. solani* and *F. oxysporum* on cluster bean. *P. aeruginosa* was more effective than *P. lilacinus* in reducing the *M. javanica* infection. Combined use of *P. lilacinus* and *P. aeruginosa* was more effective in reducing the infection of root knot nematode and *F. solani* on guar than either used alone. Use of *P. aeruginosa* and *P. lilacinus* significantly ($P<0.05$) increased plant height of guar (Perveen *et al.,* 1998) (Table 11.11).

Table 11.11. Effect of *P. aeruginosa* and *P. lilacinus* on plant height and control of root rot/wilt disease complex in cluster bean

Treatment	Plant height (cm)	Root-knot index	Infection %			
			M. phaseolina	*R. solani*	*F. solani*	*F. oxysporum*
Control	24.5	4.1	31	0	75	81
P. lilacinus (PL)	28.0	3.5	6	0	62	81
P. aeruginosa (PA)	27.7	2.0	19	0	44	50
PL + PA	32.0	1.4	6	0	37	44
CD (P = 0.05)	**2.2**	**0.34**	**6.1**	**6.1**	**6.1**	**6.1**

11.6. FIELD BEAN, *LABLAB PURPUREUS*

11.6.1. Insect Pests

11.6.1.1. Pod Borer, *Adisura atkinsoni*

(i) Damage: This is the major pod borer in field bean. The eggs are laid on tender pods. The young larvae bore into tender pods. They develop inside the bored pod and come out after attaining fourth instar, which is a migratory stage. It causes heavy damage by way of feeding the developing pods and reducing the marketable yield (Fig. 11.10).

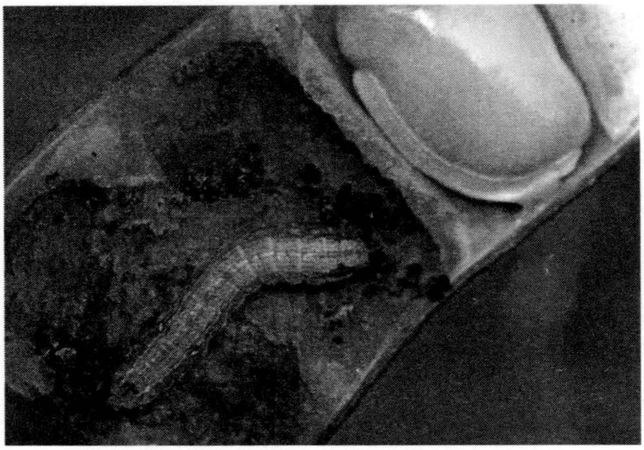

Fig. 11.10. Field bean pod infected with borer

(ii) Integrated Management

(a) Bioagents and Chemicals: Spraying of *Aa* NPV at 250 LE/ha in combination with endosulfan at 0.035%, twice at fortnightly intervals significantly reduced both larval population and pod damage by *Adisura atkinsoni* and *Sphaenanches anisodactylus* (Narayanan, 1987).

Aa NPV at 125 LE/ha along with endosulfan (0.035%) is effective in significantly reducing the pest damage to both pods and grains (Narayanan and Gopalakrishnan, 1990).

Chapter - 12

CUCURBITAECIOU VEGETABLE CROPS

12.1. PUMPKIN, *Cucurbita moschata*

12.1.1. Diseases

12.1.1.1. Gummy Stem Blight, *Mycosphaerella melonis*

(i) Symptoms: A non-descript marginal necrosis is followed by larger, wedge-shaped necrotic areas appear on leaves. Infected stems first show water-soaked lesions and later appear tan. Older stems show pycnidia within the affected tissue. Stem lesions often cause gummy, reddish-brown or black beads to exude (Fig. 12.1).

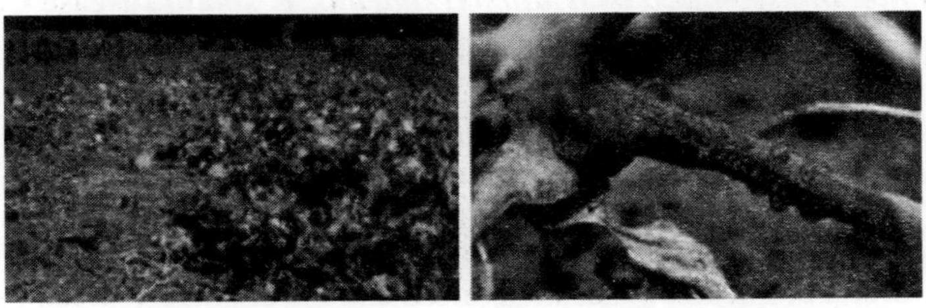

Fig. 12.1. Symptoms of gummy stem blight on leaf and stem

(ii) Integrated Management

(a) Bioagents and Botanicals: Soil application of neem cake enriched with *T. harzianum* minimized gummosis, root rot and collar rot.

12.1.2. Nematodes

12.1.2.1. Root-knot Nematodes, *Meloidogyne* spp.

(i) Symptoms: The very broad leaved plant like pumpkin which is infected with root-knot nematodes may show day time wilting and develop much larger galls. In cucurbits, the roots react to the presence of *Meloidogyne* spp. by the formation of large, fleshy galls (Fig. 12.2).

(ii) Integrated Management

(a) Bioagents and Botanicals: Addition of 250 g fresh neem cake or karanj cake enriched with *Trichoderma harzianum/Paecilomyces lilacinus* per planting pit effectively controlled the nematodes.

Fig. 12.2. Root-knot nematode on pumpkin

12.1.2.2. Root-knot Nematode, *Meloidogyne javanica* and Root Rot/Wilt, *Macrophomina phaseolina, Fusarium oxysporum, F. solani* Disease Complex

(i) Integrated Management

(a) Two Bioagents: *Pseudomonas aeruginosa* and *Paecilomyces lilacinus* when used together significantly reduced infection of the disease complex on pumpkin (Table 12.1) (Perveen *et al.,* 1998).

Table 12.1. Effect of Pseudomonas aeruginosa and Paecilomyces lilacinus on growth and control of root disease complex of pumpkin.

Treatment	Plant height (cm)	Root-knot index	M. phaseolina Infection %
Control	26.5	3.1	87
P. *lilacinus* (PL)	32.2	1.2	94
P. aeruginosa (PA)	30.0	0.6	50
PL + PA	28.2	0.5	19
CD (P = 0.05)	2.2	0.34	6.1

P. aeruginosa and *P. lilacinus* used alone or together significantly (P<0.05) reduced infection of root-knot nematode *M. javanica* and root infecting fungi viz., *M. phaseolina, R. solani, F. solani* and *F. oxysporum* on pumpkin. *P. aeruginosa* was more effective than *P. lilacinus* in reducing the *M. javanica* infection. Combined use of *P. lilacinus* and *P. aeruginosa* was more effective in reducing the infection of *M. phaseolina* and *F. oxysporum* on pumpkin than either used alone. Use of *P. aeruginosa and P. lilacinus* significantly (P<0.05) increased plant height of pumpkin.

12.2. CUCUMBER, *CUCUMIS SATIVUS*

12.2.1. Diseases

12.2.1.1. Damping-off, *Pythium* spp.

(i) Symptoms: In seedlings, a watery rot develops in the tap root and hypocotyl at or near the soil line. Damping-off or a slow decline may occur when seedling death is preceded by cotyledon and leaf chlorosis. Young seedlings wilt and die (Fig. 12.3).

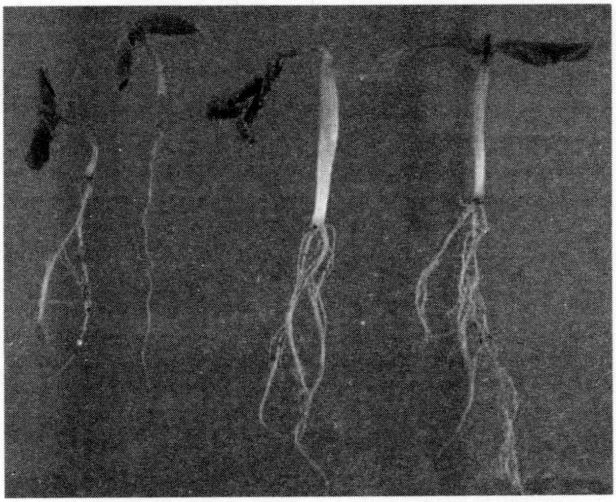

Fig. 12.3. Damping-off of cucumber seedlings. L – 3 seedlings infected, R - Healthy

(ii) Integrated Management

(a) AMF and Botanicals: Combined use of AMF *Gigaspora margarita* and 3-15% charcoal compost (which contained antagonistic microorganisms such as *Bacillus subtilis, Thermomonospora* sp. and *Thermoactinomyces* sp.) drastically reduced damping-off caused by *Pythium splendens* or *Rhizoctonia solani* in two and three-week-old cucumber seedlings. Moreover, AMF and charcoal compost stimulated rooting and increased the root volume, and hence plant growth (Kabayashi, 1989b).

12.2.1.2. Fusarium Wilt, *Fusarium oxysporum* f. sp. *cucumerinum*

(i) Symptoms: Yellowing of leaves progresses upward from the base of the plant. Wilting or yellowing may occur only on one side of a leaf or a branch or one side of the plant (Fig. 12.4). Yellow leaves wilt noticeably before they die. Wilting may occur at midday, when sunlight is bright and temperature is high. Infected plants are stunted and both fruit size and yield are reduced.

Fig. 12.4. Fusarium wilt of cucumber

(ii) Integrated Management

(a) AMF and Botanicals: Combined use of AMF *Gigaspora margarita* and 3-15% charcoal compost (which contained antagonistic microorganisms such as *Bacillus subtilis, Thermomonospora* sp. and *Thermoactinomyces* sp.) drastically reduced Fusarium wilt in cucumber. Moreover, AMF and charcoal compost stimulated rooting and increased the root volume, and hence plant growth (Kabayashi, 1989b).

(b) Two Bioagents: Combinations of non-pathogenic fusaria and fluorescent pseudomonads significantly reduced Fusarium wilt disease incidence on cucumber, although neither group alone induced significant biological control. This suggests that mixtures of biological control agents were effective in disease control strategies.

12.2.1.3. Grey Mold, *Botrytis cinerea*

(i) Symptoms: This microfungus kills cucumber plants in a day or so, rotting through the stem near the base of the plant. Under humid conditions fuzzy grey mould grows on affected buds, leaves, flowers or fruit (Fig. 12.5). Above-ground parts of many plants, particularly buds and flowers, shrivel and die. *Botrytis* infection leads to a soft brown rot, often as the fruit is ripening.

(ii) Integrated Management

(a) Bioagents and Chemicals: Elad *et al*. (1993) used a combination of *T. harzianum* and dicarboximide for successful control of grey mold in cucumbers. However, the alternation of the antagonist with fungicides was shown to be more effective than mixtures.

12.2.1.4. Rhizoctonia Fruit Rot

(i) Symptoms: Symptoms occur on the underside and blossom end of the cucumber fruit and can be observed within as little as 24 hours after the pathogen invades cucumbers that are in contact with soil. As the disease progresses, lesions become sunken and irregular in shape (Fig. 12.6). The entire fruit can rot in 72 hours.

Fig. 12.5. Grey mold on cucumber

Fig. 12.6. *Rhizoctonia* fruit rot of cucumber

(ii) Integrated Management

(a) Bioagents and Cultural: *Rhizoctonia* fruit rot of cucumber was decreased with combinations of deep ploughing to bury infested soil layers and the application of *Trichoderma* sp. (VT6) (Lewis and Papavizas, 1980).

(b) Bioagents and Chemicals: Lewis and Papavizas (1980) reported effective control of *R. solani* on cucumber with a combination of *T. harzianum* and chlorothalonil.

12.2.2. Nematodes

12.2.2.1. Root-knot Nematodes, *Meloidogyne* spp.

(i) Symptoms: Root-knot nematode causes galls or swellings on plant roots. In case of heavy attacks, galls can become very large, the root system being reduced to a swollen stump without hairs (Fig. 12.7). It restricts the uptake of nutrients from the root system to the foliage, resulting in a yellow and stunted plant.

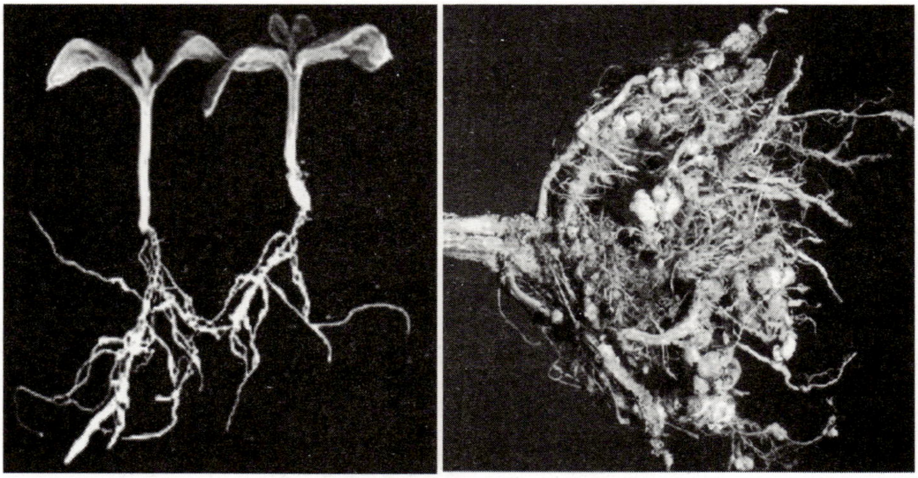

Fig. 12.7. Cucumber roots severely galled due to root-knot nematode infection.

(ii) Integrated Management

(a) Bioagents and Botanicals: Pre-plant treatment with *P. lilacinus* (20 kg/ha) in combination with organic amendments effectively reduced root-knot infection in gherkin.

(b) Bioagents and AMF: Combined inoculation of AMF and *Pseudomonas fluorescens* had positive effect on root-knot nematode control on cucumber (Jakobsen, 1999).

(c) Physical and Botanicals: Solarization alone or soil amendment alone (using poultry manure, alfalfa, cauliflower, tomato, and olive-cake residues) treatments significantly reduced densities of second stage juveniles of *M. javanica,* root galling and increased yield of cucumber. Organic amendments reduced densities of *Fusarium* spp., generally increased *Aspergillus* spp., while *Trichoderma* spp. was not affected. Combinations of solarization and addition of organic amendments substantially augmented each other, particularly with poultry manure, alfalfa hay, and to a lesser extent with cauliflower and tomato residues.

(d) Cultural and Botanicals: In commercial greenhouse trials in Spain an integrated management system was developed, including biofumigation with sheep manure and mushroom residue and the cultivation of short-cycle vegetables acting as trap crops. Using this strategy, initial very high levels of *M. incognita* were reduced to near zero in the main susceptible cucumber crop (Bello, 1998).

(e) Physical and Bioagents: In a cucumber crop in glasshouse trial, the use of solarization and *Pasteuria penetrans* had an additive detrimental effect on *M. javanica* populations (Tzortzakakis and Gowen, 1994).

(f) Bioagents and Chemicals: Oxamyl increased the efficacy of *P. penetrans* in trials against *M. javanica* infection of cucumber crop and the effects on nematode control were additive (Tzortzakakis and Gowen, 1994).

12.3. WATERMELON, *CITRULLUS LUNATUS*

12.3.1. Diseases

12.3.1.1. Root-knot Nematode, *Meloidogyne javanica* and Wilt, *Fusarium solani* Disease Complex (Fig. 12.8)

Fig. 12.8. Root-knot nematode and *Fusarium* wilt disease complex

(i) Integrated Management

(a) Two Bioagents: *Pseudomonas aeruginosa* and *Paecilomyces lilacinus* when used together significantly reduced infection of *M. javanica* and *F. solani* on watermelon. *P. aeruginosa* and *P. lilacinus* used alone or together significantly ($P<0.05$) reduced infection of root knot nematode *M. javanica* and *F. solani* and *F. oxysporum* on watermelon. *P. aeruginosa was* more effective than *P. lilacinus* in reducing the *M. javanica* root knot nematode infection (Perveen *et al.*, 1998). Combined use of *P. lilacinus* and *P. aeruginosa* was more effective in reducing the infection of *F. solani* on watermelon than either used alone. Use of *P. aeruginosa* and *P. lilacinus* significantly ($P<0.05$) increased plant height and fresh shoot weight on watermelon (Table 12.2).

Table 12.2. Effect of *Pseudomonas aeruginosa* and *Paecilomyces lilacinus* on growth and control of root disease complex of watermelon.

Treatment	Plant height (cm)	Root-knot index	*F. solani* Infection %	*F. oxysporum* Infection %
Control	35.1	3.8	69	12
P. *lilacinus* (PL)	37.5	2.8	81	0
P. aeruginosa (PA)	35.0	1.5	69	0
PL + PA	38.5	1.5	44	0
CD (P = 0.05)	**2.2**	**0.34**	**6.1**	**6.1**

12.3.2. Watermelon Pests and Diseases

(i) Integrated Management

(a) Bioagents and Chemicals: Seed treatment with thiamethoxam + Phule *Trichoderma* each at 5 g/kg was the most promising treatment against serpentine leaf miner (*Liriomyza trifolii*), thrips (*Thrips palmi*), bud necrosis disease and root rot disease in watermelon and registered highest fruit yield (17.722 MT/ha) followed by imidacloprid + Phule *Trichoderma* at 5 + 5 g/kg (12.555MT/ha), imidacloprid at 5 g/kg (14.444 MT/ha), and acetamiprid + Phule *Trichoderma* at 5 + 5 g/kg (12.555 MT/ha); while untreated plots recorded significantly lower yield (8.333 MT/ha).

12.4. MUSKMELON, *CUCUMIS MELO*

12.4.1. Diseases

12.4.1.1. Wilt, *Fusarium solani*

(i) Symptoms: On young seedlings, a hypocotyl rot and damping-off may occur. In older plants, there is marginal yellowing progressing to a general yellowing of the older leaves, and wilting of one or more runners. In some cases, sudden collapse occurs without any yellowing of the foliage. On stems near the crown of the plant, a linear, necrotic lesion may develop, extending up the plant and usually on one side of the vine (Fig. 12.9). One runner on a plant may wilt and collapse, with the rest of the runners remaining healthy. A gummy, red exudate may ooze from these lesions. Vascular discolouration should be evident and is very diagnostic (Fig. 12.9).

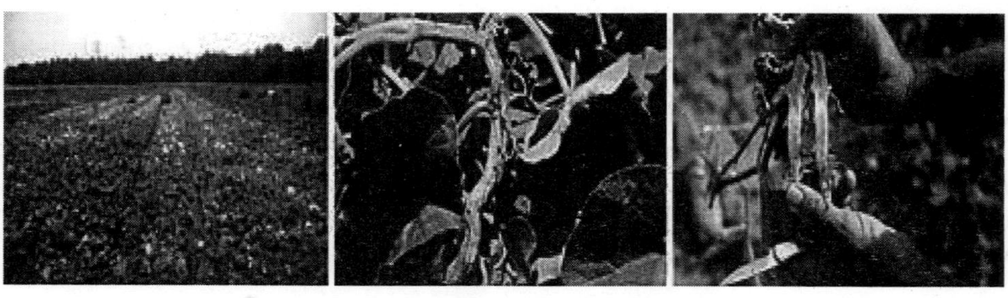

Fig. 12.9. Fusarium wilt of muskmelon

(ii) Integrated Management

(a) Bioagents and Botanicals: *T. harzianum* and *T. viride* when applied as seed treatment (each at 5 g/kg) followed by soil application along with FYM/neem cake at 1 kg/basin were effective.

(b) Bioagents and Chemicals: Integration of *T. viride* T4-1 with carbendazim and KCl was most effective in reducing wilt as well as pathogen propagules in the soil and in root tissues besides increasing shoot and root length (Chattopadhyay and Sen, 1996).

12.4.2. Nematodes

12.4.2.1. Root-knot Nematode, *Meloidogyne incognita*

(i) Symptoms: As *M. incognita* larvae enter the plant root, feed and mature, the surrounding cells of the plant root increase in size and divide causing galls on the roots (Fig. 12.10). The flow of nutrient and water is restricted, and plants wilt quickly when water becomes limiting. If plants are infected when young, they are often severely stunted and chlorotic. Infected vines rarely die, but are generally not productive.

Fig. 12.10. Root-knot nematode on musk melon.

(ii) Integrated Management

(a) Cultural and Botanicals: Combination of crop rotation (sorghum in 2005 followed by melon in 2006 and 2007), biocidal intercropping (*Eruca sativa* cv. Nemat) and biofumigant treatments (pellet based on formulated defatted seed meals of *Brassica carinata*), was effective for the management of root-knot nematodes. The results showed that sorghum cultivation as a non host crop halved the nematode population in the soil. The biofumigant treatment in spring 2006, caused a strong decrease in galling index (G.I. based on a scale from 0 to 5) on the roots of the following melon crop in rotation (G.I. = 0.3) compared with the untreated control (G.I. = 4.2). These positive results were further confirmed in 2007, when an important increase in quality and quantity was recorded in the yield of melon fruits. The autumn cultivation of *E. sativa* cv. Nemat and its incorporation into the soil brought a good amount of organic matter, with positive effects on melon yield. The intrinsic characteristics of the pellet based on *B. carinata* defatted seed-meals in rotation with *E. sativa* highlighted an excellent biofumigant effect, fully comparable to chemical nematicides, and a noticeable contribution in both organic matter and nitrogen that played an important fertilizing effect. The full effectiveness of pellet in decreasing substantially the nematode population, suggest its application in alternate years with *Eruca* green manure, in the presence of low larval infestations in the soil (Curto *et al*., 2008).

(b) Physical and Chemicals: The yield of cantaloupe was significantly increased by soil solarization with or without sulphur and by sulphur as single treatment. The thickness of the polyethylene did not affect the yields, which were significantly increased by EVA 0.150 mm, compared to EVA 0.035 mm. The addition of either 750 or 1,000 kg/ha sulphur previously to soil solarization was beneficial compared with the polyethylene tarping only. The sulphur application under EVA tarping did not statistically increased the yield with respect to 500 kg/ha sulphur alone (Table 12.3). The nematode population was significantly suppressed by either solarization or sulphur treatments: No difference was found between the two films or among the sulphur dosages and no further suppression derived by the combined use of solarization and sulphur.

Table 12.3. Effect of soil solarization and sulphur on *Meloidogyne incognita* on cantaloupe.

Sulphur (kg/ha)	Soil solarization	Crop yield (MT/ha)	Final nema popn. (eggs & J_2/ml soil)
-	-	12.6 ab	11.7 a
500	-	16.4 b	3.2 c
500	Polyethylene 0.050 mm	16.2 b	3.1 c
500	EVA 0.035 mm	19.0 bc	2.9 c
750	-	17.8 bc	4.3 c
750	Polyethylene 0.050 mm	20.4 c	2.6 c
750	EVA 0.035 mm	20.2 c	4.3 c
1000	-	17.0 bc	3.7 c
1000	Polyethylene 0.050 mm	20.4 c	3.4 c
1000	EVA 0.035 mm	20.4 c	1.6 c
-	Polyethylene 0.050 mm	17.0 bc	2.1 c
-	EVA 0.035 mm	16.4 b	4.9 bc

* Means followed by the same letters in the same column are not significantly different according to Duncan's Multiple Range Test (P = 0.01).

12.5. BITTER GOURD, *MOMORDICA CHARANTIA*

12.5.1. Insect Pests

12.5.1.1. Fruit Fly, *Bactrocera cucurbitae*

(i) Damage: Fruit fly infestation causes oozing of resinous fluid from fruits which become distorted and malformed. Maggots feed on fruit pulp (Fig. 12.11) causing premature dropping of fruits.

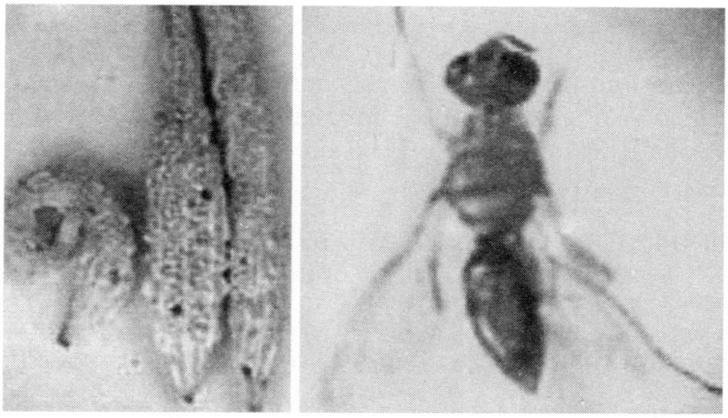

Fig. 12.11. Fruit fly damage on bitter gourd fruits.

(ii) Integrated Management

(a) Bioagents and Chemicals: Application of 0.05% acephate in combination with *Beauveria bassiana* at one ml commercial preparation/litre of solution recorded lower level of infestation of fruit fly in bitter gourd (Maicykutty and Gopalakrishnan, 2003). This botanical encouraged natural enemies namely, *Pediobius foveolatus* and *Tetrastichus ovularum*.

12.5.1.2. Leaf Hopper, *Empoasca motti*

(i) Symptoms: Small 3-4 mm long nymph and adults are destructive with the common hopper burn symptom on the leaves sometimes shiny and brownish causing premature death of leaves.

(ii) Integrated Management

(a) Pathogens and Chemicals: Application of *Bacillus thuringiensis* (Dipel) at 1 ml/litre solution along with 0.05% acephate gave satisfactory control of leaf hoppers in bitter gourd (Maicykutty and Gopalakrishnan, 2003).

12.5.1.3. Fruit Borer, Eudioptis indica (Fig. 12.12)

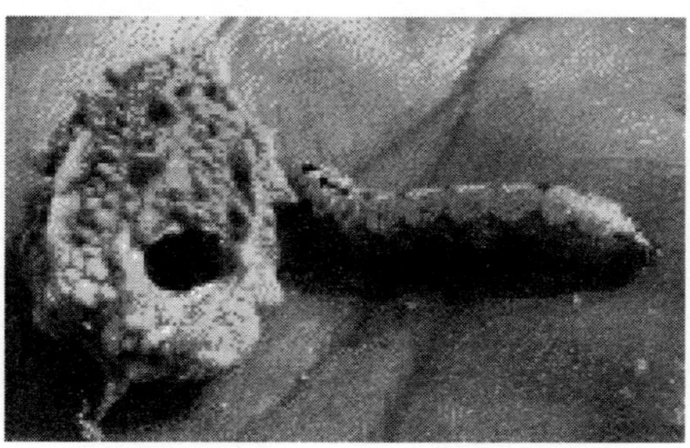

Fig. 12.12. Fruit borer infestation on bitter gourd

(i) Integrated Management

(a) Bioagents and Chemicals: Weekly application of 0.05% acephate in combination with *Beauveria bassiana* at one ml commercial preparation/litre of solution recorded lower level of infestation of fruit borer (Maicykutty and Gopalakrishnan, 2003).

12.6. POINTED GOURD, *TRICHOSANTHES DIOIDA*

12.6.1. Diseases

12.6.1.1. Stem and Fruit Rot, *Phytophthora cinnamoni*

(i) Symptoms: Infection usually starts at the middle portion of the fruits which shrinks and dries up. If there is cloudy weather or rain or soil remains moist due to application of irrigation, white mycelial growth develops in the infected portion. After starting of monsoon frequent rains and cloudy weather favours the spread of the disease, more and more vines and fruits get affected. Affected tissues become water-soaked and discoloured. In all cases the affected area of the fruits is covered with white mycelial growth. During this phase stem infection is also found mainly in the nodal region. During rainy days oozing of sticky substances from diseased stem, leaf and fruit tissues are quite common which usually takes places before formation of mycelial growth.

(ii) Integrated Management

(a) Bioagents and Botanicals: Singh *et al.* (2002) reported control of *P. cinnamoni* with certain organic amendments and *T. harzianum*.

12.6.2. Nematodes

12.6.2.1. Root-knot Nematode, *Meloidogyne incognita*

M. incognita is responsible for 30-40% loss in yield of pointed gourd.

(i) Symptoms: Root-knot nematodes attack the plant root, gall is formed in it, growth of plants retarded, leaves become chlorotic, flower and fruit form lately, and the production is significantly reduced.

(ii) Integrated Management

(a) Cultural, Bioagents and Botanicals: Integration of *Paecilomyces lilacinus* (50 g) + *Trichoderma harzianum* (100g) + neem cake (250g) + marigold as an intercrop (3 plants/pit) increased plant growth parameters and yield (6.6 kg/plant compared to 1.4 kg/plant in control), and reduced root galling (14.3/5g roots compared to 75.0/5g roots in control), number of egg masses (40/5g roots compared to 150/5g roots in control) and final nematode population (66.6/100g soil compared to 86660/100g soil in control) (Verma *et al.*, 2005) (Table 12.4).

Table 12.4. Effect of integration of bioagents, neem cake and marigold as intercrop for the management of *Meloidogyne incognita* infecting pointed gourd

Treatment	No. of galls /5g root	No. of egg masses/5g root	Yield (kg/plant)
Neem cake + *P. lilacinus* (50 g) + *T. harzianum* (100g) + (250 g) + Marigold (3 plants)/pit	10.4	30	8.5
P. lilacinus (50 g/pit)	15.3	53	6.4
Carbofuran (3 g/pit)	22.5	72	5.8
Neem cake (250 g/pit)	22.4	75	5.0
T. harzianum (100 g/pit)	25.6	45	4.2
Marigold (3 plants/pit)	35.3	113	4.0
Control	80.4	186	1.3
CD (P = 0.05)	**3.5**	**9.6**	**0.5**

(b) Bioagents, Botanicals and Chemicals: Vine dipping in monocrotophos at 1000 ppm for 6 hr + soil application of *T. viride* at 10 g/pit in 2 split doses (once at planting and another at 40 DAP) reduced root galling caused by *M. incognita*, nematode population in soil and gave fruit yield almost double of the untreated plots (Khan *et al.*, 2009) (Table 12.5).

Table 12.5. Effect of bioagents, botanicals and chemicals for the management of *Meloidogyne incognita* infecting pointed gourd

Treatment	Gall index (0-5 scale)	Nema popn. J_2/200 ml soil	Yield (kg/12m^2)
Vine dipping in 1000 ppm monocrotophos for 6 hr + Organic matter at 20 MT/ha	1.93	264	22.10
Vine dipping in 1000 ppm monocrotophos for 6 hr + Vermicompost at 2 MT/ha	1.78	232	25.53
Vine dipping in 1000 ppm monocrotophos for 6 hr + *P. lilacinus* at 10 g/pit (2 splits)	2.54	198	11.93
Vine dipping in 1000 ppm monocrotophos for 6 hr + *T. viride* at 10 g/pit in 2 split doses	1.75	162	33.80
Vine dipping in 1000 ppm monocrotophos for 6 hr + Carbofuran at 1 kg a.i./ha	3.60	386	16.90
Vine dipping in 1000 ppm monocrotophos for 6 hr + Neem cake at 500 kg/ha	2.16	280	17.43
Vine dipping in 1000 ppm monocrotophos for 6 hr alone	3.27	390	12.60
Untreated control	3.67	535	16.83
CD (P = 0.05)	**0.93**	---	6.02

(c) Physical, Cultural, Chemical and Botanicals: Integrated approaches, i.e. cultural control (deep ploughing followed by soil solarization for 15 days and stubble burning) with or without minimum synthetic chemical (2.0 or 3.0 kg a.i. carbofuran/ha) and with the inclusion of neem components like leaf, cake and azadirachtin, were highly effective in checking the nematode population build-up and infection. The eco-friendly integrated treatments recorded very low final nematode population and good yield (Chakraborti, 2000).

Chapter 13

LEAFY VEGETABLE CROPS

13.1. LETTUCE, *LACTUCA SATIVA*

13.1.1. Diseases

13.1.1.1. White Mold, *Sclerotinia sclerotiorum*

(i) Symptoms: The cool, moist conditions in the solar greenhouses in winter were perfect for white mold infection. The growing fungus would deposit hard black bodies, called sclerotia, on the soil surface. Some sclerotia would quickly sprout into tiny mushrooms (Fig. 13.1) that spread millions of spores throughout the greenhouses, infecting even more lettuce plants. Others would settle into the soil, where they might sit for years before sprouting mushrooms or infecting plant roots. The fungus can also grow directly from sclerotia, infecting plant roots.

Fig. 13.1. L- A bed of young lettuce plants infested with white mould, caused by the fungus *Sclerotinia sclerotiorum*. R - Sclerotia sprout into tiny mushrooms that spread *S. sclerotiorum* spores.

(ii) Integrated Management

(a) Biofumigation and Solarization: On-farm trails were conducted to monitor the effects of biofumigation and soil solarization on *Sclerotinia sclerotiorum*. This study suggests that a month of summer soil solarization can control populations of *S. sclerotiorum* to a depth of 15 cm in Kentucky high tunnels. The effect was seen in both at the middle and edge of solarized plots. Biofumigation, by incorporating a mixture of *Brassicae juncea* leaves and stems at a rate of 900 g/m^2, did not reduce germination of *S. sclerotiorum* sclerotia (Fig. 13.2).

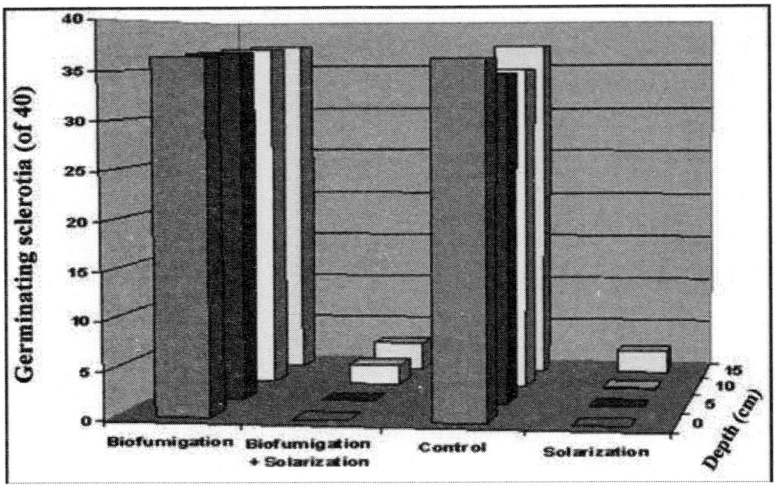

Fig. 13.2. The graph clearly shows that solarization was almost completely lethal to *S. sclerotiorum* to 15 cm, and that biofumigation had no effect on the fungus.

13.1.2. Nematodes

13.1.2.1. Root-knot Nematodes, *Meloidogyne incognita, M. hapla*

(i) Symptoms: Root knot nematodes feed within the roots and cause characteristic swelling galls on roots (Fig. 13.3). The northern root knot nematode, *M. hapla* generally occurs in cooler regions than the other three *Meloidogyne* species that prefer hot summer climates. Galls formed by *M. hapla* are spherical, distinct, and generally smaller than those caused by the three "warm-climate" species. Plants infested as seedlings may be stunted, with patches of stunted plants becoming evident by midseason. The root-knot nematode cause large decreases in yield of lettuce.

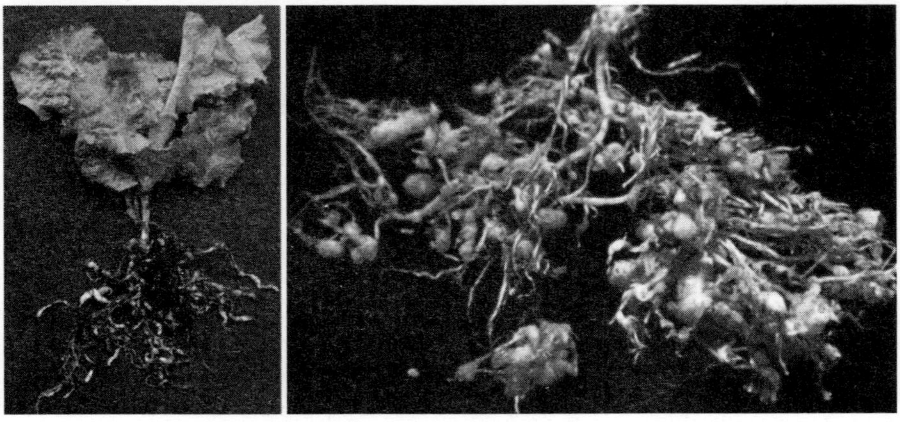

Fig. 13.3. Root-knot nematode on lettuce

(ii) Integrated Management

(a) Biofumigation, Soil Solarization and Bioagents: Biofumigation with mustard under black polythene for 30 days followed by soil application of *Pseudomonas fluorescens* at 2.5 kg/ha at the time of planting was on par with carbofuran at 1 kg a.i./ha in reducing the root-knot nematode population by 76.39% and increased the lettuce leaf yield by 23.29%.

13.2. ASPARAGUS, *ASPARAGUS OFFICINALIS*

13.2.1. Diseases

13.2.1.1. Wilt, *Fusarium oxysporum* f. sp. *asparagi*

(i) Symptoms: Mature plants infected with Fusarium gradually decline in productivity and growth. During the summer, infected plants are characterized by one to several stunted, bright yellow ferns (Fig. 13.4). A reddish brown vascular discolouration, which may extend into the crown, is present at the base of stalks infected by *F. oxysporum* f. sp. *asparagi.* Crowns and below-ground portions of stems exhibit reddish flecks or sunken brown lesions, which can be seen by cutting them open. Reddish brown, elliptical lesions occur on storage roots of infected plants. Feeder roots, most of which may be rotted off completely, show reddish brown discolouration.

Fig. 13.4. Yellowing of ferns on *Fusarium*-infected asparagus plant

(ii) Integrated Management

(a) AMF and Botanicals: Addition of charcoal or manure of coffee residue to bed soil is effective for increasing the tolerance to Fusarium root rot in AM fungus-infected asparagus plants (Matsubara *et al*., 2002).

13.3. DRUMSTICK, *MORINGA OLEIFERA*

13.3.1. Insect Pests

(i) Fruit Fly, *Gitona distigma*

(i) Damage: Fruit fly (Fig. 13.5) maggots cause drying and splitting of fruits from tip and oozing of gummy exudate from fruit.

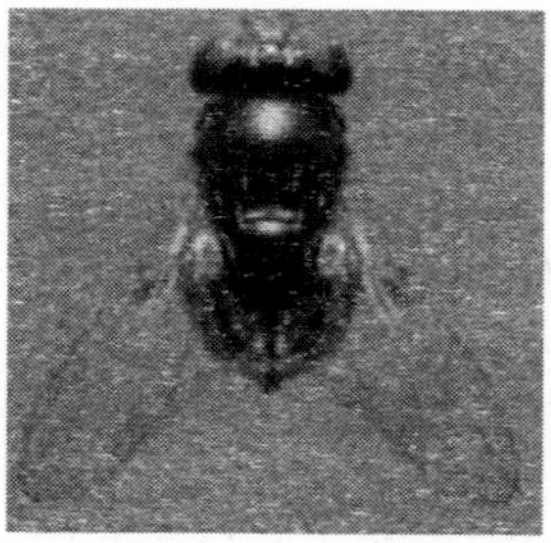

Fig. 13.5. Fruit fly on drumstick

(ii) Integrated Management

(a) Cultural, Botanicals and Chemicals: Integration of botanicals (Nimbecidine, NSKE), chemicals (Fenthion, Nuvan, Endosulfan) and cultural methods (removal of affected fruits regularly) (Modules A to F) were effective in reducing the fruit fly infestation and in increasing the fruit yield of drumstick (Raghumoorthi and Armugam, 1992) (Table 13.1).

Table 13.1. Impact of IPM components on fruit fly infestation and yield of drumstick

IPM module*	% fruit fly infestation	% reduction over control	Fruit yield (MT/ha)	% increase over control
A - Components 1,2,3,4	4.67	93.06	39.390	93.06
B - Components 2,3,4	7.33	89.11	38.291	89.11
C - Components 1,6,3,4	5.67	91.58	38.977	91.59
D - Components 5,2,7,4	7.33	89.11	38.291	89.11
E - Components 1,6,7,4	6.00	91.09	38.840	91.09
F - Components 5,2,3,4	5.00	92.57	39.254	92.58
Control-without any treatment	67.33	0.00	13.499	0.00

* Components: 1. Application of Fenthion 80 EC 0.04% at vegetative & flowering stage.
2. Application of Nuvan 76 WSC 0.04% at 50% fruit set & 35 days later.
3. Soil application of Endosulfan 4D (200 g/tree) at 50% fruit set.
4. Removal of affected fruits regularly at weekly intervals.
5. Application of Nimbecidine 0.03% at 60 ppm conc. during vegetative & flowering stage.
6. Application of Nimbecidine 0.03% at 150 ppm conc. during 50% fruit set & 35 days later.
7. Soil application of NSKE 4% at 2 litres/tree during 50% fruit set.

(b) **Cultural, Botanicals, Resistant Cultivars and Chemicals:** Application of fenthion 0.04% during vegetative and flowering stage, spraying of Nimbecidine 0.03% at 150 ppm during 50% fruit set and 35 days after, soil application of NSKE 5% at 2 l/tree in soil during 50% fruit set and removal of affected fruits are recommended along with the use of resistant accessions like MT_{18}, MT_{26} and MT_{28}.

Section - IV

BIOINTENSIVE INTEGRATED PEST MANAGEMENT IN ORNAMENTAL, MEDICINAL, AROMATIC AND TUBER CROPS

Chapter - 14

ORNAMENTAL CROPS

14.1. ROSE, *ROSA* spp.

14.1.1. Diseases

14.1.1.1. Black Spot, *Marssonina rosae* (syn. *Diplocarpon roseae)*

(i) Symptoms: This disease is more of a problem in open-field cultivation. The characteristic symptom of the disease is the appearance of leaf spots which are coal-black, circular or irregular with black margins and yellow halos (Fig. 14.1). The spots develop on either side of the leaf. Symptoms first appear on older leaves on the lower portion of plant causing severe defoliation, reducing yield and the size of flowers. On close examination with a hand lens, the spots show small, black blister-like fruiting bodies of the fungus. Continual defoliation will cause weakness, dieback or death of the plant. Some very susceptible species may have stems affected with a considerable reduction in plant vigour.

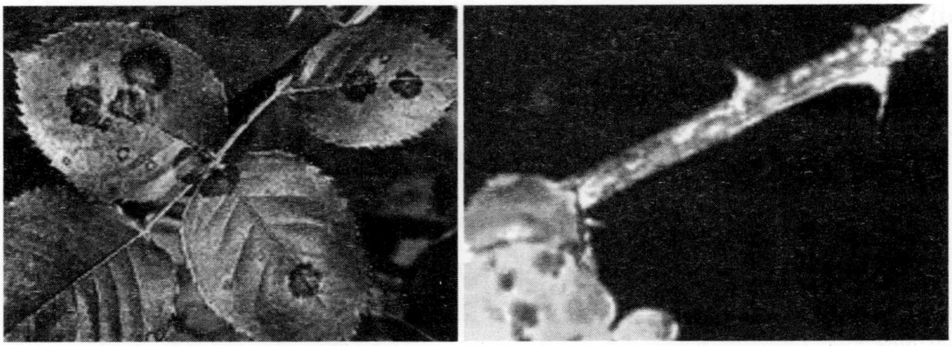

Fig. 14.1. Black spots on rose leaves and stem.

The pathogen can survive in the canes. Spores are spread by wind and water-splash.

(ii) Integrated Management

(a) Bioagents and Chemicals: Evaluation of fungal biocontrol agents (*Trichoderma harzianum, Chaetomium globosum*) and fungicides (chlorothalonil, mancozeb) either alone or in combination against black spot on rose revealed that at 100 days after first spray, lowest defoliation rating was recorded in chlorothalonil, *T.*

harzianum + chlorothalonil and *T. harzianum* + mancozeb treatments (1.00). Highest vigour index was recorded in *T. harzianum* treatment, while highest flower yield was recorded in *C. globosum* + chlorothalonil (4.33) followed by *T. harzianum* alone and *T. harzianum* + chlorothalonil treatments (4.00) (Prasad *et al.*, 2002).

14.1.2. Nematodes

14.1.2.1. Root-knot Nematodes, *Meloidogyne* spp.

(i) Symptoms: Nematodes of the genus *Meloidogyne* attack the roots of a wide range of plants including roses. The damaged root system develops root-knots (Fig. 14.2), which cause slow growth, wilting, and yellowing of leaves. Nematodes are spread by introducing infested plants or soil to the field, and also can be carried on garden tools.

Fig. 14.2. *Meloidogyne* sp. damage on rose roots.

(ii) Integrated Management

(a) Bioagents, Botanicals and Chemicals: Pre-plant treatment of soil with dazomet (at 25 g/m^2) followed by soil amendment with neem cake (at 1 kg/m^2) along with *Pochonia chlamydosporia* (at 2 x 10^{10} spores/m^2) recorded reduced root galling and increased flower yield (Nagesh and Janakiram, 2004) (Table 14.1).

Table 14.1. Integrated management of root-knot nematode on rose

Treatment Dose/m^2	Root gall index (1-5 scale)	Plant mortality (%)	% increase in flower yield
Dazomet-40 g	1.8	18	14.0
Neem cake-1kg + *P. chlamydosporia* -2 x 10^{12} spores	1.7	14	15.5
Dazomet-40 g + *P. chlamydosporia* -2 x 10^{12} spores	0.6	10	18.0
Neem cake-1kg + Dazomet-40 g + *P. chlamydosporia* -2 x 10^{12} spores	0.2	8	24.5
Control	3.0	40	---
CD (P = 0.05)	**0.19**	**3.11**	---

14.2. CARNATION, *DIANTHUS CARYOPHYLLUS*

14.2.1. Diseases

14.2.1.1. Wilt, *Fusarium oxysporum* f. sp. *dianthi*

(i) Symptoms: The initial symptoms are foliar yellowing and production of 'crook neck' or bent shoots (Fig. 14.3). The leaves and shoots wither and become brownish. Initially the symptoms are confined to a few branches or a part of the plant but ultimately the entire plant shows the symptoms. The symptoms are first noticed during mid-day when the temperature is quite high inside the polyhouses, but they are not visible when the temperature goes down. Stems when cut open show brown discolouration at the vascular region. Sometimes pith and cortex also get discoloured. Finally the stems show shredding of the internal tissue.

Fig. 14.3. Fusarium wilt on carnation.

The disease spreads through contaminated planting material as young rooted cuttings carry the pathogen without expressing the symptoms. Spread of the disease within the greenhouse occurs through root contacts, contaminated soil carried by workers and implements.

(ii) Integrated Method

(a) Bioagents and Physical/Chemical: Martinez and Pinzon (1999) reported that application of *T. harzianum* immediately after steam treatment or chemical disinfestation of soil prevented the rapid reinfestation of the soil by *F. oxysporum* f. sp. *dianthi*. By this treatment, it was possible to grow susceptible varieties of spray carnations in areas heavily infested with Fusarium wilt.

14.2.2. Nematodes

14.2.2.1. Root-knot Nematode, *Meloidogyne incognita*

The root-knot nematode is one of the serious limiting factors in commercial cultivation of carnation under polyhouse conditions.

(i) Economic Importance: Nagesh and Parvatha Reddy (2000) reported that *M incognita* was responsible for 27% loss in flower yield of carnation. Most of the highly fetching exotic cultivars of carnation from Europe have shown 40 to 60% mortality in polyhouse beds due to root-knot nematode infection in and around Bangalore (Nagesh and Parvatha Reddy, 1996a).

(ii) **Symptoms:** The root-knot infected carnation plants exhibit stunted growth, leaf yellowing and premature dropping and root galling (Fig. 14.4).

Fig. 14.4. Root galling on carnation due to root-knot nematodes.

(iii) Integrated Management

(a) Bioagents and Botanicals/Chemicals: Integration of neem cake at 0.5 kg per m^2 with *T. viride* at 100 g per m^2 was effective for the management of root-knot nematodes and increased flower yields.

Integration of *P. lilacinus/T. harzianum* at 0.5L per m^2 (aqueous spore suspension containing 2 x 10^4 spores per ml) with neem cake at 0.5 kg per m^2 or fenamiphos at 2 g a.i. per m^2 increased plant growth parameters and flower yield of carnation. The above treatments also increased root-knot egg parasitization by the parasitic fungi (Nagesh and Parvatha Reddy, 1996a) (Fig. 14.5).

Three antagonistic fungi viz., *T. harzianum*, *V. lecanii* and *P. lilacinus* at 2 x 10^4 spores/ml in combination with neem cake reduced *M. incognita* population in both soil and roots of carnation.

Application of *P. lilacinus* at 0.5 g/kg of soil along with neem cake at 1.0 MT/ha efficiently suppressed the nematode population and checked its build up, enhancing the plant growth parameters resulting in better flower production with increased flower stalk length and flower diameter in carnation. The plants also came to flowering early (Nirmal Johnson, 2000).

Fig. 14.5. Management of root-knot nematodes on carnation by integration of bioagents and botanicals. Left – Untreated; Right – Treated.

(b) Bioagents and AMF: Combined inoculation of AMF and *Pseudomonas fluorescens* had positive effect on root-knot nematode control on carnation (Anusuya and Vadivelu, 2002).

(c) Bioagents, Botanicals and Chemicals: Pre-plant treatment of beds with dazomet (at 40 g/m^2) followed by the application of neem cake (at 1 kg/m^2 15 days later) along with antagonistic fungi, *P. chlamydosporia/P. lilacinus* (at 2 x 10^{12} spores/m^2) significantly reduced root-knot nematode population (*M. incognita*), mortality of plants and suppressed the nematode infection for 2 years in carnation. The antagonistic fungi established better in the beds treated with dazomet. The above treatment also reduced root galling, nematode multiplication rate, and increased spike/stem length, flower yield and root colonization with the bioagents (Nagesh and Parvatha Reddy, 2005) (Table 14.2).

Table 14.2. Integrated management of root-knot nematode on carnation

Treatment Dose/m^2	Root gall index	Plant mortality (%)	Stem length (cm)	No. of flowers/m^2
Dazomet-40 g	2.0	32	96.0	42.5
Carbofuran-10 g	3.2	30	92.0	40.0
Carbosulfan-1 litre of 0.03%	2.2	33	96.5	47.5
Chlorpyriphos-1 litre of 0.03%	2.4	38	95.0	45.5
Neem cake-1kg + *P. lilacinus*-2 x 10^{12} spores	2.8	31	95.5	43.0
Neem cake-1kg + *P. chlamydosporia* -2 x 10^{12} spores	2.8	29	98.0	46.0
Dazomet-40 g + Neem cake-1kg + *P. lilacinus*-2 x 10^{12} spores	1.6	11	104.8	63.0
Dazomet-40 g + *P. chlamydosporia* -2 x 10^{12} spores	0.4	12	116.0	68.5
Control	4.2	52	75.5	37.5
CD (P = 0.05)	**0.18**	**1.84**	**3.43**	**1.66**

14.2.2.2. Root-knot Nematode, *Meloidogyne incognita* and Wilt, *Fusarium oxysporum* f. sp. *dianthi* Disease Complex

(i) Symptoms: In an experiment carried out to study the role of root-knot nematode, *M. incognita* in predisposing carnation to *Fusarium* wilt, it was observed that when both the pathogens were simultaneously inoculated the root galling index (RGI) was 3.1, while prior inoculation of *F. oxysporum* showed a RGI of 2.8. When *M. incognita* was inoculated alone the RGI was 3.5, while prior inoculation of *M. incognita* recorded an RGI of 4.15 twelve weeks after inoculation.

The appearance of the wilt symptoms were accelerated when *M. incognita* was inoculated two weeks prior to *F. oxysporum* f. sp. *dianthi*. The rate of wilting was observed to be 4.6, while the root galling index (RGI) value was recorded to be 4.75 during 25th week of observation. It was observed that maximum plant mortality was recorded when *M. incognita* was inoculated 2 weeks prior to *F. oxysporum* f. sp. *dianthi* which was 79.2 % at 25th week of observation. The plant growth parameters (plant height, plant weight) were also reduced significantly due to prior inoculation of *M. incognita* (Table 14.3) (Shylaja, 2004).

Table 14.3. Effect of interaction of *M. incognita* and *F. oxysporum* f. sp. *dianthi* on wilt disease complex on carnation cv. Ivonne

Treatment	Root-knot index*	Wilt disease index*	% plant mortality	Plant height (cm)	Plant weight (g)
M. incognita	8.2	0.0	14	44.0	28.56
F. oxysporum f. sp. *dianthi*	0.0	2.5	45	40.0	27.80
M. incognita + *F. oxysporum* f. sp. *dianthi* (simultaneous inoculation)	7.4	3.3	63	22.4	25.88
M. incognita 2 weeks prior to *F. oxysporum* f. sp. *dianthi*	9.5	4.6	80	32.8	28.78
F. oxysporum f. sp. *dianthi* 2 weeks prior to *M. incognita*	7.0	2.3	40	17.4	21.5
Uninoculated control	0.0	0.0	0	71.4	62.58
CD (P = 0.05)	**0.42**	**0.18**	**7.56**	**7.56**	**3.73**

* Root-knot index 1-10 scale ** Wilt disease index 1-5 scale

(ii) Integrated Management

(a) Two Bioagents: The studies carried out to evaluate combination of bio-agents for the biological control of wilt (*F. oxysporum* f. sp. *dianthi*) and root-knot nematode (*M. incognita*) disease complex in carnation, revealed that a combination of *Pochonia chlamydosporia* + *Paecilomyces lilacinus* each at 20 g/m^2 gave significant increase in plant height and flower yield (stalk length, stalk weight). The lowest root galling (1.64) and wilting index (2.0) and plant mortality (49.5%) was found in plants treated with *P. chlamydosporia* and *P. lilacinus* (Table 14.4) (Shylaja, 2004).

Table 14.4. Effect of biocontrol agents on the plant growth and flower yield of carnation infected with *M. incognita* and *F. oxysporum* f. sp. *dianthi*

Treatment	Plant height (cm)	Stalk length (cm)	Stalk weight (g)	Root galling index	Wilt disease index	Plant mortality (%)
Formulations of *P. chlamydosporia* and *P. lilacinus* each at 20 g/m^2	54.0	48.8	17.24	1.64	2	49.5
Formulations of *P. chlamydosporia* and *T. harzianum* each at 20 g/m^2	52.2	43.4	15.90	2.70	3	60.4
Formulations of *T. harzianum* and *P. lilacinus* each at 20 g/m^2	47.8	39.2	14.66	2.17	3	63.9
Control	22.8	17.6	10.82	3.66	5	95.8
CD (P = 0.05)	**5.16**	**4.54**	**2.01**	**0.82**	**0.53**	**10.46**

14.3. GERBERA, *GERBERA JAMESONII*

14.3.1. Diseases

14.3.1.1. Foot Rot, *Pythium* spp., *Sclerotium rolfsii, Rhizoctonia solani*

(i) **Symptoms:** The plants show reduced growth, smaller leaves and flowers with weak stems. Leaves show purple discolouration, with rotting of both feeder and main roots.

***(ii)* Integrated Management**

(a) **Bioagents and Botanicals:** Soil application of FYM along with *Trichoderma harzianum* is helpful for initial suppression of the disease.

14.3.2. Nematodes

14.3.2.1. Root-knot Nematode, *Meloidogyne incognita*

The root-knot nematode, *M. incognita* is one of the serious limiting factors in commercial cultivation of gerbera under polyhouse conditions.

(i) Economic Importance: Nagesh and Parvatha Reddy (2000) reported that *M incognita* was responsible for 31% loss in flower yield of gerbera. Most of the highly fetching exotic cultivars of Gerbera from Europe have shown 40 to 60% mortality in polyhouse beds due to root-knot nematode infection in and around Bangalore (Nagesh and Parvatha Reddy, 1996a).

(ii) Symptoms: The root-knot infected gerbera plants exhibit stunted growth, leaf yellowing and premature dropping and root galling.

(iii) Integrated Management

(a) Bioagents, Botanicals and Chemicals: Pre-plant treatment of beds with dazomet (at 40 g/m^2) followed by the application of neem cake (at 1 kg/m^2 15 days later) along with antagonistic fungi, *P. chlamydosporia/P. lilacinus* (at 2 x 10^{12}

spores/m^2) significantly reduced root-knot nematode population (*M. incognita*), mortality of plants and suppressed the nematode infection for 2 years in gerbera. The antagonistic fungi established better in the beds treated with dazomet. The above treatment also reduced root galling and plant mortality, and increased spike length & flower yield with the bioagents (Nagesh and Parvatha Reddy, 2005) (Table 14.5).

Table 14.5. Integrated management of root-knot nematode on gerbera

Treatment Dose/m^2	**Root gall index**	**Plant mortality (%)**	**Spike length (cm)**	**No. of flowers/ m^2**
Dazomet-40 g	1.8	28	66.5	44.0
Carbofuran-10 g	2.9	32	60.8	42.5
Carbosulfan-1 litre of 0.03%	1.8	30	65.5	44.0
Chlorpyriphos-1 litre of 0.03%	2.0	25	63.5	42.5
Neem cake-1kg + *P. lilacinus*-2 x 10^{12} spores	2.6	25	62.0	43.0
Neem cake-1kg + *P. chlamydosporia* -2 x 10^{12} spores	2.6	28	64.2	44.0
Dazomet-40 g + Neem cake-1kg + *P. lilacinus*-2 x 10^{12} spores	1.4	15	72.5	55.5
Dazomet-40 g + *P. chlamydosporia* -2 x 10^{12} spores	0.4	12	75.9	58.0
Control	3.8	40	48.5	40.5
CD (P = 0.05)	**0.11**	**4.11**	4.41	**5.11**

14.3.2.2. Root-knot Nematode, *Meloidogyne incognita* and Foot Rot, *Phytophthora parasitica* Disease Complex

(i) Symptoms: Sustainable production of gerbera is seriously hampered by the disease complex caused by *M. incognita* and *P. parasitica*. These two pathogens reduce the productivity of gerbera significantly to the tune of 40-60%.

(ii) Integrated Management

(a) Bioagents and Botanicals: Combined application of neem cake enriched with either *Trichoderma harzianum* or *Pseudomonas fluorescens* [mixing 50 g of *T. harzianum* (2 x 10^6 cfu/g) or *P. fluorescens* (2 x 10^8 cfu/g) in 1 kg of neem cake] applied at 25 g/m^2 was found effective for the management of disease complex and increased the flower yield by 26% in gerbera cv. Debora (Manoj Kumar *et al.*, 2010).

14.4. TUBEROSE, *POLIANTHES TUBEROSA*

14.4.1. Diseases

14.4.1.1. Leaf Spot/Blight, *Alternaria polyantha*

(i) Integrated Management

(a) Two Bioagents: Treatment with *Trichoderma harzianum* and *Pseudomonas fluorescens* at 4 g/kg tuber and seedling dip in *T. harzianum* suspension before planting resulted in effective control of blight and increased the plant vigour.

14.4.2. Nematodes

14.4.2.1. Root-knot Nematode, *Meloidogyne incognita*

M. incognita, M. javanica and *M. arenaria* have been reported as the major limiting factors in successful tuberose cultivation in Tamil Nadu (Sundarababu and Vadivelu, 1988b), while *M. incognita* and *M. javanica* were potential pests in Karnataka (Khan and Parvatha Reddy, 1992).

(i) Economic Importance: *M. incognita* was responsible for 13.25, 9.87, 14.30, 13.78 and 28.58% reduction in plant weight, number of flowers, spike length, spike weight and number of bulblets, respectively (Khan and Parvatha Reddy, 1994).

(ii) Symptoms: Affected plants exhibit stunting, yellowing and drying up of leaves and rotting of bulbs (Jayaraman *et al.*, 1975). Further, the emergence of side shoots from the bulbs was also affected and the numbers were conspicuously less. In severely infected plants, the emergence of spike is suppressed and 65% reduction in top growth occurred (Sundarababu and Vadivelu, 1988). Conspicuous galls can be observed on the root system (Fig. 14.6).

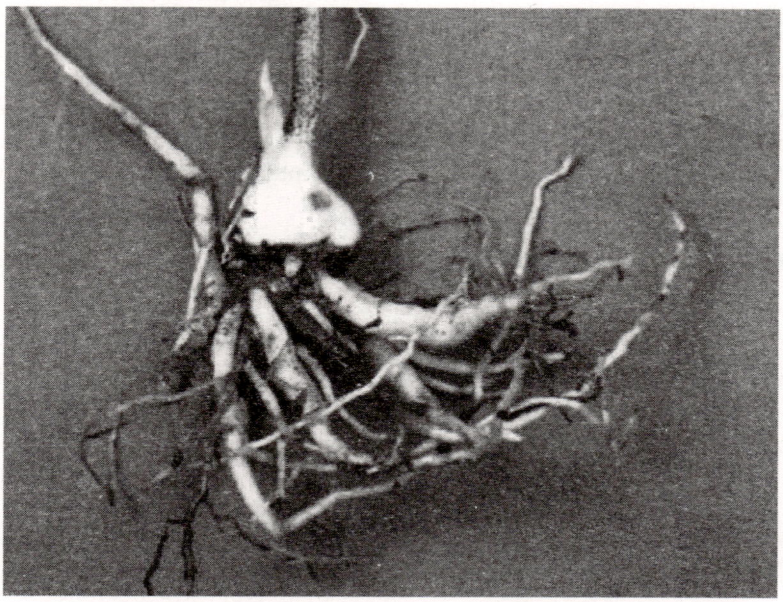

Fig. 14.6. Root-knot nematode on tuberose.

(iii) Integrated Management

(a) Bioagents and Botanicals: Split application (at planting and 45 DAP) of neem cake at $1kg/1.5m^2$ in combination with *P. lilacinus* (at $25g/1.5m^2$ containing 18 x 10^8 spores/g) significantly reduced root galling. The above treatment also significantly increased number of spikes, and parasitization of egg masses by *P. lilacinus* (Nagesh *et al.*, 1998) (Table 14.6).

Table 14.6. Effect of split application of oil cakes and *Paecilomyces lilacinus* on root galling, spike yield and propagule density on tuberose infected with *Meloidogyne incognita*

Treatment (Dose/1.5 m²)		Root-knot index	No. of Spikes/ 1.5 m²	% egg mass parasitization by *P. lilacinus*
At planting	**45 DAP**			
Control	---	4.2	34	---
P. lilacinus-25 g + Castor cake-1 kg	*P. lilacinus*-25 g + Castor cake-1 kg	3.0	55	32.3
P. lilacinus-25 g + Karanj cake -1 kg	*P. lilacinus*-25 g + Karanj cake-1 kg	2.4	56	40.3
P. lilacinus-25 g + Neem cake-1 kg	*P. lilacinus*-25 g + Neem cake-1 kg	1.8	69	48.9
Carbofuran-2 kg a.i./ha	---	3.2	45	---
CD (P = 0.05)		**0.26**	**4.82**	**5.22**

Treatment of tuberose bulbs with neem cake extract mixed with *P. lilacinus* spores significantly reduced *M. incognita* infection and multiplication besides stimulating the plant growth. Soil application of neem cake enriched with *T. harzianum* at 1 MT/ha is also effective. Treatment of tuberose bulbs with neem and calotropis leaf extracts mixed with *P. lilacinus* spores significantly reduced *M. incognita* infection and multiplication besides stimulating the plant growth (Nagesh *et al.*, 1997).

Table 14.7. Effect of integration of *Glomus mosseae, G. fasciculatum*, neem cake and aldicarb on root galling, plant growth and bulb weight of tuberose infected with *Meloidogyne incognita*

Treatment	Dose	Plant weight (g)	Root-knot index	Bulb wt. (g)
Glomus mosseae + Aldicarb	100 spores/100 g soil + 0.5 kg a.i./ha	7.99	2.0	2.3
G. fasciculatum + Aldicarb	100 spores/100 g soil + 0.5 kg a.i./ha	9.53	2.6	1.7
G. mosseae + Neem cake	100 spores/100 g soil + 0.5 kg a.i./ha	5.46	2.5	2.8
G. fasciculatum + Neem cake	100 spores/100 g soil + 0.5 kg a.i./ha	9.96	2.4	2.5
G. mosseae + Aldicarb + Neem cake	100 spores/100 g soil + 0.5 kg a.i./ha + 0.5 MT/ha	11.80	1.5	2.8
G. fasciculatum + Aldicarb + Neem cake	100 spores/100 g soil + 0.5 kg a.i./ha + 0.5 MT/ha	7.43	2.2	0.9
Control	---	3.26	4.5	1.1
CD (P = 0.05)	---	**1.35**	**0.18**	**0.50**

(b) Botanicals, AMF and Chemicals: Integration of AMF (*Glomus mosseae, G. fasciculatum*), a botanical (neem cake) and a nematicide (aldicarb) gave effective management of *M. incognita* and increased plant growth and yield of bulbs (Khan and Parvatha Reddy, 1994) (Table 14.7).

(c) Two Bioagents: Combined application of *P. chlamydosporia* + *T. harzianum* gave significantly higher control of *M. incognita* on tuberose (Shylaja *et al.*, 2004). Soil application of *T. harzianum* (10^9 cfu/g) + *P. lilacinus* (10^6 cfu/g) gave effective control of root-knot nematodes.

14.4.2.2. Bud and Leaf Nematode, *Aphelenchoides besseyi*

Foliar nematode, Aphelenchoides besseyi emerged as a serious problem in tuberose reported from Ranaghat areas of Nadia district of West Bengal. A. besseyi is becoming a major limiting factor for cultivation of tuberose in Ranaghat, Haringhata and Panskura areas of West Bengal. The 'Calcutta single' cultivar of tuberose is more vulnerable to A. besseyi than the 'Calcutta double' cultivar. Research results confirmed that A. besseyi is the primary causal agent for malformed flowers (Khan, 2001).

(i) Symptoms: Infected flower stalk initially appears rough, stalk becomes crinkled, stunted and finally distorted and in severe cases flower buds failed to bloom. Brown streaks appear on leaf bracts and petals and subsequently develop into rusty brown spots. The severely infected flower stalk becomes rotten and brittle over drying, even gets blind and the number of flowers per stalk is also reduced. The nematode forms 'nematode wool' upon dark brown spots. The ovary contains large number of nematodes. This nematode is generally more serious during rainy season from June to September and cent per cent loss of the second year crop of the 'Calcutta Single' cultivar of tuberose is encountered. However, in 'Calcutta Double' cultivar 30% to 40% flower stalk renders unsalable and individual flower stalk harbours up to 45,000 nematodes (Khan and Pal, 2001). In 'Calcutta Single', the yield loss may occur to the extent of 59% (Pathak and Khan, 2008). The presence of nematode species in the cut flower and stalk is a constraint in export of flowers to other countries of the world for pest risk.

Infested bulbs harbour nematode in coiled anhydrobiotic condition (quiescent pre-adult and adult stages) in the scaly leaves outside the bulbs. The nematode can also survive in the dried scaly leaves, stems and flowers more than 25 months; however, they cannot survive in soil for long time (Khan and Ghadipur, 2004; Khan, 2006).

(ii) Integrated Management

(a) Physical and Chemical: Pre-soaking of tuberose bulbs in water overnight followed by hot water treatment at 50°C for 20 minutes + spraying of the crop twice with monocrotophos 36 EC at 0.15% (at sprouting and 30 days after first spray) in the first year and 3 sprays in second and third year crop at 15-20 days interval was found most effective in reducing percent nematode infestation as well as disease indices (Khan *et al.*, 2004).

14.4.2.3. Root-knot Nematode, *Meloidogyne incognita* and Wilt, *Fusarium oxysporum* f. sp. *dianthi* Disease Complex

(i) Symptoms: Rao *et al.* (2002b) reported that root-knot nematodes accelerate and increase *Fusarium* wilt symptom development and ultimately increase the death rate of tuberose plants infected with both the pathogens. The fungal infection was observed to aggravate in the presence of *M. incognita*.

The root-knot nematode, *M. incognita* reduced flower yield considerably and the tuberose plants became highly susceptible to the attack by *F. oxysporum* f. sp. *dianthi* (Rao *et al.*, 2002b).

In an experiment undertaken to assess the impact of different densities of root-knot nematodes in inducing *Fusarium* wilt-root knot disease complex on the flower yield of tuberose, it was observed that the spike emergence was significantly delayed (65.07 and 68.33 days) due to the high population levels of nematodes at 100 and 150/100 cm^3 soil. The longest delay in spike emergence (81 days) was recorded when *Fusarium* was present along with the highest density of nematodes at 150/100 cm^3 soil. It was observed that the disease complex drastically reduced the yield of flowers in tuberose, thereby bringing down the production of flowers in this commercial crop. It was found that the damage was maximum in the presence of both pathogens *viz; M. incognita* and *F. oxysporum* f. sp. *dianthi*, when compared to the presence of either one of these pathogens (Table 14.8) (Shylaja, 2004).

Table 14.8. Effect of interaction of *Meloidogyne incognita* with *Fusarium oxysporum* f. sp. *dianthi* on flower yield of tuberose

Treatment	Days for spike emergence	No. of spikes/ plot	Spike length (cm)	Days for I^{st} flowering
F. oxysporum f. sp. *dianthi* at 10 ml (10^2 spores/ ml)	63.00	17.33	74.33	74.00
M. incognita at 50 J_2/100 cm^3 soil	62.33	18.00	69.33	72.33
M. incognita at 100 J_2/100 cm^3 soil	65.07	17.00	63.67	75.33
M. incognita at 150 J_2/100 cm^3 soil	68.33	13.67	62.67	77.67
F. oxysporum f. sp. *dianthi* at 10 ml + *M. incognita* at 50 J_2/100 cm^3 soil	74.67	13.33	61.33	85.33
F. oxysporum f. sp. *dianthi* at 10 ml + *M. incognita* at 100 J_2/100 cm^3 soil	76.00	12.50	59.33	89.33
F. oxysporum f. sp. *dianthi* at 10 ml + *M. incognita* at 150 J_2/100 cm^3 soil	81.00	11.00	56.67	89.67
Control	60.67	24.00	80.67	72.00
CD (P = 0.05)	**4.00**	**2.31**	**5.85**	**4.33**

(ii) Integrated Management

(a) Two Bioagents: In course of the experiments carried out to evaluate combination of bio-agents for the control of wilt and root-knot nematode disease complex in tuberose, the best result was obtained in plants treated with *P. chlamydosporia* + *T. harzianum* (Rao *et al.*, 2003). The above treatment gave significant reduction in root galling, nematode population both in soil and roots, disease index, and increased egg parasitization and root colonization of bioagents and flower yield (Table 14.9).

Table 14.9. Effect of integration of bioagents for the management of disease complex and flower yield in tuberose

Treatment/Dose (g/4m²)	**Gall index (1-10 scale)**	**Disease index (1-5 scale)**	**No. of Spikes/ 4m²**	**Root colonization (cfu/g)**		**% egg parasitization**	
				Pc	***Th***	***Pc***	***Th***
P. chlamydosporia – 20g	6.21	3.76	24	30458	---	40.98	---
P. chlamydosporia – 40g	4.26	3.11	26	38943	---	54.69	---
T. harzianum – 20g	6.56	2.82	20	---	40369	---	50.68
T. harzianum – 40g	5.21	2.33	23	---	45653	---	55.84
P. chlamydosporia – 20g + *T. harzianum* – 20g	4.79	2.18	29	30879	38789	37.49	51.69
P. chlamydosporia – 40g + *T. harzianum* – 40g	4.20	1.59	24	36278	44926	53.96	52.38
Control	8.49	4.42	18	---	---	---	---
CD (P = 0.05)	**1.76**	**0.76**	**2.33**	**4563.28**	**5219.75**	**6.47**	**7.42**

Shylaja (2004) found that integration of *P. chlamydosporia* with *T. harzianum* gave maximum increase in plant height. Integration of *P. chlamydosporia* with *P. lilacinus* gave least root galling and wilt disease index. Integration of *P. chlamydosporia* with *T. harzianum* gave maximum flower yield characteristics (spike length, number of flowers/spike and number of spikes/plot) (Table 14.10) (Shylaja, 2004).

Table 14.10. Effect of bioagents on plant growth parameters and disease complex of tuberose infected with *M. incognita* and *F. oxysporum* f. sp. *dianthi*

Treatment	**Plant height (cm)**	**Root galling index**	**Wilt disease index**	**Spike length (cm)**	**No. of flowers/ spike**	**No. of spikes/ plant**
Formulations of *P. chlamydosporia* and *P. lilacinus* each at 20 g/m²	33.01	2.18	1.50	70.63	29.87	22.80
Formulations of *P. chlamydosporia* and *T. harzianum* each at 20 g/m²	33.39	2.40	2.28	75.73	45.00	24.0
Formulations of *T. harzianum* and *P. lilacinus* each at 20 g/m²	32.01	3.36	3.10	72.89	29.90	21.60
Control	20.60	4.10	4.28	45.69	23.65	18.40
CD (P = 0.05)	**4.94**	**0.36**	**0.40**	**5.09**	**6.22**	**2.64**

14.5. GLADIOLUS, *GLADIOLUS* spp.

14.5.1. Diseases

14.5.1.1. Wilt, *Fusarium oxysporum* f. sp. *gladioli*

(i) Symptoms: This disease is often referred to as Fusarium yellows or Fusarium Wilt. The symptoms are yellowing of the leaves starting with the older outside ones which is accompanied by the apparent stunting of newer leaves. In addition, the spike itself will often be stunted and faded in colour. The plant may also have blackened areas at the base which spread onto the corm (Fig. 14.7). Eventually the whole plant will wilt. The interior of the corm when opened will appear marbled with a brown colour.

Fig. 14.7. Fusarium infection on gladiolus.

(ii) Integrated Management

(a) Bioagents and Botanicals: *T. harzianum*/*T. viride* either alone or in combination with compost (mushroom, vermin, paddy straw) reduced wilt effectively (58 to 92%).

(b) Bioagents, Botanicals and Chemicals: Under field conditions, the disease can be effectively controlled by application of neem cake at 1.25 MT/ha along with *T. harzianum* before planting, followed by soil drenching with 0.1% carbendazim one month after planting.

(c) Bioagents and Chemicals: Mishra *et al.* (2002) reported significant reduction of gladiolus corm rot or wilt by integration of corm treatment with *T. virens* and carboxin.

14.5.2. Nematodes

14.5.2.1. Root-knot Nematode, *Meloidogyne incognita*

(i) Symptoms: The root-knot nematode causes stunting of plants and reduction in leaf count. The threshold population of 1000 juveniles of *M. incognita* per plant reduced the growth of plant markedly and there was no emergence of plants at 10,000 juveniles of *M. incognita* per plant (Chandel *et al.*, 1997).

Severe galling on roots results in yellowing of leaves which subsequently leads to stunted growth. The nematode invades roots, daughter corms and cormels which develop after flowering.

(ii) Integrated Management

(a) Botanicals and Bioagents: Corms of gladiolus treated with neem suspension enriched with *Pseudomonas fluorescens* strains 1 and 2 (5 g/litre of water) and application of neem cake enriched with *P. fluorescens* strains to beds at 20g/m^2 was found to be effective in reducing the root population of *M. incognita* by 68% and increased the flower yield by 28%.

Soil application of neem cake enriched with *T. harzianum* at 1 t/ha is also effective.

(b) Botanicals and AMF: Both the neem cake and the AM fungus (*Glomus fasciculatum*) increased the spike length, number of florets/spike and the floret diameter whereas the root-knot nematode (*Meloidogyne incognita*) reduced these floral growth characters. The neem cake at higher doses of 1 and 2 % and the AM fungus significantly increased the floral growth characters and suppressed root galls. These effects were significantly greater in the presence of neem cake/AM fungi is around nematode infected plants. Lower doses of cake in combination with *G. fasciculatum* is suggested as a means to protect gladiolus from root-knot damage (Hasan and Nehal Khan, 2004).

14.5.2.2. Root-knot Nematode, *Meloidogyne incognita* and Wilt, *Fusarium oxysporum* f. sp. *gladioli* Disease Complex

(i) Symptoms: The commercial production of gladiolus is limited by soil-borne pathogens like root-knot nematodes and *Fusarium* wilt. Their combined occurrence in cultivated soils aggravated the wilt problem causing high plant mortality in gladiolus fields.

(ii) Integrated Management

(a) Bioagents and Botanicals: Gladiolus plants treated with *P. lilacinus* + *T. harzianum* + neem cake and *P. lilacinus* + *T. viride* + neem cake combinations not only controlled *M. incognita* infection but also *Fusarium* wilt till the harvest of flower spikes. The corms and cormels obtained from plants treated with these combinations were free from *Fusarium* infection. Bioagent colonization of galled roots was maximum in *P. lilacinus* + *T. viride* combination (94%) followed by *P. lilacinus* + *T. harzianum* combination (69%). Similarly, the parasitization of eggs was maximum in *P. lilacinus* + *T. viride* combination (44%) followed by *P. lilacinus* + *T. harzianum* combination (38%), while egg mass parasitization was same in both the combinations (58%) (Nagesh *et al.*, 1998) (Table 14.11).

Table 14.11. Effect of integration of antagonistic fungi with neem cake on root-knot and wilt disease complex on gladiolus.

Treatment/Dose/plant	% Healthy plants	% Redn. in nema multipln.	% Infected corms & cormels	% Root coloni-zation	% Parasitization	
					Egg masses	eggs
P. lilacinus-8 x 10^{10} spores	29	38	52	48	49	46
T. harzianum-8.8 x 10^{10} spores	36	16	30	50	44	32
T. viride-8.8 x 10^{10} spores	40	24	20	53	47	39
P. lilacinus + Neem cake 8 x 10^{10} spores + 20 g	34	64	32	62	61	58
P. lilacinus + *T. harzianum* + Neem cake ½ dose each	78	26	4	69	58	38
P. lilacinus + *T. viride* + Neem cake + ½ dose each	88	47	0	74	58	44
CD (P = 0.05)	**4.36**	**6.22**	**8.54**	**3.88**	**4.11**	**3.56**

14.6. CHRYSANTHEMUM, *DENDRATHEMA GRANDIFLORA*

14.6.1. Nematodes

14.6.1.1. Root-knot Nematodes, *Meloidogyne* spp.

M. arenaria and *M. javanica* have been reported on chrysanthemum (Sen and Dasgupta, 1977; Chandwani and Reddy, 1967). Root-knot nematodes have been reported to cause economic losses in cut-flower plantations.

(i) Symptoms: The root-knot nematode, Meloidogyne javanica cause severe plant stunting, chlorosis, and extensive root galling on chrysanthemum cv. Yellow Vero in a commercial cut-flower production facility.

(ii) Integrated Management

(a) Botanicals and Bioagents: Field application of two formulations of *P. lilacinus* (talc and pesta granules) in combination with neem cake reduced root gall index, nematode populations and enhanced flower yield of chrysanthemum by 23 to 28% and increased the *P. lilacinus* spore viability in the rhizosphere for longer time.

14.7. CROSSANDRA, *CROSSANDRA UNDULAEFOLIA*

14.7.1. Diseases

14.7.1.1. Foot Rot or Root Rot, *Phytophthora nicotianae*

(i) Symptoms: This disease leads to sudden death of plants. In young seedlings, symptoms appear as lesions on rootlets followed by rotting of the entire rootlet. On the collar region, peculiar berry root can be seen. Leaves show pink discolouration and drooping. In advanced stage of infection, wilting of whole plants can be noticed.

(ii) Integrated Management

(a) Bioagents, Botanicals and Chemicals: Application of neem cake along with *Trichoderma harzianum* followed by soil drenching with 0.2% aliette gave effective control of the disease.

14.7.2. Nematodes

14.7.2.1. Root-knot Nematode, *Meloidogyne incognita*

(i) Economic Importance: The nematode is responsible for 25.62% and 21.64% loss in number of flowers and weight of flowers, respectively (Khan and Parvatha Reddy, 1994).

(ii) Symptoms: Infected plants are stunted with dried peripheral branches bearing smaller chlorotic leaves almost turning to white at later stages (Rajendran *et al.*, 1976). Roots exhibit severe galling. Inflorescences are small and sometimes fail to produce flowers.

(iii) Integrated Management

(a) Botanicals and Bioagents: Combinations of *Verticillium lecanii* and *Paecilomyces lilacinus* (2 x 10^4 spores/ml each) with 5% neem leaf extract resulted in significantly higher plant growth parameters and crossandra flower yield (Nagesh and Parvatha Reddy, 1995). Root galling, nematode multiplication factor were least and the parasitization of eggs and egg masses was highest in *V. lecanii* + 5% neem leaf extract (Table 14.12).

Table 14.12. Effect of integration of *Paecilomyces lilacinus, Verticillium lecanii* and leaf extracts on root galling and flower yield of crossandra infected with *Meloidogyne incognita*

Treatment	Dose/plant	Rot-knot index	Flower yield (g/plant)
Control	---	4.21	18
Castor leaf extract	5%	3.8	24
Neem leaf extract	5%	3.0	28
P. lilacinus	10^4 spores/ml	3.0	25
V. lecanii	10^4 spores/ml	3.4	24
P. lilacinus + Castor leaf extract	10^4 spores/ml + 5%	2.8	32
P. lilacinus + Neem leaf extract	10^4 spores/ml + 5%	2.8	33
V. lecanii + Castor leaf extract	10^4 spores/ml + 5%	2.5	28
V. lecanii + Neem leaf extract	10^4 spores/ml + 5%	2.0	38

Incorporation *V. lecanii* with neem cake facilitated the effective management of *M. incognita* on crossandra. Application of neem cake enriched with *T. harzianum* at 2 kg/m^2 (2 MT/ha) in nursery beds gave effective control.

(b) Cultural and Botanicals: Incorporation of FYM in soil and intercropping of crossandra with marigold or pangola grass would reduce the root-knot nematode infection.

(c) AMF and Botanicals: *G. fasciculatum* and *G. mosseae* in combination with neem cake gave better control of nematodes over carbofuran treatment. The root colonization of AM fungi increased significantly in the presence of neem cake which in turn improved their efficacy in reducing *M. incognita* population in crossandra roots (Nagesh *et al*., 1998).

Integration of neem, karanj and castor cakes with *G. mosseae* significantly enhanced plant growth parameters and flower yield of crossandra, root colonization and sporulation of AMF. The above treatments also reduced root-knot nematode multiplication and root-galling (Nagesh and Parvatha Reddy, 1996b).

(d) Bioagents, AMF and Botanicals: Integration of a bioagent (*V. lecanii*), endomycorrhiza (*G. mosseae*) with botanicals improved the growth of crossandra and reduced the population of *M. incognita*. *G. mosseae* reduced the requirement of phosphatic fertilizer and favoured the antagonistic potential of *V. lecanii* against *M. incognita*.

(e) Two or More Bioagents: Combined application of *P. lilacinus* and *Pasteuria penetrans* enhanced plant growth and flower yield of crossandra besides reducing root galling due to *M. incognita* (Nagesh *et al*., 1995).

Integration of *V. lecanii, P. lilacinus* and *P. marguandi* gave effective control of *M. incognita* infecting crossandra (Khan and Parvatha Reddy, 1994) (Table 14.13).

Table 14.13. Integration of *Verticillium lecanii, Paecilomyces lilacinus* and *P. marguandi* on final nematode population of *Meloidogyne incognita* infecting crossandra

Treatment	Dose (g)/kg soil	Final nematode population
V. lecanii	1	2737
V. lecanii	2	2198
P. lilacinus	1	6165
P. lilacinus	2	1929
P. marguandi	1	3688
P. marguandi	2	714
V. lecanii + P. marguandi	1 + 1	1563
V. lecanii + P. lilacinus	1 + 1	3501
P. lilacinus + P. marguandi	1 + 1	1389
V. lecanii + P. lilacinus + P. marguandi	1 + 1 + 1	453
Control	---	12145
CD (P = 0.05)	---	**1176**

Treatment of the nursery bed with the formulations of *P. chlamydosporia* and *P. fluorescens* each at 50 g/m^2 and seed treatment with later was significantly effective in reducing number of nematodes in roots and soil, increasing the per cent parasitization or per cent suppression of eggs by bio-control agents and also flower yield of crossandra. The seedlings were colonized by both the bio-agents and when transplanted in the field the bio-agents reached the field soil as they were recovered from root and soil samples at harvest of the crop. Individual effect of bio-agents was maximized when both these organisms were integrated in the nursery bed stage. This could be due to the combined effect of both organisms on root-knot nematode. Combined use of *P. fluorescens* and *P. chlamydosporia* did not affect the colonization of each other on root (Rao, 2007b).

(f) Bioagents, Botanicals and Chemicals: Integrated management of root-knot nematodes infecting crossandra was achieved by rational combination of a biocontrol agent (*T. harzianum* or *P. chlamydosporia*) with a nematicide (aldicarb) or oil cake (neem cake) (Khan and Parvatha Reddy, 1994) (Table 14.14).

Table 14.14. Effect of integration of *Trichoderma harzianum*/*Pochonia chlamydosporia,* neem cake and aldicarb on plant growth and root galling in crossandra infected with *Meloidogyne incognita*

Treatment	Dose	Plant wt (g)	Root-knot index
P. chlamydosporia	4 g/kg soil	2.37	2.5
T. harzianum	4 g/kg soil	4.10	2.8
Aldicarb	1 kg a.i./ha	2.53	2.0
Neem cake	1 MT/ha	2.12	1.8
P. chlamydosporia + Aldicarb	2 g/kg soil + 0.5 kg a.i./ha	4.70	1.5
T. harzianum + Aldicarb	2 g/kg soil + 0.5 kg a.i./ha	5.06	1.7
P. chlamydosporia + Neem cake	2 g/kg soil + 0.5 MT/ha	5.20	1.7
T. harzianum + Neem cake	2 g/kg soil + 0.5 MT/ha	8.26	1.6
P. chlamydosporia + Aldicarb + Neem cake	2 g/kg soil + 0.5 kg a.i./ha + 0.5 MT/ha	5.16	1.3
T. harzianum + Aldicarb + Neem cake	2 g/kg soil + 0.5 kg a.i./ha + 0.5 MT/ha	2.90	1.7
Control	---	0.50	4.5
CD (P = 0.05)	---	**1.23**	**0.42**

14.7.2.2. Root-knot, *Meloidogyne incognita* and Foot Rot, *Phytophthora nicotianae* Disease Complex

(i) Symptoms: Infection by the root-knot nematode, *Meloidogyne incognita* was shown to make the plants more prone to foot rot disease, *Phytophthora nicotianae*.

(ii) Integrated Management

(a) Bioagents and Chemicals: Combined application of *Trichoderma harzianum* (20 g/plant) and Aliette (0.3%) did not show any mortality of plants and was also effective for the management of root rot disease and root-knot nematode disease complex (Table 14.15).

Table 14.15. Effect of *Trichoderma harzianum*, Aliette and neem cake on root-knot nematode and foot rot disease complex in crossandra

Treatment	Mortality*	Collar rot*	Root rot	Root-knot index
Nematode alone	Nil	Nil	1.60	2.80
Nematode + Neem cake + *P. nicotianae*	1	Nil	1.75	1.25
Nematode + *T. harzianum* + *P. nicotianae*	Nil	Nil	1.20	1.20
Nematode + Aliette + *P. nicotianae*	Nil	Nil	0.40	1.40
Nematode + *P. nicotianae*	2	2	3.50	1.70

* Denotes number of plants

Scale for root rot index: 0 = No root rot, 1 = 1-25%, 2 = 25-50%, 3 = 50-75%, 4 = 75-100%

Scale for root-knot index: 0 = No galls, 1 = 1-10%, 2 = 10-20%, 3 = 20-50%.

Chapter - 15

MEDICINAL PLANTS

15.1. COLEUS, *COLEUS FORSKOHLII*

15.1.1. Diseases

15.1.1.1. Wilt, *Fusarium chlamydosporum*

(i) Symptoms: In the field, the infected plants were characterized by gradual yellowing marginal necrosis and withering of leaves followed by loss in vigour and premature death. Such plants showed discolouration of roots and complete decaying of tap and lateral root system. The bark of such plants easily peeled off. There was extensive sloughing off and shredding of affected bark. Such affected plants were finally killed due to severe root and collar rot. The infected tubers showed rotting and emitting bad odour (Fig. 15.1) (Shyla, 1998).

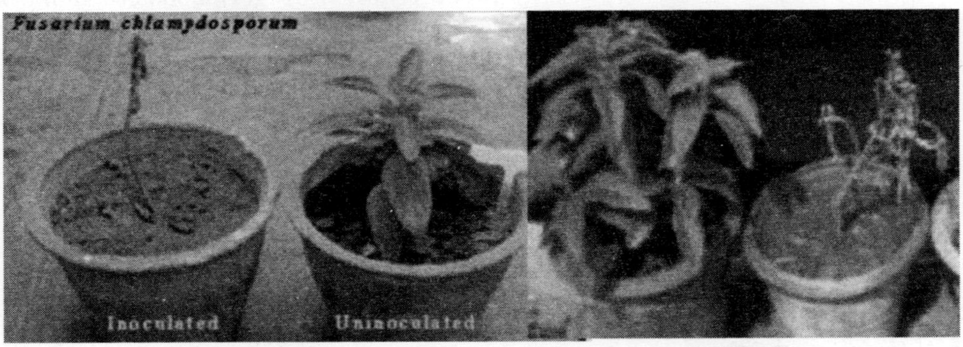

Healthy Infected

Fig. 15.1. Symptoms of *Fusarium* wilt on coleus

(ii) Integrated Management

(a) Bioagents and AMF: Inoculation with *T. viride* + *G. mosseae* gave the best result in controlling the disease. The same treatment also resulted in maximum plant growth, yield and root forskohlin concentration of coleus. The next best treatment was *P. fluorescens* + *T. viride* followed by *G. mosseae* + *P. fluorescens* and *T. viride* alone (Boby and Bagyaraj, 2003).

(b) Two Bioagents: Paramasivan *et al.* (2007) reported that the use of bioinoculants like *T. viride* and *P. fluorescens* reduced the disease incidence by 20-21%.

(c) Bioagents and Botanicals: Combination of *T. viride* + Neemto (neem based product applied at 500g/5m^2) resulted in lowest wilt incidence (12.76%) (Kulkarni *et al.*, 2007).

15.1.1.2. Blight, *Rhizoctonia bataticola*

(i) Symptoms: Blight disease is common during monsoon or during period of high humidity. The disease initially expressed as water soaked areas and the affected tissues soon turned into a soft, black, watery mass at the collar region of the plant. The infection was also found on roots and caused decay, which ultimately resulted in collapse of the plant. The infected plant roots showed discolouration followed by rotting of root hairs. Extensive sloughing off of affected bark was also observed. Under conditions of high humidity, the disease was found to spread rapidly. Severe infection results in defoliation and death of the plants (Fig. 15.2) (Ramaprasad Shresti, 2005).

Fig. 15.2. Blight symptoms on coleus

(ii) Integrated Management

(a) Botanicals and Bioagents: The blight incidence and colony forming units of *R. bataticola* were significantly minimum in the plots where *T. viride* (10 ml/plant spore suspension) combined with neemto (500 g/5 m2) were applied compared to other treatments (Table 15.1) (Ramaprasad Shresti, 2005).

Table 15.1. Management of blight of *Coleus forskohlii* using different biocontrol agents, organic amendments and chemicals (Ramaprasad Shresti, 2005).

Treatment	% blight incidence*
Trichoderma viride at 10 ml / plant (8x10^3 cfu/ml)***	21.09 (27.33)**
Trichoderma harzianum at 10 ml/plant (8x10^3 cfu/ml)	18.87 (25.74)
Pseudomonas fluorescens at 10 ml/plant (24x10^5 cfu/ml)	19.98 (26.51)
Pronto at 5% soil drench	23.31 (28.84)

Neemto at 500 g/5 m^2	21.09 (27.24)
Carbofuran 3G at 15 g a.i./5 m^2	24.42 (29.57)
Farm yard manure at 5 kg/5 m^2	25.53 (30.38)
Trichoderma viride at 10 ml/plant ($8x10^3$ cfu/ml) + Neemto at 500 g/5 m^2	12.7 (20.93)
Carbendazim at 0.1% soil drench	21.19 (27.33)
Propiconazole at 0.1% soil drench	23.31 (28.84)
Control	35.52 (36.59)
CD at 5%	**3.48**

*Observations recorded at harvest (150 days after planting).

**Figures in parentheses are arc sin angular transformed values.

***Cfu – Colony forming units

15.1.1.3. Root Rot, *Sclerotium rolfsii*

(i) Symptoms: The earliest symptom of the disease was darkening of the stem at collar region of the plant. The leaves became flaccid and dropped off. White, fan shaped mycelial strands crept over the stem portion, developing small light to dark brown sclerotia on the infected portion. The sclerotial initials were white at first, later turned brown with age. Finally the plant wilted and dried (Fig. 15.3) (Ramaprasad Shresti, 2005).

Fig. 15.3. Root rot symptoms on coleus

(ii) Integrated Management

(a) Cultural, Botanicals, Bioagents and Chemical: Treatment involving field sanitation + dipping stem cuttings in carbendazim (0.1%) + one more drench with 0.1% carbendazim 30 days after planting (DAP) recorded maximum reduction of disease incidence over control (76.50 and 72.72% at 45 and 90 DAP, respectively). This was on par with treatment soil application of zinc sulphate at 20 kg/ha + soil drenching with neem cake and *Trichoderma viride* mixture at 50 g/plant, where reduction in disease incidence was 73.73 and 70.11% over control at 45 and 90 DAP, respectively. The tuberous root yield was also maximum in the above treatments (5787 and 5759 kg/ha).

15.1.1.4. Bacterial Wilt, *Ralstonia solanacearum*

Bacterial wilt is the major disease of *C. forskohlii* causing heavy losses (>50%) in south India (Chandrashekara and Prasannakumar 2010).

(i) **Symptoms:** Disease show various symptoms like yellowing and wilting of leaves, brown to black roots, oozing, putrefaction and decaying of roots and unhealthy plants (Fig. 15.4).

Water soaked patches with linear streaks on collar region of the infected plants were observed. Leaves became flaccid and drooped quickly, wilting and drying of the plants. The leaves showed roll up symptoms and the whole plant dried up (Fig. 15.4). Wilted plants came off easily with a gentle pull and vascular discolourations were observed. Such tubers when pressed exhibited oozing of bacterial exudates.

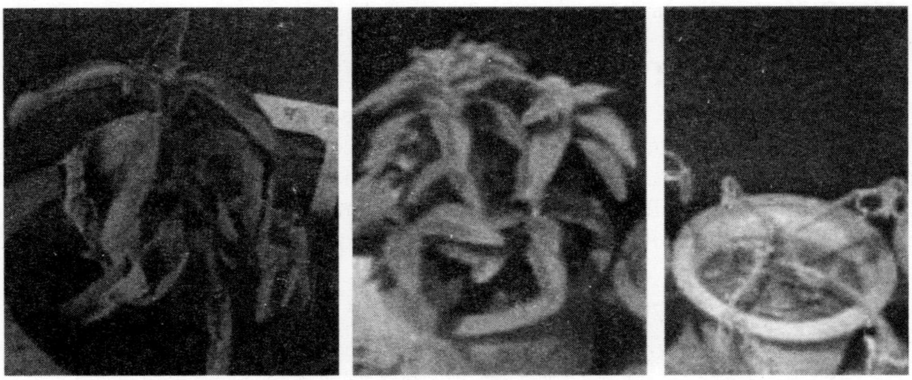

Wilting of Plant Healthy Infected

Fig. 15.4. Symptoms of bacterial wilt on coleus

(ii) Integrated Management

(a) Bioagents and Botanicals: Application of 2 kg of *Pseudomonas fluorescens* mixed with 300 kg of compost is effective.

15.1.2. Nematodes

15.1.2.1. Root-knot Nematode, *Meloidogyne incognita*

The root-knot nematode infestation was reported on coleus from Kerala and Orissa. The dry weight of the tubers was reduced by 20% due to root-knot nematodes. The percentage of starch on fresh weight basis showed drastic reduction (16%) in the infested tubers. *M. incognita* was responsible for 70.2% loss in tuber yield of coleus.

(i) Symptoms: The galls on coleus roots are very big and pronounced. The root-knot nematode damage often leads to crop failure. The infested tubers swell in size with irregular surface and cracking of the skin (Fig. 15.5). When the infestation is severe, rotting sets in even before harvest. Infested tubers rot after harvest and rarely reach market.

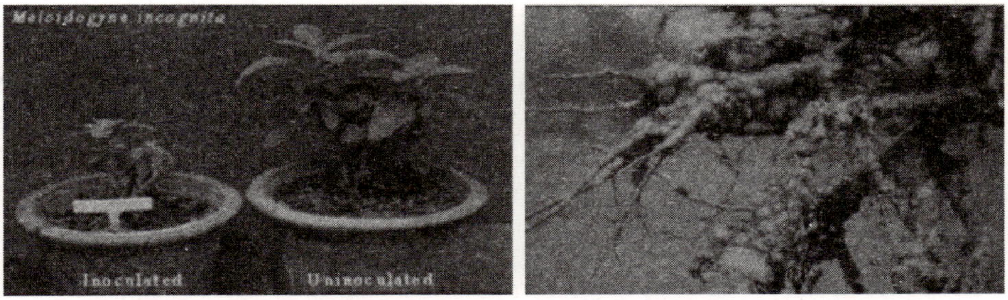

Fig. 15.5. Root-knot nematode on *Coleus forskohlii* (Mallesh, 2008)

(ii) Integrated Management

(a) Cultural, Botanicals and Bioagents : Integration of dipping stem cutting in 0.1% *Pseudomonas fluorescens* + soil application of neem cake at 400 kg/ha + growing marigold as an intercrop followed by their biomass incorporation during earthing up increased the root tuber yield by 22.7 to 30.0% and reduced the root-knot nematode (*M. incognita*) population by 71.2 to 73.8%. However, integration of *P. fluorescens* + marigold intercrop proved to be more economical (with benefit: cost ratio of 6.4 to 8.8) and effective management practice for the management of root-knot nematodes in medicinal coleus (Seenivasan and Devarajan, 2008) (Table 15.2).

Table 15.2. Effect of integration of bioagents, neem cake and marigold intercrop for the management of *Meloidogyne incognita* infecting coleus

Treatment	**Gall index (1-5 scale)**		**Yield (MT/ha)**		**Benefit: cost ratio**	
	Kharif	***Rabi***	***Kharif***	***Rabi***	***Kharif***	***Rabi***
Cutting dip in 0.1% *P. fluorescens* + neem cake at 400 kg/ha	4.3	3.6	10.41	9.86	1.8	1.3
P. lilacinus at 2.5 kg/ha + neem cake at 400 kg/ha	4.3	3.6	10.21	9.81	1.4	1.2
Cutting dip in 0.1% *P. fluorescens* + marigold intercrop	2.3	2.1	12.08	11.07	8.8	6.4
P. lilacinus at 2.5 kg/ha + marigold intercrop	3.6	3.3	10.92	10.20	4.6	3.2
P. fluorescens + neem cake + marigold	2.3	2.1	12.11	11.12	3.4	2.5
P. lilacinus + neem cake + marigold	3.6	3.3	11.06	10.26	2.0	1.4
Carbofuran at 1 kg a.i./ha	4.3	3.6	10.36	9.85	2.1	1.6
Control	4.6	4.3	9.31	9.06	---	---
CD (P = 0.05)	**0.36**	**0.36**	**0.83**	**0.80**	---	---

(b) Physical, Botanicals and Bioagents: Integration of soil solarization in the nursery for 15 days with 150 gauge LDPE film and application of *Paecilomyces lilacinus* + neem cake or *P. lilacinus* + *Bacillus macerans* in the main field are the best treatments in increasing plant height (64.3 and 60.3 cm compared to 40.0 cm in control), number of leaves (593.3 and 583.3 compared to 310.0 in control), weight of tubers/plant (560.0 and 546.6 g compared to 350.0 g in control), yield (11.5 and 11.3 kg/plot compared to 6.9 kg/plot in control) and in reducing root galls (0.3 and 1.0 compared to 50.6 in control), nematode population in soil (25.0 and 30.0/100 ml soil compared to 196.6/100 ml soil in control) and roots (1.0 and 1.6/5 g roots compared to 79.0/5 g roots in control) (Nisha and Sheela, 2006).

15.1.2.2. Root-knot Nematode *(Meloidogyne incognita)* and Wilt *(Fusarium chlamydosporum)* Disease Complex

(i) Symptoms: Among the different diseases affecting coleus, root-knot and wilt disease complex caused by *Meloidogyne incognita* and *Fusarium chlamydosporum* was observed in severe form (Kumar, 2008). In the interaction studies, *M. incognita* was the most aggressive pathogen compared to *F. chlamydosporum*. However simultaneous inoculation of *M. incognita* and *F. chlamydosporum* caused greater reduction in plant growth as well as nematode multiplication (Fig. 15.6). In case of sequential inoculation of *M. incognita* 7 days prior to *F. chlamydosporum* caused reduction in plant growth parameters. The effect of simultaneous inoculation of *M. incognita* with *F. chlamydosporum* on coleus was additive in nature. However, when *M. incognita* was inoculated with *F. chlamydosporum* the resultant effect was almost equal to sum of individual effect. These results suggest that the nematode can predispose the coleus to infection by *F. chlamydosporum* and can aggravate the disease. Reduction in number of galls per plant and final nematode population was observed in simultaneous inoculation of nematode and fungus.

Fig. 15.6. Complete rotting of root system due to root-knot and wilt disease complex

(ii) Integrated Management

(a) Bioagents and Botanicals: Combined application of plant products (Neem seed kernel powder at 5 g/kg of soil) with biocontrol agents (*Paecilomyces lilacinus,*

Trichoderma viride + *Pseudomonas fluorescens* at 10 g/kg of soil) was found effective in reducing the number of galls, nematode population, number of egg masses, root-knot index, root rot index and improving the plant growth parameters as compared to inoculated control.

15.1.2.3. Root-knot Nematode and Collar Rot Disease Complex

Rhizoctonia bataticola, Sclerotium rolfsii, F. chlamydosporum and *M. incognita* were found to be the most commonly associated fungi with collar rot disease complex (Ramaprasad Shresti, 2005).

(i) Symptoms: Inoculation with *M. incognita* seven days prior to inoculation of all three fungal pathogens or vice versa, the resultant effect on plant growth parameters was more than simple additive effect. Wilt symptoms were first recorded at 45 days after inoculation when *M. incognita* was inoculated seven days prior to inoculation of all the fungal pathogens simultaneously (*F. chlamydosporum* + *R. bataticola* + *S. rolfsii*). The highest root-knot index (Fig. 15.7) and nematode population were recorded in the treatment inoculation with *M. incognita* seven days prior to inoculation of *F. chlamydosporum* + *R. bataticola* + *S. rolfsii* (Ramaprasad Shresti, 2005).

Fig 15.7. Root-knot and collar rot disease complex of coleus. Left – Infected, Right –Healthy (Mallesh, 2008)

(ii) Integrated Management

(a) Botanicals, Bioagents and Chemicals: The wilt incidence and number of galls were significantly minimum in the plots where *T. viride* (10 ml/plant spore suspension) combined with neemto (500 g/5 m^2) were applied compared to other treatments. Colony forming units of *F. chlamydosporum* and *R. bataticola* were significantly minimum in the plots treated with carbendazim and propiconazole (Table 15.3) (Ramaprasad Shresti, 2005).

Table 15.3. Management of collar rot complex of *Coleus forskohlii* using different biocontrol agents, organic amendments and chemicals.

Treatment	% wilt incidence*	No. of galls/5 g roots	Cfu***	
			Fusarium chlamydosporium	***Rhizoctonia bataticola***
Trichoderma viride @ 10 ml/plant (8x10^3 cfu/ml)	21.09 (27.33)**	21.13	7.60	12.20
T. harzianum @ 10 ml/plant (8x10^3 cfu/ml)	18.87 (25.74)	19.53	8.00	12.60
Pseudomonas fluorescens @ 10 ml/plant (24x10^5 cfu/ml)	19.98 (26.51)	18.27	8.00	14.20
Pronto @ 5% soil drench	23.31 (28.84)	17.33	10.60	15.60
Neemto @ 500 g/5 m^2	21.09 (27.24)	16.07	12.60	16.40
Carbofuran 3G @ 15 g a.i./5 m^2	24.42 (29.57)	14.93	16.20	17.60
Farm yard manure @ 5 kg/5 m^2	25.53 (30.38)	25.67	15.20	18.80
Trichoderma viride at 10 ml/plant (8x10^3 cfu/ml) + Neemto @ 500 g/5 m^2	12.76 (20.93)	10.13	6.20	9.60
Carbendazim @ 0.1% soil drench	21.19 (27.33)	23.33	3.60	6.80
Propiconazole @ 0.1% soil drench	23.31 (28.84)	23.00	3.80	7.40
Control	35.52 (36.59)	28.40	19.60	21.60
CD at 5%	**3.48**	**5.38**	**2.49**	**2.72**

*Observations recorded at harvest (150 days after planting).

**Figures in parentheses are arc sin angular transformed values.

***Cfu – Colony forming units x10^{-3}/g of soil (Average of 5 replications).

15.1.2.4. Root-knot Nematode, *Meloidogyne incognita* and Root Rot, *Macrophomina phaseolina* Disease Complex

The productivity of coleus has been hampered by its susceptibility to root knot nematode (*Meloidogyne incognita*) and root rot disease (*Macrophomina phaseolina*) complex. Due to this disease complex the yield loss ranged from 50 to 60 per cent.

(i) Symptoms: Simultaneous inoculation of *M. incognita* and *M. phaseolina* as well as nematode inoculation followed by fungus 15 days later caused significant reduction in tuber yield and 100% root rot disease in medicinal coleus (Table 15.4) (Senthamarai *et al.*, 2008).

Table 15.4. Effect of *Meloidogyne incognita* and *Macrophomina phaseolina* on root galling and yield of coleus

Treatment	Tuber yield/ plant (g)	No. of galls/ plant	% disease incidence
M. incognita (1 J2/g soil)	44.00	768	0
M. phaseolina (5 g/kg soil)	61.50	0	50
M. incognita (prior) + *M. phaseolina* (15 days later)	34.50	373	100
M. phaseolina (prior) + *M. incognita* (15 days later)	52.50	110	50
M. incognita + *M. phaseolina* (simultaneously)	13.00	316	100
Uninoculated control	84.38	0	0
CD (P = 0.05)	**10.07**	**2.08**	**43.99**

(ii) Integrated Management

(a) Cultural and Bioagents Integrated nematode management strategy includes dipping of stem cuttings in *Pseudomonas fluorescens* (strain Pf1) + growing marigold as intercrop and their biomass incorporation during earthing up. The above treatment increased the tuber yield by 40.6% and reduced nematode infestation in terms of number of juveniles per 100 cc soil (73.2%), number of adult females per g of root (82.4%), number of egg mass/g root (85.9%), number of eggs per g of root (87.9%) with least gall index (1.6). This treatment also decreased the incidence of *M. phaseolina* root rot disease up to 50.4%. Hence it could be concluded that dipping of stem cuttings in 0.1% *P. fluorescens* + marigold intercropping can be commercially exploited for the management of *M. incognita* and *M. phaseolina* disease complex in medicinal coleus (Seenivsan, 2010).

15.2. ASHWAGANDHA, *WITHANIA SOMNIFERA*

15.2.1. Insect Pests

15.2.1.1. Mealy Bug, *Ferrisia virgata*

(i) Damage: Mealy bugs suck sap from lower surface of leaves and also from pods. Infested leaves showed yellowish discolouration followed by drying symptoms (Fig. 15.8).

Fig. 15.8. Mealy bug infestation on ashwagandha leaf

(ii) Integrated Management

(a) Botanicals and Chemicals: Soil application of FYM at 12.5 t/ha + azophos at 2 kg/ha + neem cake at 1 t/ha and need based foliar spray of neem oil (3%) was found effective.

15.2.2. Diseases

15.2.2.1. Seedling Blight

(i) Symptoms: Seedling blight is known to be the major disease of ashwagandha. The disease reduced the plant population drastically by seedling mortality, which ultimately reduced the yield.

(ii) Integrated Management

(a) Bioagents and Chemicals: Seed treatment with *Trichoderma viride* at 4 g/kg of seed along with metalaxyl at 6 g/kg of seed showed minimum seedling mortality (18.7%).

15.2.3. Nematodes

15.2.3.1. Root-knot Nematode, *Meloidogyne incognita*

(i) Symptoms: The nematode infected plants typically show chlorotic, stunted, less branched with fewer and smaller leaves and poor response to fertilizer and irrigation. Such symptoms usually are not noticeable until severe damage to root system by the nematodes. Roots of such plant were severely galled with reduced alkaloids. When stem touches the soil it was also found to be infested with root knot nematode. The root-knot nematode infected plants are more likely to be killed early than healthy non-infected plants (Fig. 15.9).

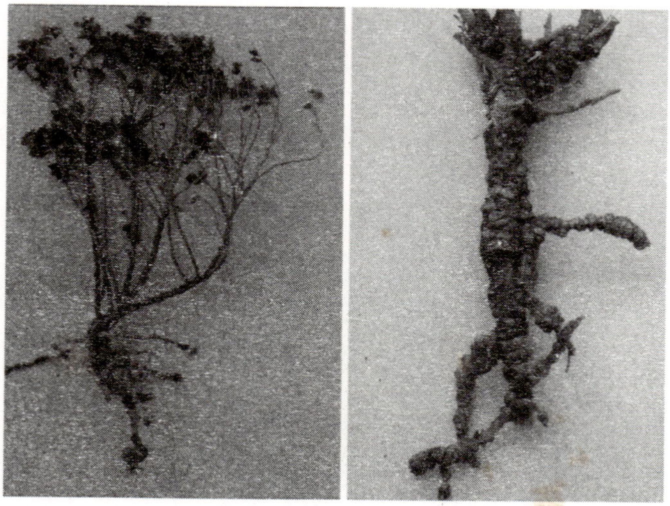

Fig. 15.9. Root knot infested plants and roots of *Withania somnifera*

(ii) Integrated Management

(a) Botanicals and Bioagents/AMF: Integration of neem cake + *T. harzianum,* vermicompost + *T. harzianum*, and cow urine + *T. harzianum* considerably reduced the root-knot nematode development and enhanced plant growth and yield. Maximum root-knot suppression was noticed in vermicompost + *T. harzianum* followed by mentha distillate + *Glomus aggregatum*. Highest increase in plant yield was recorded when the soil was amended with mentha/curry leaf distillates along with *T. harzianum/G. aggregatum* (Pandey and Kalra, 2003) (Table 15.5). Vermicompost apart from providing a complete nutrition to the plant may also support the growth *T. harzianum* thus helping it proliferate in the rhizosphere and improve the soil health and thus sustainability to restrict the nematode population.

Table 15.5. Effect of organic materials and bioagents on plant growth of Ashwagandha infected with *Meloidogyne incognita*

Treatment	Dry weight of plant (g)	Root-knot index	Total nema popn (soil + roots)
Untreated – Uninoculated	8.5	---	---
Untreated – Inoculated	5.2 (-38.8)	4.00	6460
Carbofuran	8.0 (-5.9)	1.66	3012
Neem cake + Davana	23.8 (+62.4)	1.66	2820
Neem cake + Curry leaf	7.4 (-12.9)	1.33	2261
Neem cake + Vermicompost	12.0 (+41.2)	1.66	2649
Neem cake + *T. harzianum*	12.8 (+50.6)	1.66	2674
Neem cake + *G. aggregatum*	14.3 (+68.2)	3.00	2964
Neem cake + *Mentha* distillate	14.2 (+67.1)	1.66	2409
Davana + Curry leaf	10.8(+27.1)	1.33	2030
Davana + Vermicompost	14.1 (+65.9)	3.00	2594
Davana + *T. harzianum*	9.6 (+12.9)	1.66	2110
Davana + *G. aggregatum*	10.3 (+21.2)	1.33	2050
Davana + *Mentha* distillate	13.2 (+55.3)	1.33	2150
Curry leaf + Vermicompost	14.0 (+64.7)	1.66	2440
Curry leaf + *T. harzianum*	15.0 (+76.5)	3.33	3252
Curry leaf + *G. aggregatum*	13.4 (+57.6)	3.00	2635
Curry leaf + *Mentha* distillate	15.3 (+80.0)	1.66	2404
Vermicompost + *T. harzianum*	14.4 (+69.4)	0.66	1400
Vermicompost + *G. aggregatum*	13.6 (+60.0)	1.99	2480
Vermicompost + *Mentha* distillate	13.9 (+63.5)	1.33	2000
G. aggregatum + *Mentha* distillate	15.6 (+83.5)	1.00	1784
CD (P = 0.05)	**0.71**	**0.01**	**465.70**

(b) Three AMF: Root-knot nematode infection was drastically impaired and plant growth biomass improved when plants were inoculated with three AMF simultaneously as compared to their inoculation alone (Table 15.6).

Table 15.6. Effect of integration of AMF for the management of *Meloidogyne incognita* infecting ashwagandha

Treatment	Fresh wt (g)	Dry wt. (g)	Oil yield (%)	% AMF colonization	Root-knot indices
Untreated-uninoculated	395.2	107.1	0.51	---	---
Untreated-inoculated	285.0	77.9	0.38	---	3.6
Glomus aggregatum (*Ga*)	322.0	91.6	0.40	42.3	2.3
G. fasciculatum (*Gf*)	333.0	93.8	0.41	58.2	2.0
G. mosseae (*Gm*)	388.0	105.9	0.48	63.5	1.6
Ga + *Gf* + *Gm*	360.0	99.8	0.46	78.5	1.3

(c) Bioagents and Botanicals: Soil application of super *Pseudomonas* (consortia of *Pseudomonas fluorescens* strains with chitinase, ACC deaminase and neem mixture) at 2.5 kg/ha was effective to suppress the root-knot nematode incidence by 78% and to enhance the biomass of economic parts of aswagandha plants by 26.87% (dry root) and 43.19% (seed) over untreated control.

Substantial reduction in root galling was noticed in neem oil seed cake + *T. harzianum*, vermicompost + *T. harzianum*, cow urine + *T. harzianum* treated plots as compared to untreated control. A significant and marked improvement in plant growth and yield was also noticed in plots treated with vermicompost + *T. harzianum*, cow urine + *T. harzianum*, neem oil seed cake + *T. harzianum* with maximum root yield obtained from vermicompost + *T. harzianum* treated plots (Table 15.7).

Table 15.7. Influence of organic and biological amendments on plant growth, yield and root knot index in *Withania somnifera* infected with *Meloidogyne incognita*

Treatments	Shoot dry weight (kg/m^2)	Root dry weight (kg/m^2)	Root-knot index (RKI)
Control (zero fertilizers)	1.3f	0.15h	3.33a
Farmyard manure	1.8e	0.20g	1.66b
Trichoderma harzianum	2.3d	0.25e	0.66cd
Cow urine	2.7b	0.28d	0.83cd
Vermicompost	2.3d	0.29bc	1.33bc
Neem oil seed cake	2.5c	0.23f	1.16bc
Farmyard manure + *T. harzianum*	2.5c	0.23f	1.16bc
Cow urine + *T. harzianum*	2.8ab	0.30b	0.33d
Vermicompost + *T. harzianum*	2.9a	0.32a	0.66cd
Neem oil seed cake + *T. harzianum*	2.8ab	0.29bc	0.33d

Mean in each column followed by same letters do not differ significantly (P=0.05) according to Duncan's multiple range test.

15.3. SARPAGANDHA, *RAUVOLFIA SERPENTINA*

15.3.1. Diseases

15.3.1.1. Foliar Blight/Spot, *Alternaria tenuis*

(i) Symptoms: The pathogen attacks the leaves, resulting in minute, brownish or dark coloured circular spots with a yellowish margin on the ventral side of the leaves. The fungus also affects the flowers and fruits.

(ii) Integrated Management

(a) Bioagents, Botanicals and Chemicals: Application of two split doses of neem cake at 50 g/plant at 30 days interval plus two sprays of *B. subtilis* in September and October along with three foliar sprays of mancozeb at 15 days interval plus two foliar sprays of carbendazim at 0.15% at 21 days interval gave 201.5 g fresh root/plant, and 75.5% protection against foliar blight pathogens (*Alternaria alternata, Cercospora rauvolfia, C. serpentina, Colletotrichum gloeosporioides, Corynespora cassiicola, Curvularia lunata, Macrophomina phaseolina* and *Rhizoctonia solani*).

15.3.2. Nematodes

15.3.2.1. Root-knot Nematodes, *Meloidogyne* spp.

The root-knot nematodes are recognized as the major limiting factors in successful cultivation of sarpagandha crop.

(i) Integrated Management

(a) Botanicals and Bioagents: Soil application of neem cake enriched with *T. harzianum* at 1 MT/ha is recommended.

15.4. HENBANE, *HYOCYAMUS MUTICUS, H. NIGER, H. ALBUS*

15.4.1. Nematodes

15.4.1.1. Root-knot Nematode, *Meloidogyne incognita*

Henbanes (*Hyoscyamus niger* and *H. muticus*) were severely infested with *M. incognita* and *M. javanica*. Even 3-4 root-knot larvae/g of soil caused significant damage to the crop (Pandey, 1990).

(i) Symptoms: In the field, up to 60-70% henbane plants were chlorotic and stunted showing a patchy appearance with fewer smaller leaves and flowers. The roots of infested plants were severely galled to various degrees. Varying sizes of galls were found in the root system (Fig. 15.10).

(ii) Integrated Management

(a) Bioagents and AMF: Application of bio-inoculants have not only enhanced the total biomass yield of *Hyoscyamus niger* but also significantly decreased the multiplication of nematodes. However, a significantly higher reduction was recorded in the treatment where all bio-inoculants are combined. This may be attributed to the fact that these bioagents are secreting potential chemicals which are either non-favourable for multiplication of *M. incognita* or inducing tolerance in the plant against the attack of root-knot nematodes. Hence, mixed inoculation of rhizo-bacterium with AMF could be considered for biomanagement for reducing the deleterious effect of root-knot nematodes in black henbane. Similar results were also obtained with different bioagents in *H. muticus* plants against *M. incognita* (Pandey *et al.*, 2000) (Table 15.8).

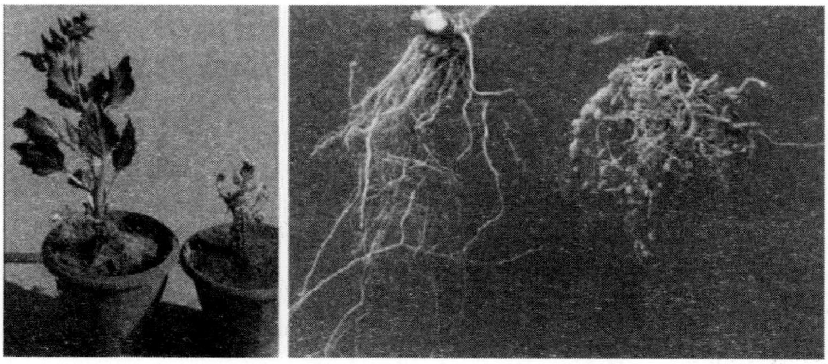

Fig. 15.10. Healthy & root-knot infested plants and roots of Egyptian henbane (*H. muticus*)

Table 15.8. Effect of bioagents for the management of *Meloidogyne incognita* infecting henbane

Treatment	Fresh biomass wt. (g)	Reproduction factor	Root-knot index
Control	120.6	8.14	4.00
Pseudomonas fluorescens (*Pf*)	171.0	4.24	1.33
Glomus aggregatum (*Ga*)	172.7	4.52	1.66
G. fasciculatum (*Gf*)	144.6	5.78	2.66
G. mosseae (*Gm*)	142.0	6.05	3.00
Pf + *Ga* + *Gf* + *Gm*	202.0	4.03	1.00
CD (P = 0.05)	**12.01**	---	**1.16**

15.5. ALOE, *Aloe indica*

15.5.1. Diseases

15.5.1.1. Black Spot/Rust

(i) Symptoms: Aloe rust causes round brown or black spots on aloe leaves.

(ii) Integrated Management

(a) Bioagents and AMF: Among four selected bioinoculants namely, *Bacillus subtilis, Glomus aggregatum, G. intraradices* and *Trichoderma harzianum* evaluated against black spot disease either alone or in different combinations; maximum plant height attained was 56 cm in *G. intraradices* treated plants followed by 53 cm in *B. subtilis* and *B. subtilis* + *G. aggregatum* treatments. Maximum plantlets production (14) was recorded in the treatments of *B. subtilis* + *T. harzianum* + *G. aggregatum* followed by other treatments (12). Maximum herb yield obtained was 3.65 kg/plant in *G. aggregatum* treated pots, followed by 3.35 kg/plant in *G. aggregatum* + *B. subtilis* and *G. intraradices* treated pots.

15.6. BABCHI, *PSORALIA CORYLIFOLIA*

15.6.1. Diseases

15.6.1.1. Collar Rot

(i) Integrated Management

(a) Bioagents, Botanicals and Chemicals: Soil application of neem cake at 400 kg/ha + seed treatment with *T. viride* at 4 g/kg + spraying of carbendazim at 0.05% in December-January significantly reduced the collar rot incidence.

15.7. SODA APPLE, *SOLANUM VIARUM*

15.7.1. Diseases

15.7.1.1. Wilt, *Fusarium oxysporum* f. sp. *lacimiti*

(i) Symptoms: The disease generally appears after first showers and is characterized by wilting and sudden death of plants. The disease spreads very fast and within a fortnight majority of the plants in a field show wilting symptoms which ultimately die irrespective of age of the plant.

(ii) Integrated Management

(a) Bioagents and Botanicals: Mixing 2.5 kg *Trichoderma viride* in 50 kg of FYM and applying along plant lines gave effective management of Fusarium wilt.

Chapter 16

AROMATIC PLANTS

16.1. JASMINE, *JASMINUM* SPP.

16.1.1. Diseases

16.1.1.1. Collar Rot and Root Rot, *Sclerotium rolfsii*

(i) Symptoms: Plants at all stages are infected. First the older leaves become yellow followed by younger leaves and finally death of the plant. In the root black discolouration can be seen. White strands of mycelia and mustard like sclerotia are seen on the infected tissues and stem surface.The young leaves turn yellow and the twigs start drying from tip downwards (Fig. 16.1). The plants wilt in patches.

Fig. 16.1. Collar rot and root rot symptoms on jasmine

(ii) Integrated Management

(a) Bioagents and Botanicals: Heavy application of FYM with *Trichoderma viride* is helpful.

16.1.2. Nematodes

16.1.2.1. Root-knot Nematode, *Meloidogyne incognita*

M. incognita was reported on *J. sambac* and *J. flexile* from Tamil Nadu by Rajendran and Rajendran (1979).

(i) Symptoms: The roots exhibited very small swellings and enlarged rootlets. The minute galls were more in *J. sambac*. Pale coloured leaves and dieback symptoms were associated (Rajendran and Rajendran, 1979).

(ii) Integrated Management

(a) Botanicals and Chemicals: Application of phorate at 4 g a.i. per plant during May and September months and incorporation of 20 kg FYM per plant increased flower yield of jasmine by 50% and reduced the root-knot nematode population by 70% (Sundarababu, 1992).

(b) Cultural, Bioagents and Botanicals: Cuttings or planting material should be raised in nematode-free or treated nursery beds. Application of neem cake enriched with *T. harzianum* at 1 t/ha is recommended.

(c) Botanicals and AMF: Nursery beds should be treated with neem or pongamia cakes enriched with mycorrhizal spores at 1 kg/m^2.

16.2. MINTS, *MENTHA* SPP.

16.2.1. Nematodes

16.2.1.1. Root-knot Nematodes, *Meloidogyne* spp.

Root-knot disease of Menthol/Japanese mint, Spearmint, Scotch Spearmint, Peppermint and Bergamot mint caused by *M. incognita* and *M. javanica* was observed for the first time in Lucknow and Terai region of Uttar Pradesh which reduce the herbage yield and oil content (Haseeb and Pandey, 1989). *M. incognita* dominates over *M. javanica* in mixed infection. The quality of mint oil was also adversely affected due to nematode infection. The root-knot nematode, *M. incognita* caused 40.2% loss in herbage yield and 46.6% loss in oil yield in menthol mint (Pandey, 2001).

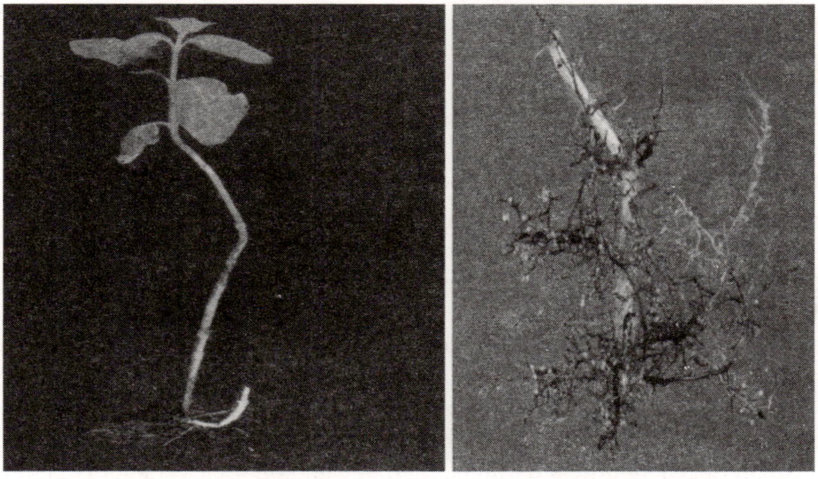

Fig. 16.2. L - Root-knots on suckers of menthol mint. R - Infested runners of menthol mint

(i) Symptoms: Initial symptoms of the disease include occasional yellowing of leaves which in a month's time spread to a large portion of the foliage. Growth ceases soon after yellowing. Leaves turn yellow and thin, scorching easily and eventually turning brown. Initially symptoms appear in patches as the reduced plant growth with smaller leaf size and temporary wilting under slight stress of water during hot sun. As the crop grows, especially after first harvest, symptoms become more severe and appear as stunted growth with yellowing of leaves, while veins remain green.

The below-ground symptoms are the large numbers of small sized galls with large egg masses on the roots (Fig. 16.2). Under severe infection, initiation of lateral roots and rootlets on suckers is checked. As a result, uptake of nutrients is inhibited, which in turn produces deficiency symptoms on the aerial portion of the plants (Haseeb and Pandey, 1989).

(ii) Integrated Management

(a) Three AMF: Application of arbuscular mycorrhizal fungi such as *G. aggregatum, G. fasciculatum* and *G. mosseae* improved plant growth, enhanced herbage and oil yield and effectively inhibited root-knot nematode infection (Pandey *et al*., 1997) (Table 16.1).

Table 16.1. Effect of arbuscular mycorrhizal fungi on the productivity of mint infected with *Meloidogyne incognita*

Treatment	Fresh herbage weight (g)	% oil yield	% mycorrhizal colonization	Root-knot index
Untreated – Inoculated	285.0	0.38	---	3.6
Glomus aggregatum (Ga)	322.0	0.40	42.3	2.3
G. fasciculatum (Gf)	333.0	0.41	58.2	2.0
G. mosseae (Gm)	388.0	0.48	63.5	1.6
Ga + Gf + Gm	360.0	0.46	78.5	1.3

(b) Botanicals and Chemicals: Application of carbofuran and neem cake in combination improved the growth of Japanese mint and oil yield and reduced root-knot nematode population very effectively.

(c) Botanicals, Bioagents and AMF: Pandey (2005) conducted field trial to determine the efficacy of *Trichoderma harzianum* isolate U, *Glomus aggregatum*, oil seed cakes of neem (*Azadirachta indica*), and mustard (*Brassica campestris*) in the management of *Meloidogyne incognita* and their impact on yield of menthol mint (*Mentha arvensis*) cv. Kosi. Significant reductions in nematode populations and root-knot indices were noticed in plots receiving oil seed cakes and bioagents, whose effects were equal to that of carbofuran. The maximum reduction in *M. incognita* population was recorded in beds treated with mustard cake along with *T. harzianum* which also produced significantly higher herbage and oil yields (Pandey, 2005).

The maximum enhancement of plant dry weight was recorded in *T. harzianum* + *G. intraradices* (20.0%) followed by *T. harzianum* + *P. fluorescens* (14.4%), *G. intraradices* + *P. fluorescens* (5.0%), *G. intraradices* + *B. megaterium* (1.2%), *B. megaterium* + *P. fluorescens* (1.5%) as compared to untreated uninoculated control (Table 16.2). Root galling was minimum in T. *harzianum* + *G. intraradices* (1.33) followed by *T. harzianum* + *P. fluorescens* (1.66). Root colonization by *G. intraradices* was maximum in *T. harzianum* + *G. intraradices* (78%) followed by *P. fluorescens* + *G. intraradices* (74%) (Pandey *et al.*, 2011).

Table 16.2. Effect of mutualistic endophytes and PGPRs on root-knot and mycorrhizal population development and on yield of menthol mint.

Treatment	**Plant dry weight (g)**	**Root-knot index**	***G. intraradices* % root colonization**
Control - Uninoculated	34.0	---	---
Control - Inoculated	25.7 (-24.4)*	3.66	---
Carbofuran	31.4 (-7.6)	2.33	---
Trichoderma harzianum	34.8 (-2.4)	2.00	---
Glomus intraradices (*Gi*)	33.3 (-2.1)	2.66	65
Bacillus megaterium (*Bm*)	30.9 (-9.1)	3.00	---
Pseudomonas fluorescens (*Pf*)	33.4 (-1.8)	2.66	---
T. harzianum + *Gi*	40.8 (+20.0)	1.33	78
T. harzianum + *Bm*	36.0 (+5.9)	2.00	---
T. harzianum + *Pf*	38.9 (+14.4)	1.66	---
G. intraradices + *Bm*	34.4 (+1.2)	2.00	70
G. intraradices + *Pf*	35.7 (+5.0)	2.00	74
B. megaterium + *Pf*	34.5 (+1.5)	2.66	---
LSD at 5%	**---**	**0.7433**	**4.321**

*Percent increase (+) or decrease (-) over untreated-uninoculated control,
Note: Inoculated with 5000 freshly hatched juveniles of *Meloidogyne incognita* (Mi)/pot.

16.3. PATCHOULI, *POGOSTEMON PATCHOULI*

16.3.1. Nematodes

16.3.1.1. Root-knot Nematodes, *Meloidogyne* spp.

Root-knot nematodes (*Meloidogyne incognita*, *M. javanica*, *M. hapla*) have become most important constraints for the successful cultivation of patchouli in India. Prasad (1978) and Prasad and Reddy (1979, 1984) reported 47.0 and 86.7 per cent loss in shoot weight and shade dried leaf yield, respectively in patchouli by *M. incognita*. They further reported that the multiplication of *M. incognita* was more in sandy soil on patchouli than in clay soil.

(i) **Symptoms:** Infection of root-knot nematodes occurs when plants are in their early stage of development. Root-knot infested plants are weak and grow slowly. Heavy galling on root system with root-knot nematode on patchouli results in stunting, wilting, defoliation and chlorosis of the plant. Sometimes root galls are very small or the surrounding galls coalesce to form large one up to 2-5 cm or even more (Fig. 16.3).

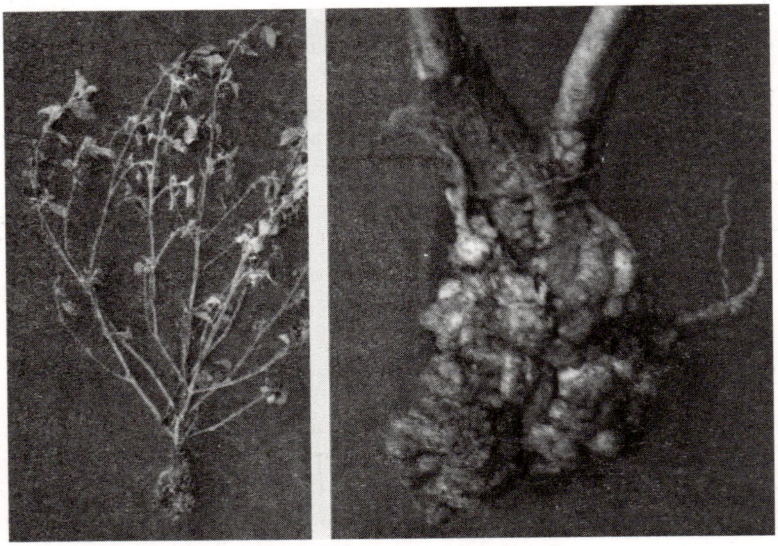

Fig. 16.3. Root-knot nematode infested plant of patchouli and infested root

(ii) Integrated Management

Treat nursery beds with neem/pongamia cake at 1 kg/m^2 along with *Glomus fasciculatum* at 50 g/m^2.

16.3.1.2. Root-knot Nematode and Root Rot Disease Complex

(i) Integrated Management

(a) Bioagents and Botanicals: Combined mortality due to root rot and root-knot nematode could be minimized by the application of *Trichoderma harzianum* + karanj cake at 5 MT/ha.

16.4. CHAMOMILE, *MATRICARIA CHAMOMILLA*

16.4.1. Nematodes

16.4.1.1. Root-knot Nematode, *Meloidogyne incognita*

(i) Symptoms: *M. incognita* causes considerable reduction in growth, flower buds and essential oil yield of chamomile (Pandey *et al.*, 1999).

(ii) Integrated Management

(a) Botanicals and AMF: Integration of *G. mosseae* and neem cake reduced the severity of root-knot disease and enhanced the growth, biomass and flower yield by 12 to 37% of chamomile.

16.5. DAVANA, *ARTEMISIA PALLENS*

16.5.1. Nematodes

16.5.1.1. Root-knot Nematode, *Meloidogyne incognita*

The root-knot nematode is a major problem in the cultivation of davana. *M. incognita* is responsible for more than 30% reduction in oil yield of davana (Haseeb and Pandey, 1990).

(i) Symptoms: The main symptoms are chlorotic and stunted plants with less number of flower buds (which are the major source of essential oil) showing patchy appearance in the field. Their roots are severely galled by root-knot nematodes (Fig. 16.4). One larva/2 g soil has been found as an economic threshold level of *M. incognita* on this crop.

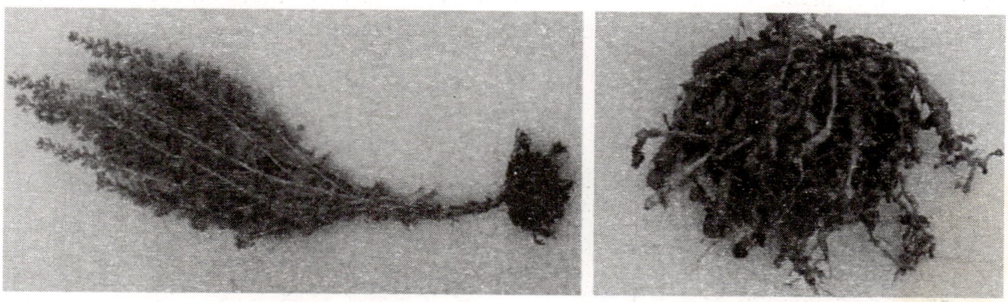

Fig. 16.4. Root-knot infested plant and roots of davana

(ii) Integrated Management

(a) Two Botanicals: Increase in nematicidal efficacy of neem cake along with FYM for 90 to 120 days in comparison to 60 days for fenamiphos has been noticed (Anitha and Vadivelu, 1997).

Chapter 17

TUBER CROPS

17.1. SWEET POTATO, *IPOMEA BATATAS*

17.1.1. Insect Pests

17.1.1.1. Weevil, *Cylas* spp.

(i) Damage: Tubers and vines are tunneled by small white, legless larvae. The tunnels may be partially filled with frass. Ant-like weevils, either having glassy blue black elytra, reddish brown legs and thorax and a black head (*C. formicarius*) or entirely shiny black (*C. punctcollis* or *C. brunneus*) may be found on the leaves or in the tunnels (Fig. 17.1). Yield, storage life and plant vigour are reduced.

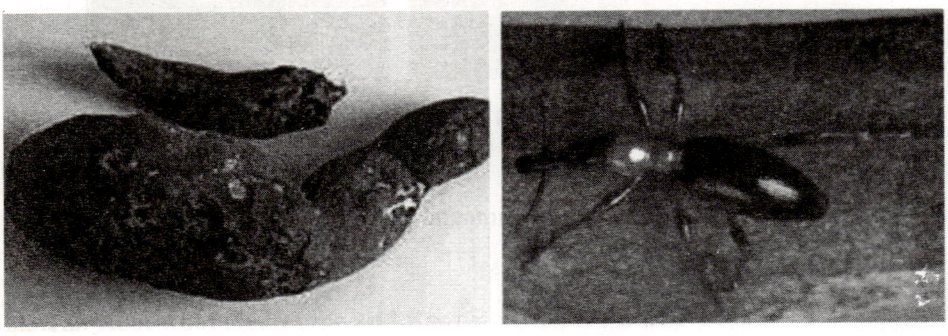

Fig. 17.1. L – Sweet potato weevil damage, R – Adult beetle

(ii) Integrated Management

(a) Cultural and Bioagents: The solitary ectoparasitoid, *Rhaconotus menippus* (at 10 pairs/5 m^2) and the green muscardine fungus, *Metarrhizium anisopliae* (at 3 x 10^9/ml) in combination with reridging (putting an additional layer of soil around the plant) at 65 days after planting is highly effective in reducing the weevil damage. This method is equally effective as that of chemical method. The fungus and parasitoid are recovered from the treated plots.

(b) Cultural, Biological and Chemical: The following IPM technology was developed at Central Tuber Crops Research Institute, Tiruvanthapuram, Kerala (Palaniswami, 2002):

- Removal and destruction of *Ipomoea* weeds.

- Selection of weevil-free planting material.
- Dipping the vines in 0.05% fenthion or fenitrothion or monocrotophos for 10 min.
- Installation of traps with septa (sex pheromone impregnated in 4 mm rubber tube uniformly in such a way, that one cm bits of such tube contains 1 mg sex pheromone) at 100 traps/ha. The traps are placed with the commencement of planting and continued a fortnight after the harvest. The male weevils trapped inside the bin trap (bottom of the trap contains water with a pinch of detergent) are removed on alternate days.
- Reridging (putting an additional layer of soil around the plant) the crop at 30 and 50 days after planting.
- Conserving the naturally occurring braconid solitary ectoparasitoids of weevil by not spraying insecticides especially on the vines.
- Harvesting the crop at 100-110 days maturity.
- Removal and destruction of crop residues like infested tubers and vines.
- Practicing crop rotation.

The above IPM technology was assessed, refined and validated at National level in ten different centres under All India Coordinated Research Project on Tuber Crops. The farmers from different sweet potato growing areas were convinced about the benefits of this technology in enhancing tuber production and reducing weevil infestation.

17.2. COLOCASIA, *COLOCASIA ESCULENTA*

17.2.1. Diseases

17.2.1.1. Leaf Blight, *Phytophthora colocasiae*

The leaf blight of colocasia is a major and widespread disease causing 50% loss in tuber yield. This disease occurs widely in India, Indonesia, Malaysia, Sarawak and the Pacific. It has also been recorded from parts of Africa and the Caribbean.

(i) Symptoms: Lesions are initially small, dark and round but rapidly enlarge to 2.5-5.0 cm in diameter and become purplish to brownish in colour (Fig. 17.2). Drops of a clear liquid exude from the spots, and turn yellow, orange or purple when dry. There is usually chlorotic halo around the spots. As the disease progresses, the spots coalesce and have characteristic rings of yellow or brown colour. Eventually the whole leaf may be affected and may die. Leaf blight is common in wetland and is favoured by humid, cloudy conditions and poor soil fertility. Losses up to 50% have been reported. Spores are produced on the leaf spots and are readily spread by rain.

(ii) Integrated Management

(a) Bioagents and Chemicals: Tuber treatment with *T. viride* + 0.25% mancozeb spray after the first appearance of the disease + 0.2% ridomil spray 20 days after mancozeb spray was effective in reducing the percent leaf area damaged and increased the yield.

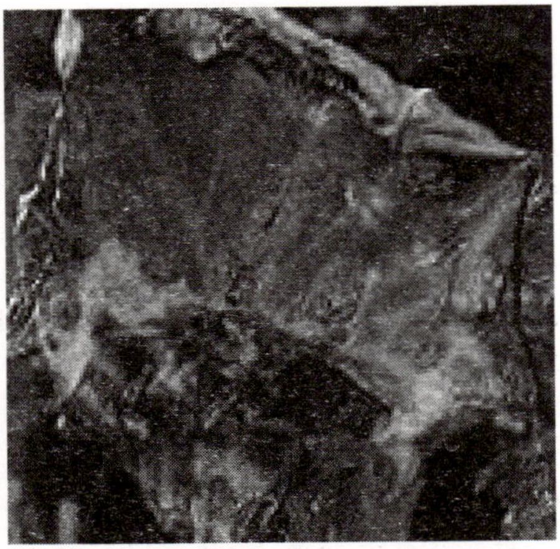

Fig. 17.2. Leaf blight damage on colocasia.

17.3. ELEPHANT FOOT YAM, *AMORPHOPHALLUS PAEONIIFOLIUS*

17.3.1. Diseases

17.3.1.1. Collar Rot, *Sclerotium rolfsii*

(i) Integrated Management

(a) Botanicals and Bioagents: Corm treatment with *Trichoderma viride* at 5 g/litre along with soil application of neem cake at 250 g/pit effectively controlled collar rot disease.

Section - V

BIOINTENSIVE INTEGRATED PEST MANAGEMENT IN PLANTATION AND SPICE CROPS

Chapter - 18

PLANTATION CROPS

18.1. COFFEE, *COFFEA ARABICA, COFFEA CANEPHORA*

18.1.1. Diseases

18.1.1.1. Brown Root, *Fomes noxius;* Red Root, *Poria hypolaterita;* Black Root and Santavery Root Disease, *Corticium koleroga*

(i) Symptoms: Stump or brown root-rot affected bushes show a gradual yellowing of leaves and defoliation and death. Affected roots are brittle and show dark brown wavy lines of the fungus. Stem near the ground level becomes soft and spongy. Brown fungal encrustation can be seen on the affected roots (Fig. 18.1).

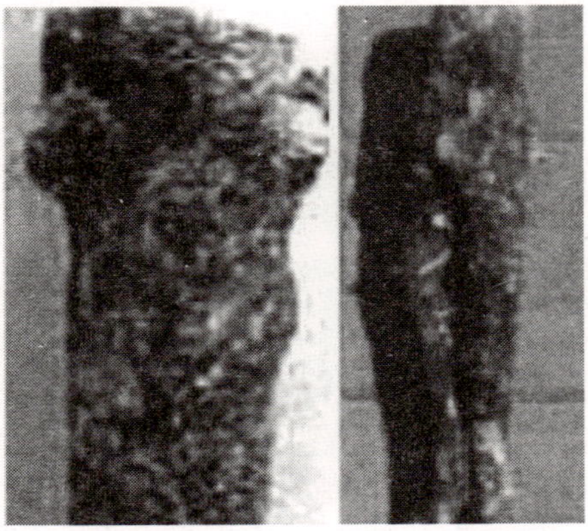

Fig. 18.1. Black root and red root on coffee.

The pest infestation causes blackening and rotting of affected leaves, young twigs and berries. Affected leaves get detached and hang down by means of slimy fungal strands. Defoliation and berry drop occur.

(ii) Integrated Management

(a) Bioagents and Botanicals: Application of 5-10 kg/plant of well decomposed FYM/compost fortified with *Trichoderma* sp. to the surrounding healthy plants is effective.

Soil application of 2 kg *T. harzianum* or *T. viride* in compost/plant in drip circle during June and October gave good control of brown root and red root diseases

(Nirmala Kannan *et al.*, 1997) (Table 18.1). Mix 2.5 kg of *Trichoderma* spp. in 100 kg of FYM. Leave it overnight. Spread these equally around plant basins twice a year (September-October and May-June).

Table 18.1. Effect of *Trichoderma viride* and *T. harzianum* on red and brown root diseases of coffee

Disease	Antagonistic fungi	Type of application	Dose/plant	% Disease control
Red root	*Trichoderma viride* & *T. harzianum* (10^{-8} - 10^{-9})	Soil application along with 5 kg FYM	0.5 – 1.0 kg twice yearly (June & Oct.)	83-91
Brown root	*Trichoderma viride* & *T. harzianum* (10^{-8} - 10^{-9}) in compost	Soil application	2.0 kg	87

18.1.1.2. Pink Disease, *Corticium salmonicolor*

(i) Symptoms: The symptoms include pink encrustation on infected branches and development of longitudinal cracks through which pink encrustation bursts. Cobweb like mycelial branches develop on affected branches. Infected branches lose leaves and die (Fig. 18.2).

Fig. 18.2. Pink disease in coffee

(ii) Integrated Management

(a) Bioagents and Botanicals: Mix 2.5 kg of *Trichoderma viride* in 100 kg of FYM. Leave it overnight. Spread these equally around plant basins twice a year.

18.2. TEA, *CAMELLIA SINENSIS*

18.2.1. Insect Pests

18.2.1.1. Diaspine Scale, *Fiorinia theae; Purple Scale, Chrysomphalus aonidum*

(i) Damage: The most common insect pests of tea are scales. Scale insects feed on plants by piercing plant tissue and sucking sap. Scales do not look like typical insects. They are small, immobile and have no visible legs. They vary in appearance depending on species and sex. Some look like small fish scales attached to the plant. As a result of their unusual appearance, populations can reach damaging levels before they are noticed.

On tea, scales usually attach to leaves but some species also attach to stems. Their feeding weakens the plant. With a heavy infestation, symptoms include yellowing of the upper leaf surface (Fig. 18.3), fewer and smaller blossoms, leaf drop, twig dieback and sometimes death.

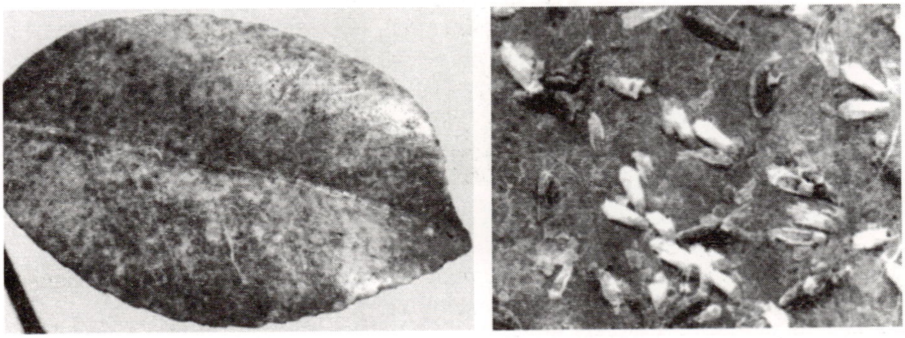

Fig. 18.3. L - Tea scale damage to top of leaf, R - Adult tea scales on underside of leaf (males are snowy white and females are dark)

(ii) Integrated Management

(a) Two Bioagents: Combined action of reduviid bug (predator) and parasitoids (*Aphytis* sp.) keeps the diaspine scale (*Fiorinia theae*) population very low.

18.2.2. Diseases

18.2.2.1. Blister Blight, *Exobasidium vexans, E. camelliae*

(i) Symptoms: The blister blight is the major disease affecting the tender harvestable shoots of tea resulting in enormous crop loss. Translucent spots appear on tender leaves. Tender stem also gets affected. Leaf galls are most often observed during the spring flush of growth. New shoots and leaves become enlarged, thickened and fleshy, and appear abnormal (Fig. 18.4). The colour of the affected areas turns from light green to nearly white or pink. The galls later rupture on the undersides of the leaves revealing a whitish mass of spores. The galls eventually harden and become brown. Plants are seldom severely damaged.

(ii) Integrated Management

(a) Bioagents and Chemicals: Field evaluation of *T. harzianum, T. virens, P. fluorescens* and *B. subtilis* revealed that they could provide moderate control of the disease; but when they were supplemented with external nutrients (salicylic acid and vermiwash), their efficacy was improved (Balasuriya and Kalaichelvan, 2000) (Table 18.2).

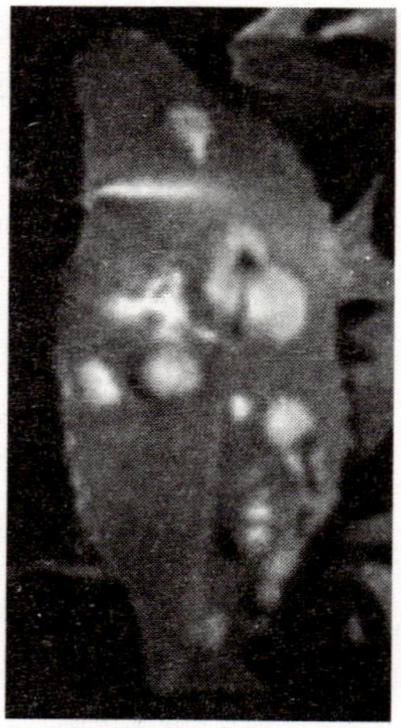

Fig. 18.4. Tea leaves infected by blister blight disease.

Table 18.2. Impact of nutritional supplements on the efficacy of biocontrol agents in controlling blister blight disease of tea

Biocontrol agents	Disease incidence (%)	Protection (%)
Trichoderma virens	52.4	38.7
T. virens + Vermiwash	52.7	38.4
Trichoderma harzianum	60.5	29.2
T. harzianum + Vermiwash	57.8	32.4
Pseudomonas fluorescens	60.4	29.3
P. fluorescens + Salicylic acid & Ammonium sulphate	52.3	38.8
Bacillus subtilis	57.0	33.3
B. subtilis + Salicylic acid & Ammonium sulphate	47.1	44.9
Control (unsprayed)	85.5	---

18.2.2.2. Grey Blight and Dieback, *Pestalotia theae/Pestalotiopsis theae*

(i) Symptoms: Mature leaves (Fig. 18.5), young shoots and bare stalks are affected by this pathogen. Infection on young shoot results in dieback of shoots. Dieback of shoots became a major problem mainly due to continuous shear harvesting. Grey blight adversely affects the health of the bushes, which in turn affects yield while dieback of young shoots directly leads to substantial crop loss. Studies indicated that the disease incidence is in its peak during July to December. The crop loss due to the disease is 17%. The economic threshold level of the disease is fixed as 18% at a sale price of Rs. 50/- per kg of made tea.

Fig. 18.5. Grey blight infection on tea leaf.

(ii) Integrated Management

(a) Bioagents and Chemicals: Field testing of biocontrol agents like *T. harzianum* and *P. fluorescens* could check grey blight disease. Efficacy of these organisms was improved when they were supplemented with nutrients (Table 18.3) (Chakraborty *et al.*, 1994).

Table 18.3. Effect of biocontrol agents on grey blight incidence in tea

Treatment	Disease incidence (%)	
	Pre-treatment	Post-treatment
Pseudomonas fluorescens	28.1	20.3 (-27.8)
P. fluorescens + **Vermiwash**	20.5	10.6 (-48.4)
Trichoderma harzianum	22.1	16.3 (-50.8)
T. harzianum + **Salicylic acid & Ammonium sulphate**	20.4	10.0 (-50.8)
Mancozeb	20.7	8.4 (-59.7)
Carbendazim	20.6	6.6 (-68.0)
Control (unsprayed)	34.3	39.2 (+14.1)

Values in parenthesis indicate the % decrease (-) or increase (+) in disease incidence

18.2.2.3. Collar Canker, *Phomopsis theae*

(i) Symptoms: A wound pathogen, Phomopsis theae is responsible for the disease (Fig. 18.6). It is prevalent in young tea and clones are more susceptible than seedlings. *In vitro* studies indicated that the pathogen completes its life cycle in 10-13 days. Impact of predisposing factors on disease development indicated that the nature of soil is an important factor. Disease incidence was more in gravelly soil.

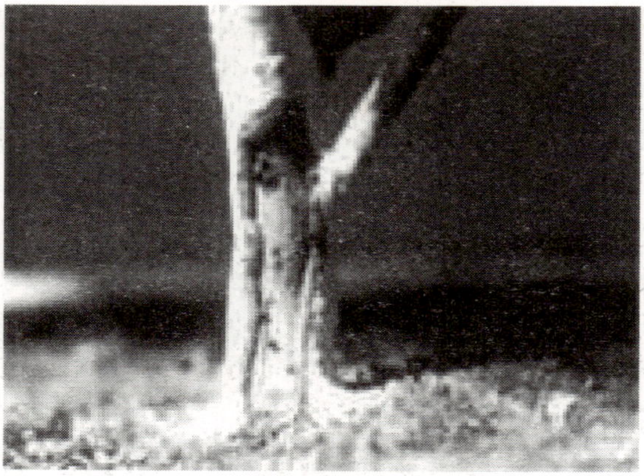

Fig. 18.6. Collar canker on tea.

(ii) Integrated Management

(a) Bioagents and Chemicals: The biological control agents like *Trichoderma harzianum* and *T. virens*, when applied both to the soil around the bush and used for wound dressing were superior for the management of collar canker compared to chemical treatments. These plants were more healthy and vigorous (Table 18.4) (Ponmurugan and Baby, 2005).

Table 18.4. Effect of chemicals and biocontrol agents on collar canker disease of tea

Treatment	**Canker size (cm^2)**	
	Pre-treatment	**Post-treatment**
T. harzianum soil application + wound dressing	11.3	5.6 (-50.4)
T. virens soil application + wound dressing	17.0	5.5 (-67.6)
T. harzianum wound dressing + carbendazim soil drenching	39.9	27.9 (-30.1)
T. virens wound dressing + carbendazim soil drenching	12.1	7.1 (-41.3)
Carbendazim soil drenching	13.8	10.2 (-26.1)
Copper oxychloride wound dressing	3.0	4.8 (+60.0)
Untreated control	45.2	97.2 (+115.0)

18.2.2.4. Black Root, *Rosellina arcuata*

The black root disease generally occurs in soils that are rich in organic litter, in cool or cold habitat. It is a major problem in southern India at elevations of 1500 m above MSL and Darjeeling (West Bengal). Occasionally, it is also noticed at mid and low elevations.

(i) Symptoms: The fungal mycelium is cob-webby in appearance, whitish grey or black in colour, spreading rapidly on the surface of the soil on organic debris.

When the mycelium comes in contact with tea roots, it penetrates the bark and produces typical star-shaped mycelial structures on the surface of the wood beneath the bark. Very rarely black spherical perithecia are seen at the collar region of the plant. Apart from tea, the disease also affects certain green manure and shade plants commonly grown in tea plantations.

(ii) Integrated Management

(a) Cultural and Bioagents: Black root disease can be controlled by avoiding burial of prunings in infested field and incorporation of *Trichoderma virens / T. viride* at 200 g/plant at the time of planting. Rehabilitating the soil with Guatemala grass and use of biocontrol agents at 200 g/plant of *T. viride* effectively check the disease (Baby and Chandramouli, 1996).

18.2.2.5. Brown Root, *Phellinus noxius*

(i) Symptoms: The brown root is the most common root disease of tea. This pathogen mainly spreads by root contact. Hence, the disease occurs in distinct concentric patches in tea fields. The fungal mycelium is fawn in colour. *Phellinus noxius* affected roots become soft and spongy. The infected root disintegrates in advanced stages of the disease and becomes spongy. When such roots are split longitudinally, typical brownish honeycomb-like reticulations are seen. The surface of the root is heavily encrusted with soil. The fructifications are typical bracket-shaped (Fig. 18.7), but occur rarely.

Fig. 18.7. Brown root of tea

(ii) Integrated Management

(a) Cultural and Bioagents: Rehabilitating the soil with Guatemala grass and use of biocontrol agents at 200 g/plant of *Trichoderma harzianum/T. viride/T. hamatum/T. resei/T. koningii/T. virens* effectively check the disease (Baby and Chandramouli, 1996).

18.2.2.6. Red Root, *Poria hypolateritia*

(i) Symptoms: The red root disease is extremely common in all tea areas of southern India. *Poria hypolateritia* produces blood red mycelium on affected roots. The fungal mycelium is initially white, then turns pink to pale red and in later stages it is dark red, almost black in colour. The mycelium is generally present as a sheet on the root surface and the typical red colour is invariably seen, when the root is rubbed vigorously under running water and held to bright light. The disease spreads both by root contact as well as by free mycelial spread to a limited extent. Roots of the affected bushes are encrusted with soil. In advanced stages, the affected root becomes spongy. It takes almost 5 years for a mature bush to succumb following the attack.

(ii) Integrated Management

(a) Cultural and Bioagents: Rehabilitating the soil with Guatemala grass and use of biocontrol agents at 200 g/plant of *Trichoderma harzianum/T. virens* effectively check the disease (Baby and Chandramouli, 1996).

18.2.2.7. Root Rot or Collar Crack, *Armillaria mellea*

(i) Symptoms: The collar crack disease occurs in isolated areas, mainly in South India. It derives its name from longitudinal fissures observed on the bark of the invaded tea roots (Fig. 18.8). It can also be observed readily by the characteristic black rhizomorphs which resemble a shoe lace and are seen in large numbers. The fungal mycelium which is white in colour grows densely under the bark and due to the pressure the bark splits. The fungus invades the host plant either by root contact or rhizomorphs which are capable of traversing long distances through the soil from the food base. The fruiting body is typical mushroom-like basidiocarp, and although it is rarely recorded. When it occurs, it is found in abundance around the collar of the diseased plant.

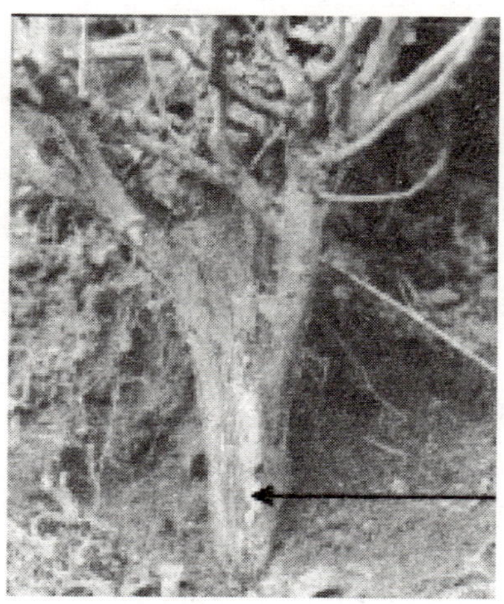

Fig. 18.8. Collar crack on tea

(ii) Integrated Management

(a) Cultural and Bioagents: Rehabilitating the soil with Guatemala grass and use of biocontrol agents at 200 g/plant of *Trichoderma harzianum/T. virens* effectively check the disease.

(b) Physical and Bioagents: Talam and Otieno (2002) reported that soil solarization and application of *T. harzianum* were effective in controlling root splitting disease. They also found that the control achieved with *T. harzianum* was excellent when applied after soil solarization.

18.2.3. Nematodes

18.2.3.1. Root-knot Nematodes, *Meloidogyne incognita* (in nurseries)

The first report of root-knot nematode in young tea was from South India, where large number of seedlings were found infected (Barber, 1901). *Meloidogyne incognita, M. javanica* and *M. hapla* have been found associated with tea in India.

(i) Symptoms: Young nursery plants, both seedlings and vegetatively propagated clonal plants, are severely damaged by root-knot nematodes. Seedling plants in which both the tap-root and lateral rots are severely attacked suffer greater damage than do clonal tea plants of equivalent age, probably because seedling tea plants possess less than half the root bulk of clonal plants of similar age. A marked increase in resistance is observed between 8 and 15 months.

(ii) Integrated Management

(a) Botanicals and Chemicals: Combination of neem cake at 5 g/seedling + carbofuran 3G at 1.5 g/seedling improved plant growth characters and reduced number of galls, egg masses and eggs/egg mass (Kalita and Bora, 2006).

(b) Cultural and Bioagents: The rehabilitation crops such as Mana grass (*Cymbopogon confertiflorus*) or Guatemala grass (*Tripsacum laxum*) (which help to improve the soil) and use of biocontrol agents at 200 g/plant of *Trichoderma harzianum/T. virens* effectively check the disease. The above grasses are resistant to *Meloidogyne* spp. The soil population of parasitic nematodes decline rapidly when these grasses are grown for one or more years before replanting tea.

18.3. COCONUT, *COCOS NUCIFERA*

18.3.1. Insect Pests

18.3.1.1. Rhinoceros Beetle, *Oryctus rhinoceros*

(i) Damage: Rhinoceros beetle feeds on unopened fronds and spathes of coconut. Characteristic fan like geometric cuts in the newly emerged frond are observed (Fig. 18.9). Infestation on spathe often destroys the inflorescence and prevents the production of nuts.

Fig. 18.9. Rhinoceros beetle damage on coconut and adult beetle.

(ii) Integrated Management

(a) Physical, Bioagents and Chemicals: Integrated pest management includes hooking out of beetle, placement of perforated polythene sachets containing phorate in leaf axil and adoption of biological control measures using baculovirus.

18.3.2. Diseases

18.3.2.1. Basal Stem Rot or Thanjavur Wilt, *Ganoderma lucidum, G. applanatum*

(i) Symptoms: In Tamil Nadu and Andhra Pradesh, Thanjavur wilt, otherwise called Ganoderma wilt is a serious problem. The symptoms are presence of bleeding patches at the stem base (Fig. 18.10), premature yellowing and drooping of outer whorl of leaves and gradual drying of spindle. The other symptoms include decay of root system, flaccid spindle leaves, browning of outer leaves, and appearance of bleeding patches on the basal region of the stem.

Fig. 18.10. Basal stem rot in coconut.

(ii) Integrated Management

(a) Bioagents and Botanicals: Application of *T. harzianum/T. viride* along with neem cake at 5 to 10 kg/plant has been found to control the multiplication of the pathogen in sick soils. When combined with phosphobacteria or plant growth promoting rhizobacteria, synergistic effects have been noticed. *T. harzianum* applied along with neem cake reduced the disease index and increased the yield of coconut.

Raising seedlings in *T. harzianum/P. fluorescens* amended compost and soil application of *T. harzianum/P. fluorescens* along with neem cake/compost in the main field was useful in minimizing the incidence of basal stem rot of coconut.

18.3.2.2. Stem Bleeding, *Theilaviopsis paradoxa*

(i) Symptoms: The characteristic symptom of stem bleeding is the dark gummy exudation from the trunk. Exudation of reddish brown liquid is observed through longitudinal cracks in the trunk, generally at the base of the trunk (Fig. 18.11). Bleeding patches spread through out as the disease advances. The liquid oozing out dries up and turns black. Tissues below the lesions rot and turn yellow and then black. Premature yellowing of leaves is observed in the outer whorl. Trunk gradually tapers at apex and crown size becomes reduced (Fig. 18.11).

Fig. 18.11. Stem bleeding on coconut.

Growth cracks on trunk, severe summer followed by sudden wetting, imbalanced nutrition, and excess salinity are the predisposing factors for the disease.

(ii) Integrated Management

(a) Bioagents and Botanicals: Control measures include application of neem cake along with antagonistic fungi like *Trichoderma*. Neem cake enriched with *Trichoderma virens, T. hamatum* and *T. harzianum* has been found to be very effective in reducing the population of the pathogen in the soil. These antagonistic fungi thrive well in neem cake supplemented with a small quantity of rice/wheat bran thus effecting their multiplication an easy task. *Trichoderma harzianum, T. viride, T. hamatum* and *T. virens* grow very well in rice bran and neem cake (1:1 w/w) and reduced stem bleeding when applied to the soil. Soil application of neem cake and FYM mixed with *T. virens* showed reduction of stem bleeding up to 31.3%, least disease index and the highest yield.

Raising seedlings in *T. harzianum/P. fluorescens* amended compost and soil application of *T. harzianum/P. fluorescens* along with neem cake/compost in the main field was useful in minimizing the incidence of stem bleeding disease of coconut.

(b) Bioagents, Botanicals and Chemicals: Treatment with 4% tridemorph root feeding and wound dressing + coal tar sealing + soil application of *T. virens*, neem cake, FYM and NPK fertilizers showed the lowest disease index and the highest yield followed by treatment with carbendazim root feeding and wound dressing + coal tar sealing + soil application of *T. virens*, neem cake, FYM and NPK (Ramanujam *et al.*, 1997).

18.3.3. Nematodes

18.3.3.1. Burrowing Nematode, *Radopholus similis*

The burrowing nematode was reported from coconut palms in Kerala, India by Weischer (1967). *R. similis* is the most important nematode pest of coconut and is responsible for considerable amount of root rotting (Koshy *et al.*, 1978). It causes 30% yield loss in coconut (Koshy and Geetha, 1992). The threshold inoculum density required to cause significant reduction in various growth parameters of coconut is 100 nematodes/625 ml sandy loam soil over a period of 5 years under field conditions (Koshy, 1986).

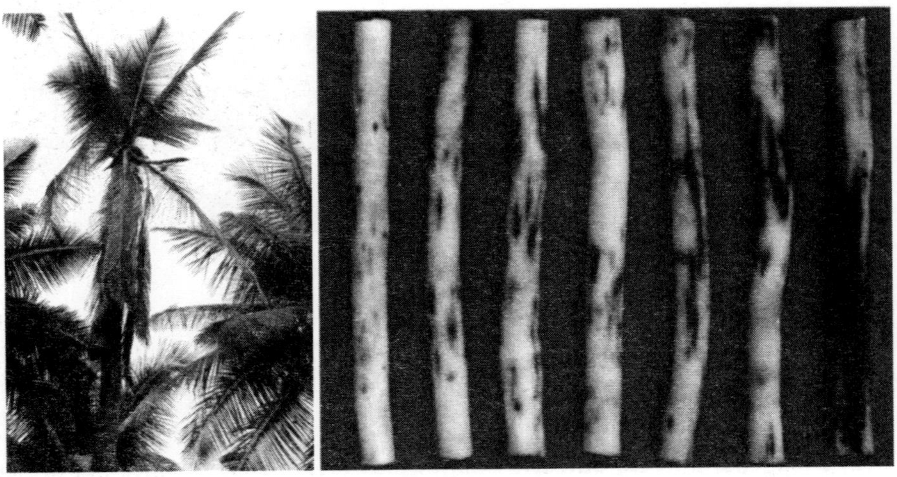

Fig. 18.12. Burrowing nematode symptoms on coconut

(i) Symptoms: The burrowing nematode infested coconut palms exhibit general decline symptoms like yellowing, button shedding, reduction in number and size of leaves and leaflets (Fig. 18.12), delay in flowering and reduced yield which are non-specific. Symptoms on the root are more specific. *R. similis* on infestation produces isolated elongate orange coloured lesions on tender and semi-hard roots (Fig. 18.12). Consequent to nematode parasitization and multiplication, these lesions enlarge and coalesce to cause extensive rotting of roots. Tender roots on heavy infestation become spongy in texture. On semi-hard orange coloured roots, surface cracks are commonly seen. As high as 4,000 nematodes were recovered from one gram (one inch length) of main roots. The drastic reduction in the number and mass of tertiary feeder roots on parasitization by the nematode limits plant growth (Koshy and Sosamma, 1987).

(ii) Survival and Spread: Under field conditions, *R. similis* survives for 6 months in moist soil (27°-36°C) and one month in dry soil (29°-39°C). In roots of stumps of felled coconut palms, the nematode survives for up to 6 months. The infested coconut roots yielded maximum number of *R. similis* during October to November and minimum during March to July. A mean soil temperature below 25°C and light rain fall coupled with availability of tender fleshy roots are the factors favourable for *R. similis* multiplication (Koshy and Sosamma, 1978).

Infested coconut seedlings help in the dissemination of the nematode to distant places. Apart from coconut, infested planting materials of intercrops such as areca nut, banana, pepper, ginger, turmeric also serve as sources of inoculum.

(iii) Integrated Management

(a) Botanicals and Chemicals: Thirty per cent increase in yield and 5 to 10% decrease in disease indices of palms affected with root wilt disease was recorded by the application of *Hydnocarpus* oil cake and phorate granules at 10g a.i./palm in June-July and in October- November.

(b) Cultural, Botanicals and Chemicals: Application of farmyard manure or oil cakes, cultivation of *Crotolaria juncea* in the basin and interspaces (and used as green manure), and application of phorate 10G at 100 g/palm/twice a year (during May-June and Sep-Oct.) is effective against burrowing nematode.

18.4. ARECANUT, *ARECA CATECHU*

18.4.1. Nematodes

18.4.1.1. Burrowing Nematode, *Radopholus similis*

(i) Symptoms: Plants infested with *R. similis* induces 'yellow leaf' disease. Infestation produces small, elongate, orange lesions in young, succulent, creamy-white to light-orange portions of the main and lateral roots. The adjoining lesions coalesce and cause extensive rotting.

Large number of crops like banana, coconut, black pepper, ginger, turmeric, betel vine, etc. are known hosts of *R. similis*.

(ii) Integrated Management

(a) Botanicals and Chemicals: Integration of neem oil cake application at 1 kg/plant with phorate at 15 g a.i./plant gave effective control of *R. similis* in areca nut. In areca nut + banana + black pepper cropping system, integration of phorate at 15 g a.i./plant with neem oil cake at 500 g/plant was found most effective in reducing the nematode population (Sudha and Sundararaju, 1998).

(b) Botanicals and Bioagents: Soil amended with glyricidia leaves and bioagents (*Paecilomyces lilacinus*, *Pasteuria penetrans* and AMF) was very effective in increasing plant height, leaf area and root growth; and in reducing burrowing nematode population (95%) and root lesion index.

18.5. BETEL VINE, *PIPER BETEL*

18.5.1. Diseases

18.5.1.1. Collar/Basal Rot, *Sclerotium rolfsii*

(i) Symptoms: The plants are usually attacked at ground level (collar region). Dense white cottony mass of threads (mycelium) are seen on stems and soon many small mustard like sclerotia appear in the soil near collar region. Bark shredding is seen on infected stem. This causes rotting of affected portion causing wilting and ultimate death of plants (Fig. 18.13).

Fig. 18.13. Collar rot symptoms on betel vine

(ii) Integrated Management

Bioagents and Chemicals: Maiti and Sen (1988) integrated *T. harzianum* with nitrogenous fertilizers for managing *S. rolfsii*.

Bioagents and Botanicals: Application of *T. viride* at 5 kg mixed in 125 kg FYM per ha is useful.

Two Bioagents: Field trials demonstrated that strain *P. fluorescens* NBRI-N6 was better than *P. fluorescens* NBRI-N in increasing the yield of betel vine significantly, whereas a consortium of the two strains controlled the collar rot disease more than either of the strains (Singh *et al*., 2003).

18.5.2. Nematodes

18.5.2.1. Root-knot Nematode, *Meloidogyne* spp.

Root-knot disease caused by *M. incognita* and *M. javanica* is the most common disease affecting betel vine plant in more than 90 per cent of the fields. *M. incognita* was responsible for 21-38% loss in leaf yield of betel vine (Saikia, 1992; Jonathan *et al.*, 1990).

(i) Symptoms: The affected plants show growth reduction and yellowing and abnormal thickening of leaves with necrosis commencing from the tip and margins of leaf and extending inwards. The disease causes reduction in quality and quantity of leaves, sometimes leading to serious wilt disease that greatly affect the growth of plants and produce heavy losses to the farmers. Blackening and drooping of the growing tip and yellowing of leaves occurs due to root-knot nematode infestation. Galls of varying sizes and shapes develop in roots leading to quick root decay.

(ii) Integrated Management

(a) Botanicals and Bioagents: *Trichoderma viride* mixed with mustard oil cake at 100 g/10 kg cake applied in three split doses reduced the nematode infestation and enhanced the leaf yield.

18.5.2.2. Root-knot Nematode, *Meloidogyne incognita* and Foot Rot, *Phytophthora capsici* Disease Complex

(i) Symptoms: The population of root-knot nematode had been found to be positively correlated with *Phytophthora* wilt disease incidence in diseased betel vine gardens.

(ii) Integrated Management

(a) Cultural, Bioagents, Botanicals and Chemicals

- Summer ploughing and exposing the field to sunlight during May prior to sowing of *Sesbania* sp. (live standard) minimizes the initial load of inoculum of both the nematode and the fungus.
- Selection of healthy seed vines from nematode-free and disease-free mother plants for planting.
- Dipping seed vines in 0.25% Bordeaux mixture solution for 5 min before planting.
- Application of FYM at 30 MT/ha to promote multiplication of antagonistic microbes which in turn kills nematodes.
- Spot application of neem cake enriched with *P. lilacinus* at 3 MT/ha in 3 split doses -1st dose at 45 DAP and remaining 2 splits at 45 days interval during North-East monsoon season (Oct-Dec).
- Rotation of betel vine crop with rice.

(b) Two Bioagents: Combined application of *Pseudomonas* sp. (Pfbv 22) and *Bacillus* sp. (Bbv 57) gave significant reduction in nematode infestation (gall index, number of egg laying females and soil population), wilt disease incidence and increase in leaf yield. The treatment also enhanced the biochemical markers responsible for induced systemic resistance such as peroxidase, polyphenol oxidase and phenylalanine ammonia lyase (Jonathan *et al.*, 2006) (Table 18.5).

Table 18.5. Effect of rhizobacterial formulations on leaf yield, nematode and wilt incidence in betel vine under glasshouse conditions

Treatment/Dose (2.5 x 10^8 cfu/g)	**No. of leaves /vine**	**Gall index (0-5 scale)**	**Wilt index (0-5 scale)**
Pseudomonas spp. - Pfbv 22	195	3.0	2.4
Bacillus spp. - Bbv 57	201	3.0	2.6
P. fluorescens - Pf 1	190	3.5	2.4
Pfbv 22 + Bbv 57	225	2.5	1.7
Pfbv 22 + Pf 1	197	3.0	2.2
Bbv 57 + Pf 1	190	3.5	2.4
Pfbv 22 + Bbv 57 + Pf 1	191	3.0	2.4
Metalaxyl (0.2%) + Carbofuran (2 g/vine)	211	2.4	1.9
Control	163	5.0	3.3
CD (P = 0.05)	**6.5**	**0.5**	**0.4**

(c) Bioagents and Botanicals: Spot application of neem cake enriched with *P. lilacinus* at 3 MT/ha in 3 split doses-1st dose at 45 DAP and remaining 2 splits at 45 days interval during North-East monsoon season (Oct-Dec).

18.6. COCOA, *THEOBROMA CACAO*

18.6.1. Diseases

18.6.1.1. Black Pod Rot, *Phytophthora palmivora*

(i) Symptoms: Black pod disease is the major disease of cocoa prevalent during monsoon season. The affected pods become pale brownish spoiling the quality of beans. Infection appears as chocolate brown spot which spreads very rapidly and soon occupies the entire surface of the pod (Fig. 18.14). As disease advances, a whitish growth of fungus consisting of fungal sporangia is produced over the affected pod surface. Affected pods become dark brown or black.

(ii) Integrated Management

(a) Bioagents and Chemicals: Application of *Trichoderma* at 100 g/tree to the tree base during June and September and spraying of fungicides (Ridomil/Akomin) to the trunk and canopy pods at monthly intervals from June to September gave effective control of black pod disease.

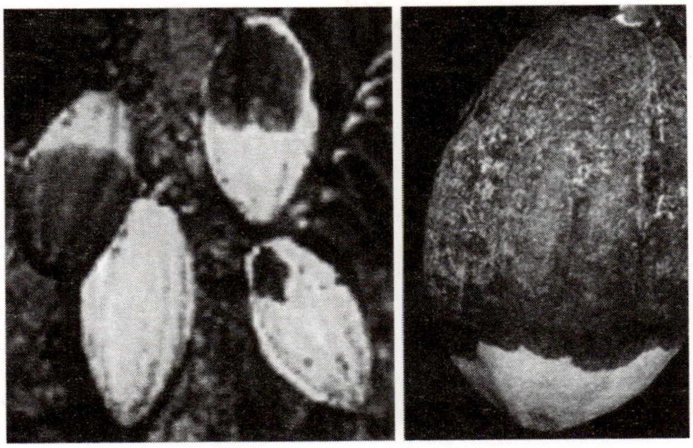

Fig. 18.14. Black pod on cocoa.

18.7. RUBBER, *HEVEA BRASILIENSIS*

18.7.1. Diseases

18.7.1.1. Brown Root, *Phellinus noxius*

(i) Symptoms: The incidence of the disease can be detected by the discolouration of the foliage along with cessation of growth. Roots become encrusted with a mass of soil, sand and small stones which can not be washed off easily. Wood also shows brown discolouration and in advanced stages, honey combing is seen.

(ii) Integrated Management

(a) Bioagents and Chemicals: An integrated approach using both fungicide (Tridemorph) and the antagonistic fungus (*Trichoderma*) has been observed to improve the disease control (Hashim, 1990).

Chapter - 19

SPICE CROPS

19.1. BLACK PEPPER, *PIPER NIGRUM*

19.1.1. Diseases

19.1.1.1. Quick Wilt/Foot Rot, *Phytophthora capsici, P. palmivora*

The disease is more severe in Karnataka and Kerala. The pathogen is responsible for 25-70% loss in yield in Cannanore of Kerala. The mortality of pepper vines due to this disease ranged from 5.15 to 61.75% in Cannanore and Kasargod districts. If management practices are not initiated at early stage, 7.0 to 78.8% loss in yield has been reported (Jahagirdar and Siddaramaiah, 2000).

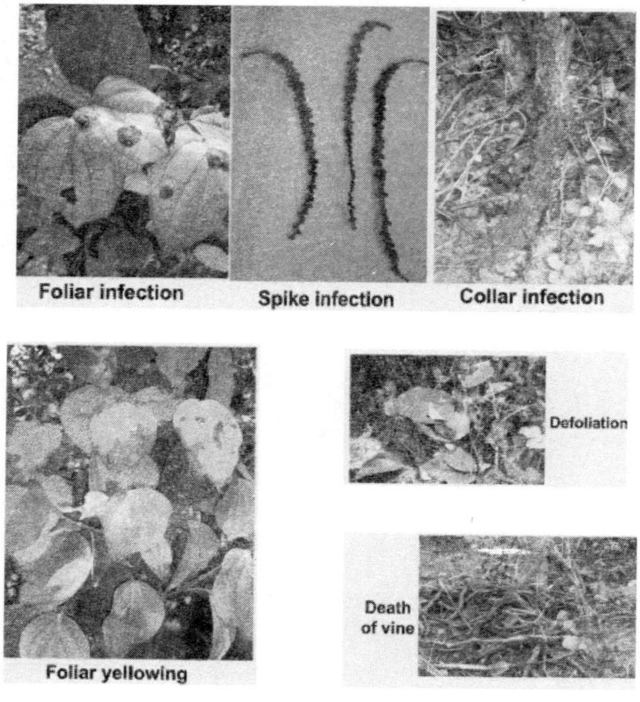

Fig. 19.1. Symptoms of foot rot on different parts of black pepper (Courtesy: Shashidhara, 2007).

(i) Symptoms: The pathogen causes black spots on leaves with fine fibre like projection from advancing margins which rapidly enlarge and cause defoliation (Fig. 19.1). Tender leaves and succulent shoot tips of new runner shoots trailing on the soil turn black. When the main stem is affected, the entire vine wilts followed by shedding of spikes with or without black spots.

(ii) Epidemiology: In areca nut-black pepper mixed cropping system, high rainfall (1000 mm) and microclimatic conditions like high relative humidity (84-89%), low temperature (22.7-29.6°C) and short sunshine hours (2.8-3.5 hr/day) favoured the disease development (Ramachandran *et al.,* 1988). Soil and infected plant debris in a plantation are the main sources of inoculum. The pathogen spreads through rain splashes. The survival of the pathogen in the infected plant debris was recorded up to 240 days and on alternate hosts up to 140 days. The incidence of quick wilt was maximum during July and minimum during January to May. A significant positive correlation has been established between total rainfall, number of rainy days and relative humidity; and a negative correlation with maximum temperature and sunshine hours.

(iii) Integrated Management

(a) Bioagents and Botanicals: Application of *Trichoderma* spp. at 75 g/vine along with neem cake at 2 kg/vine during May-June (pre-monsoon) with suitable carrier medium such as coffee husk and well rotten cow dung followed by second round of application during August-September (post-monsoon) gives effective control of the disease.

In disease affected gardens, plants treated with *T. harzianum* (IISR 1369 strain with 10^{10} cfu/g) at 50g/vine along with 10 kg of FYM or 1 kg neem cake during May – June period immediately after the receipt of first early monsoon shower and second dose of inoculum applied during August-September around the base of the vine gave effective control of foot rot and slow decline (*Meloidogyne incognita, Radopholus similis*).

Pre-monsoon application of *T. harzianum* culture mixed with pre-wetted neem cake or FYM at 1 kg/100 kg and incubated for 2 days applied at 5 kg/vine below 10 years and 10 kg/vine above 10 years gave good control of wilt disease.

Effective package for integrated management of pepper wilt with soil application of *T. viride* at 75 g/vine or combination of *T. viride* + modified panchagavya-3 (10 dilution at 2.3 litre/vine) applied twice, once during pre-monsoon (April-May) and another during post-monsoon (August-September) was developed by Jahagirdar and Siddaramaiah (2003).

(b) Bioagents and Chemicals: *T. harzianum* can be integrated with potassium phosphonate (Akomin) to get effective management of foot rot since the latter was compatible with *T. harzianum* and *T. virens* (Rajan and Sarma, 1997a).

(c) Bioagents, Botanicals and AMF: Fortification of nursery mixture with *G. fasciculatum* (AMF) (100 g containing 10^{11} cfu/g), *T. harzianum* or *T. virens* (containing 10^{11} cfu/g) and using this nursery mixture to raise the nursery stock, ensured production of healthy and robust plants. Pre-plant application of the bioagent to the planting pit along with FYM and also field application of neem cake (1 kg/vine) mixed with 50g of the bioagent to the standing crop, during May-June and subsequently one more dose of the bioagent at 50g/vine during August decreased the foliar yellowing and vine death (Sarma *et al.*, 1996).

(d) Chemicals and AMF: Integration of chemicals (metalaxyl or potassium phosphonate) with AMF was effective in reducing the plant mortality and increasing the yield (Table 19.1).

Table 19.1. Effect of AMF and agrochemicals on foot rot and yield of black pepper

Treatment	Mortality of vines (%)			Yield (kg/vine)		
	AMF	Non AMF	Mean	AMF	Non AMF	Mean
Control	44.4	60.9	52.7	1.323	1.226	1.275
AMF	10.8	27.7	19.7	6.313	1.828	4.070
Copper oxychloride, Bordeaux mixture	27.6	33.8	30.4	8.920	2.973	2.460
Metalaxyl 100 ppm (Ridomil mancozeb)	21.9	38.8	30.4	2.821	0.456	1.639
Potassium phosphonate (Akomin)	10.8	21.9	16.4	4.745	3.633	4.181
CD (p=0.05)	**NS**	**NS**	**18.12**	**2.122**	**---**	**1.294**

(e) Bioagents and AMF: Integration of *P. fluorescens* (IISR 16) with AMF was found to be highly beneficial for the growth and vigour of black pepper rooted cuttings. The rooted cuttings raised in nursery mixture fortified with the biocontrol inoculum showed field establishment as high as 90-98% (Sarma and Saju, 2004).

(f) Two Bioagents: Integration of *T. harzianum* (IISR 1369) with *Pseudomonas fluorescens* (IISR-41) gave very effective control of foot rot (10% disease incidence as compared to 90% in control) and increased plant height under field conditions. Integration of *T. harzianum* (IISR 1369) with *Pseudomonas fluorescens* (IISR-11 or 6) suppressed the root rot to the extent of 63% over control and improved the vigour of pepper vines (Saju, 2004; Saju *et al.*, 2003). Combined formulation of *T. harzianum* and *P. fluorescens* (IISR-6) was more effective for production of healthy black pepper rooted cuttings (Thankamani *et al.*, 2003; 2005).

(g) Bioagents, Botanicals and Chemicals: Reduction in intensity of leaf infection, yellowing, defoliation, collar infection and wilting was maximum in treatment metalaxyl spray and soil drench (1.25ml/L) + soil application of *T. harzianum* (50 g/vine) + soil application of *P. fluorescens* (100 ml/vine) + soil application of neem cake (1 kg/vine) (T1) followed by Potassium phosphonate spray and soil drench (3 ml/L) + soil application of *T. harzianum* (50 g/vine) + soil application of *P. fluorescens* (100 ml/vine) + soil application of neem cake (1 kg/vine) (T2) (Table 19.2) (Shashidara, 2007).

Table 19.2. Effect of bioagents, botanicals and chemicals on the management of foot rot on black pepper

Treatment	Leaf infection (%)	Leaf yellowing (%)	Defoliation (%)	Collar infection (%)	Wilting (%)
T1	2.44	1.65	0.62	5.56	0.00
T2	5.97	6.58	5.56	10.07	5.56
T3 - Control	52.92	62.76	52.67	77.77	50.00
CD (P=0.05)	**1.229**	**3.733**	**4.646**	**2.567**	**2.966**

T1 - Metalaxyl spray and soil drench (1.25ml/L) + soil application of *T. harzianum* (50 g/vine) + soil application of *P. fluorescens* (100 ml/vine) + soil application of neem cake (1 kg/vine).

T2 - Potassium phosphonate spray and soil drench (3 ml/L) + soil application of *T. harzianum* (50 g/vine) + soil application of *P. fluorescens* (100 ml/vine) + soil application of neem cake (1 kg/vine)

19.1.1.2. Leaf Rot, *Rhizoctonia solani*

(i) Symptoms: The disease is often serious in nurseries during April-May when warm humid conditions prevail. The fungus infects both leaves and stems. Greyish sunken spots and mycelial threads appear on the leaves. The infected leaves are attached to one another with the mycelial threads. On stem, the infection occurs as dark brown lesions, which spread both upwards and downwards. The new flushes subtending the points of infection gradually droop and dry up.

(ii) Integrated Management

Bioagents and Chemicals: Foliar spray with liquid formulation of *Pseudomonas fluorescens* and drenching at both pre- and post-monsoon treatment with 1% Bordeaux mixture provided good protection. The treated plants had 35% disease index as compared to 61% in control and the yield was 2.34 kg/vine as compared to 0.78 kg/vine in control.

19.1.2. Nematodes

19.1.2.1. Root-knot Nematodes, *Meloidogyne* spp.

Butler (1906) reported root-knot nematodes in black pepper from Wynad, Kerala, India. *M. incognita* and *M. javanica* have been reported from India. The root-knot infestation is a serious problem in Government nurseries in Kerala. Up to 91% root-knot infestation was reported from Para, Brazil (Ichinhoe, 1975) and Kerala, India (Ramana *et al.*, 1987).

An initial inoculum level of 10 J_2 per rooted cutting was found to reduce growth by 15%, while at 1,00,000 J_2 level 50% reduction in growth was observed over one year period. More than 50% death of transplants occurs in field planting of infected cuttings. *M. incognita* was responsible for 46% loss in yield of black pepper (Mohandas and Ramana, 1991).

(i) Symptoms: Prominent symptoms of root-knot infestation on black pepper are unthrifty growth and yellowing of leaves. Inter-veinal yellowing of the foliage is also noticeable. The leaves exhibit dense yellowish discolouration of the interveinal areas making the leaf veins prominent with deep green colour. Heavy galling of the root system is also present (Fig. 19.2).

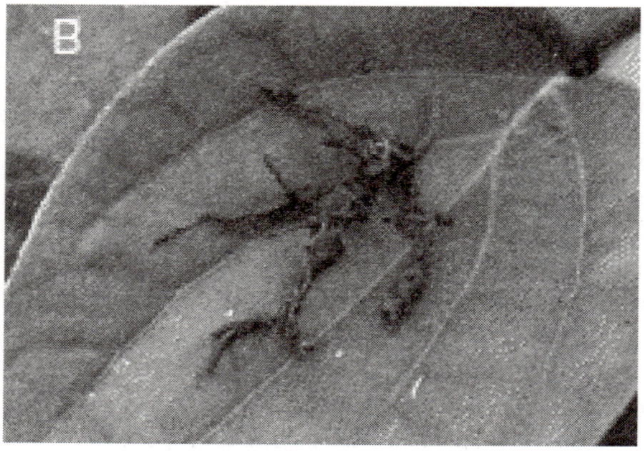

Fig. 19.2. Black pepper roots infected with root-knot nematode

(ii) Integrated Management

(a) Two Bioagents: Black pepper vines combinedly inoculated with *P. lilacinus* and *P. penetrans* had put out 23-112 per cent increase in plant growth over control and were very effective in the management of root-knot nematodes (Sosamma and Koshy, 1995).

Under field conditions, plants treated with the consortial formulation of *Pseudomonas fluorescens* Pf 123 and *Bacillus subtilis* Bs 214 significantly enhanced the yield parameters and reduced nematode infestation both in soil and roots (Table 19.3) (Devapriyanga *et al.*, 2012).

Table 19.3. Efficacy of talc formulations of *Pseudomonas* and *Bacillus* isolates on root-knot nematode and yield parameters of black pepper cv. Paniyur 1 under field conditions

Treatments	No. of spikes/vine	Wt. of spikes/ vine (g)	No. of yellow leaves/vine	No. of egg masses/ 5g roots	No. of eggs/egg mass	Gall index
Pf 123	270.30	1928.00	13.60	30.80	155.00	2.02
Bs 214	251.00	1898.20	17.80	37.20	169.60	2.36
Pf 123 + Bs 214	294.20	2192.40	9.40	25.60	143.60	1.10
Carbofuran	247.00	1838.60	17.40	37.60	176.60	2.16
Control	172.20	906.80	26.20	48.60	260.00	4.08
CD (P=0.05)	**3.7298**	**57.9689**	**1.6746**	**1.8914**	**1.1727**	**0.2523**

(b) Botanicals and Chemicals: Integration of neem cake application at 1 kg/vine along with phorate/carbofuran at 3g a.i./vine during May-June and again during September-October gave effective control of nematodes infesting black pepper.

(c) Cultural and Chemicals: Application of aldicarb at 1g a.i./vine twice a year (May/June and October/November) integrated with fertilizers (N-100g, P- 40g, K-140g/vine) in two equal splits, earthing up to 50 cm radius at the base of the vines and mulching the base of the vines with leaves reduces foliar yellowing by 83% and *M. incognita* population by 33-38% (Venkitesan and Jacob, 1985).

(d) Physical, Bioagents, Botanicals and AMF: Integrated management of foot rot (*P. capsici*) and nematodes (*M. incognita* and *R. similis*) on black pepper was achieved by (i) mixing AMF and *Trichoderma harzianum* in solarized nursery mixture to raise healthy and robust seedlings, (ii) application of *T. harzianum* and FYM in planting pit, (iii) field application of neem cake at 1 kg/vine mixed with 50 g of *T. harzianum* during August.

19.1.2.2. Burrowing Nematode, *Radopholus similis*

(i) Symptoms: *R. similis* on black pepper is associated with pepper yellows (slow-wilt) disease, which appears as pale yellow or whitish-yellow drooping leaves on the vines. The number of such leaves increases gradually until large numbers of leaves, or even the entire foliage, become yellow (Fig. 19.3). Yellowing is followed by shedding of leaves, cessation of growth and dieback symptoms. The symptoms are well pronounced when soil moisture is depleted. Within 3-5 years of initiation of yellowing, all the leaves are shed and death of the vine takes place; hence the name slow-wilt disease.

Fig. 19.3. Yellows disease of black pepper plants caused by *R. similis* destroying roots.

In bearing vines, shedding of inflorescences is a major symptom. Large numbers of shed inflorescences are seen at the base of affected vines. In large plantations, affected patches become conspicuous initially as yellowed plants, and later with large numbers of barren standards that have lost the vines, or standards supporting dead vines without any leaves. Young and old plants are affected and the replanted vines normally die within 2 years.

The tender, thin, white feeding roots show typical orange- to purple-coloured lesions (Fig. 19.3). Lesions are not clearly seen on older roots, being brown in colour. The root system exhibits extensive rotting and this results in a lack of fine feeder roots from the main roots. Extensive necrosis of larger lateral roots develops subsequently.

(ii) Integrated Management

(a) Cultural, Botanicals and Chemicals: Integrated methods of the burrowing nematode management that can be suggested are:

- Planting of nematode-free rooted cuttings.
- Uprooting of affected vines and replanting after a period of 9-12 months.
- Use of non-living supports or standards.
- Exclusion of *R. similis*-susceptible trees as standards for trailing black pepper vines.
- Exclusion of susceptible intercrops such as banana, ginger and turmeric.
- Application of organic amendments, such as neem oil cake, green foliage, or farm-yard manure.
- Earthing-up after application of nematicides, NPK fertilizers and organic amendments.

19.1.2.3. Nematodes, *Meloidogyne incognita, Radopholus similis* and Foot Rot, *Phytophthora capsici* Disease Complex

(i) Symptoms: Increased susceptibility of *M. incognita* and *M. javanica* infested cultivars of black pepper to *Phytophthora* infestation has been reported.

(ii) Integrated Management

(a) Bioagents, Botanicals, AMF and Physical Methods: Integrated management of foot rot (*Phytophthora capsici*) and nematodes (*Meloidogyne incognita* and *Radopholus similis*) on black pepper was achieved by:

- Mixing VAM and *Trichoderma harzianum* in solarized nursery mixture to raise healthy and robust seedlings.
- Application of *T. harzianum* and FYM in planting pit.
- Field application of neem cake at 1 kg/vine mixed with 50 g of *T. harzianum* during August (Sarma, 2003).

19.2. CARDAMOM, *ELETTARIA CARDAMOMUM*

19.2.1. Diseases

19.2.1.1. Damping-off, *Pythium vexans, Rhizoctonia solani*

(i) Symptoms: Infection is observed at the collar region. Infected leaves become pale, yellow and ultimately the young leaves die. Older leaves die prematurely and new shoots that arise are weak, decay and the rhizomes rot at the base of the stem. The diseased shoot can be pulled out easily.

(ii) Integrated Management

(a) Physical and Bioagents: Solarization of nursery beds before sowing seeds and incorporation of *T. harzianum* has resulted in prevention of damping-off and production of pathogen-free healthy seedlings.

(b) Physical, Bioagents and AMF: Soil solarization can be done for sterilizing the nursery mixture. To the sterilized mixture, biocontrol agents such as arbuscular mycorrhizal fungus at 100 g/kg and *T. harzianum* (10^{10} cfu/g) may be added at the time of filling of nursery mixture in polybags.

19.2.1.2. Capsule Rot/Azhukal, *Phytophthora* spp.

(i) Symptoms: This disease occurrs during the rainy season. It affects the leaves, tender shoots, panicles and capsules. On the infected leaves, water soaked lesions appear first and rotting and shedding of leaves along the veins occur there after (Fig. 19.4). The infected capsules become dull greenish brown and decay. This emits a foul smell and subsequently capsules are shed. Infection spreads to the panicles also.

Fig. 19.4. Capsule rot of cardamom. L – Rotting and shredding of leaves, R – Rotting of capsules

(ii) Integrated Management

(a) Bioagents and Botanicals: *Trichoderma* spp. can be used along with cow dung for controlling this disease. For field application, *T. viride* and *T. harzianum* inoculum in decomposed coffee pulp and FYM in 1: 1 ratio at 1 kg/plant was found to be the best (Suseela Bhai *et al*., 1994). Application of *T. harzianum* along with neem cake at the base of the clump reduced *Phytophthora* propagules and consequent reduction in the disease incidence.

(b) Two or more Bioagents: Soil application of *T. viride, T. harzianum, Bacillus subtilis* and *Latiseria aravalis* reduced *Phytophthora* population and suppressed the disease to the extent of 30-50% (Suseela Bhai *et al*., 1993, Joseph Thomas *et al*., 1993).

(c) Bioagents and Chemicals: Spraying of Bordeaux mixture followed by two applications of *T. harzianum* at 1 kg/plant (28 x 10^8 cfu/g) during May and July significantly reduced the disease incidence. Spraying of *T. harzianum* followed by two applications of Bordeaux mixture reduced the disease potential index. Three applications of *T. harzianum* combined with Akomin recorded maximum population of *T. harzianum* (Table 19.4) (Suseela Bhai, 1998).

Table 19.4. Integrated management of capsule rot of cardamom

Treatment			Disease incidence (%)	Disease potential index	Population of *Trichoderma* x 10^3
BM	*Th*	*Th*	0.30	7.56	18.76
Th	BM	BM	0.72	2.00	17.94
Ak+*Th*	Ak+*Th*	Ak+*Th*	0.28	16.17	29.69
Ak	Ak	Ak	0.10	15.89	3.03
BM	BM	BM	0.13	4.28	2.47
Th	COC	COC	0.37	15.78	15.47
COC	*Th*	*Th*	0.20	21.17	27.10
Th	*Th*	*Th*	0.20	34.44	28.99
COC	COC	COC	0.03	5.11	2.08
Control			6.61	96.89	1.50
CD (P = 0.05)			**2.96**	**32.45**	**9.24**

BM: Bordeaux mixture; *Th*: *Trichoderma harzianum* (1 x 10^7); Ak: Akomin: COC: Copper oxychloride.

(d) Botanicals, Bioagents and Chemicals: Management of capsule rot (*Phytophthora* spp.) of cardamom was achieved by two applications of *T. harzianum* at 1 kg/plant (grown on decomposed coffee pulp and FYM in 1:1 ratio) during May and July integrated with foliar spray of Akomin (potassium phosphonate) (Anandraj and Eapen, 2003).

(e) Physical and Bioagents: Solarization of nursery beds and subsequent soil application of *T. harzianum* reduced root rot infection and also nematode infection caused by *Meloidogyne incognita* (Eapen and Ramana, 1996).

19.2.1.3. Rhizome Rot or Clump Rot, *Pythium vexans, Fusarium oxysporum, Rhizoctonia solani*

(i) Symptoms: The disease occurs during South West monsoon. The symptoms include pale yellow colour on the foliage and premature death of older leaves. Collar portion of the aerial shoots become brittle and tiller breaks easily from the rhizome at the bulbous base (Fig. 19.5). Water soaked lesions appear first on leaves. The pathogen causes rotting and shredding of leaves along the veins. Rotting develops in the collar region emitting a foul smell. Capsules turn dull greenish brown emitting a foul smell.

Fig. 19.5. Clump rot of cardamom.

(ii) Integrated Management

(a) Physical and Bioagents: Solarization of nursery beds before sowing seeds and incorporation of *T. harzianum* has resulted in prevention of damping off and production of pathogen-free healthy seedlings (Eapen and Ramana, 1996).

(b) Bioagents, AMF, Chemicals and Physical: Soil solarization can be done for sterilizing the nursery mixture. To the sterilized mixture, biocontrol agents such as arbuscular mycorrhizal fungus at 100 g/kg and *T. harzianum* (10^{10} cfu/g) may be added at the time of filling of nursery mixture in polybags. Since the biocontrol agents protect the root system only, the aerial portion may be protected with 1% Bordeaux mixture spray.

(c) Bioagents, Botanicals and Chemicals: Incorporation of *Trichoderma* spp. multiplied in suitable organic medium (1 kg/clump) prior to the onset of monsoon season is a prophylactic operation. Spraying of 1% Bordeaux mixture at 3 litres/plant with an adhesive by the commencement of monsoon and continuing the spraying operation up to November-December is effective.

(d) Two Bioagents: Integration of *T. harzianum* (IISR 1369) with *Pseudomonas fluorescens* (IISR-11 or 6) suppressed the clump rot to the extent of 36% over control.

19.2.2. Nematodes

19.2.2.1. Root-knot Nematodes, *Meloidogyne* spp.

M. incognita and *M. javanica* have been reported as widely occurring in the cardamom nurseries and plantations in Kerala, Karnataka and Tamil Nadu (Kumar *et al.*, 1971; Koshy *et al.*, 1976; Ali, 1982, 1986). An yield loss of 32-47% due to root-knot infestation has been reported (Ali, 1984, 1986). An initial population of 100 nematodes/ plant causes discernible damage to cardamom (Eapen, 1987).

(i) Symptoms: Heavy infestation of root-knot on mature plants in a plantation causes stunting, reduced tillering, yellowing, premature drying of leaf tips and margins, narrowing of leaf blades, delay in flowering, immature fruit drop and reduction in yield. Infested roots are characterized by excessive branching. Galling of roots is not conspicuous on mature plants (Fig. 19.6).

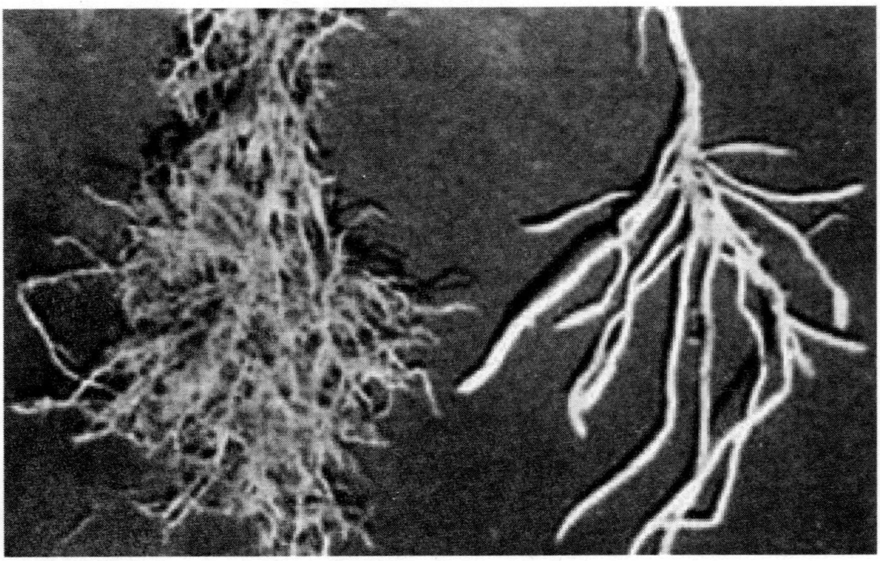

Fig. 19.6. Cardamom roots damaged by root-knot nematode. L – healthy, R - infested

In primary nursery, the second stage juveniles infest the radical and plumule as a result of which even 50% of the germinating seeds do not emerge. At the two leaf stage itself the infested seedlings show marginal yellowing and drying of leaves and severe galling of roots. Up to 40% of such seedlings do not establish in the secondary nursery. The infested secondary nursery plants exhibit stunting, yellowing, poor tillering, drying of leaf tips and margins and heavy galling of roots (Ali and Koshy, 1982).

(ii) Integrated Management

(a) Physical and Bioagents: Soil solarization of nursery beds and application of bioagents such as *P. lilacinus* or *Trichoderma* spp. improved growth of cardamom seedlings by suppressing root-knot nematode population (Eapen and Venugopal, 1995).

Besides soil solarization of nursery beds, subsequent soil application of *T. harzianum* reduced root rot infection and also infection caused by the root-knot nematode, *M. incognita* (Eapen and Ramana, 1996).

(b) Botanicals and Bioagents: Application of *T. harzianum* multiplied on decomposed coffee husks (7 day old) at the time of sowing at 2.5 kg/bed (4.5 m x 1.0 m) and repeated after 3 months is recommended for the control of root-knot nematodes and damping-off in nurseries.

19.2.2.2. Root-knot Nematode, *Meloidogyne* sp. and Rhizome Rot, *Rhizoctonia solani* Disease Complex

(i) Symptoms: *M. incognita* was found to predispose cardamom seedlings to *R. solani* infection, which causes damping-off and rhizome rot in the primary nursery (Ali and Venugopal, 1992, 1993).

(ii) Integrated Management

(a) Physical and Bioagents: *P. lilacinus* in combination with *Trichoderma* spp. suppressed *Meloidogyne* spp. and rhizome rot disease (*R. solani*) complex when incorporated in solarized cardamom nursery beds (Eapen and Venugopal, 1995).

(b) Botanicals and Chemicals: Application of carbofuran along with neem cake was found to reduce immature fruit-drop and to increase capsule yield.

(c) Physical, Bioagents and Chemicals: Soil solarization alone enhanced the germination by 25.5% and suppressed weed growth by 82.0%. Solarization also enhanced the growth and vigour of cardamom seedlings. The disease complex was suppressed by incorporation of *P. lilacinus/T. harzianum* and phorate into the solarized nursery beds. This approach is being adopted on a large scale for the production of nematode-free cardamom seedlings.

19.3. GINGER, *Zinziber officinale*

19.3.1. Insect Pests

19.3.1.1. Shoot Borer, *Dichcrosic punctiferalis*

(i) Damage: The larvae bore into the shoots. They can cause them to wilt and die. Caterpillar bores the rhizomes and pseudo stem causing dead heart. Extrusion of frass through holes can be observed (Fig. 19.7).

Fig. 19.7. Shoot borer infestation on ginger.

(ii) Integrated Management

(a) Cultural and Chemical: Pruning and destroying freshly infested pseudo stems (at fortnightly intervals) during July-August and spraying 0.1% malathion (at monthly intervals) during September-October is effective against the pest.

19.3.2. Diseases

19.3.2.1. Rhizome Rot, *Pythium aphanidermatum, P. myriotylum*

(i) Symptoms: Soft rot is the most serious disease of ginger in India and in some other countries. It is caused by *Pythium* spp., of which *P. aphanidermatum* is the principal species in India, although *P. butleri*, *P. gracile*, *P. myriotylum*, *P. nigriotilum*, and *P, vexans* have also been recorded. The bases of the aerial shoots become soft, watery and then rot. The affected plants become pale; the tips of the leaves turn yellow, followed by complete yellowing and drying up of the leaves (Fig. 19.8). The shoots fall and cease to produce rhizomes. The infection extends to the rhizomes; the inner tissues being reduced to a soft and black, putrefying mass. Losses can be high. The disease is favoured by high moisture content of the soil with insufficient drainage.

(ii) Integrated Management

(a) Bioagents and Botanicals: Application of *Trichoderma harzianum* along with neem cake at 1 kg/bed helps in preventing the disease. *T. harzianum, T. viride, T. hamatum* and *T. virens* as seed treatment and also as soil application along with neem cake at 1 kg/m^2 reduced the disease incidence and increased the yield. The treatment was even superior to mancozeb treatment. The population stability of biocontrol agent was maintained up to 60 days. In Rajasthan, *T. viride* applied along with wood saw dust or karanj or neem cake has been effective against rhizome rot caused by *Pythium myriotylum* and *Fusarium solani* (Lodha *et al.*, 1994).

Fig. 19.8. Rhizome rot of ginger.

Raising seedlings in *T. harzianum/P. fluorescens* amended compost and soil application of *T. harzianum/P. fluorescens* along with neem cake/compost in the main field was useful in minimizing the incidence of rhizome rot of ginger.

(b) Physical and Bioagents: Soil solarization prior to planting followed by application of biocontrol agents showed synergistic effects in disease reduction and increased yield (Sarma *et al.*, 1996) (Table 19.5).

Table 19.5. Effect of soil solarization and biocontrol agents on germination, rhizome rot incidence and yield of ginger

Biocontrol agent	Solarized soil			Non- solarized soil		
	Germination (%)	Disease index (%)	Yield (kg/plot)	Germination (%)	Disease index (%)	Yield (kg/plot)
T. viride	77.69	13.70	2.846	71.52	35.25	1.225
T. harzianum 1	79.61	15.80	3.552	79.43	32.07	2.673
T. harzianum 2	77.87	19.38	2.910	72.91	43.61	1.439
T. hamatum	75.00	16.16	2.744	73.23	39.26	1.602
T. virens	74.48	18.25	2.641	74.43	36.63	1.705
Mancozeb	79.23	36.22	2.260	73.89	46.70	0.818
Control	81.53	57.58	1.692	79.86	53.80	0.992
Mean	77.92	25.30	2.278	74.90	41.08	1.485
CD (P=0.05)	**NS**	**6.26**	**0.430**	**NS**	**6.26**	**0.430**

Usman *et al.* (1996) successfully utilized the native strains of *Trichoderma* and *Gliocladium* against rhizome rot of ginger caused by *P. aphanidermatum, P. myriotylum* and *Fusarium solani* complex. The disease suppression was enhanced when biocontrol was succeeded by soil solarization.

As a seed disinfection procedure, a novel technique called rhizome solarization has been devised at the Indian Institute of Spices Research, Calicut. The disinfected rhizomes when treated with *T. harzianum* and rhizobacterial strain consortia as seed treatment and soil application resulted in higher yields and growth promotion and soft rot suppression.

(c) Bioagents and Chemicals: The biocontrol was further integrated with chemical control using metalaxyl as seed treatment since it is compatible with *Trichoderma* spp. (Balakrishnan *et al.*, 1996; Sarma, 1997; Lodha and Mathur, 1997).

Integration of soil application of bioagents with fungicidal rhizome treatment using bavistin (carbendazim) + ridomil MZ (metalaxyl + mancozeb) increased the efficiency of disease control as compared with their individual treatments. Soil application of *T. harzianum* and rhizome treatment with *Pseudomonas* sp. and fungicides was the most effective among all the tested treatments (Ram *et al.*, 1999).

(d) Two Bioagents: Combined application of *Pseudomonas fluorescens* (IISR-11) and *T. harzianum* (IISR-1369) imparted 66.2% survival of ginger tillers, reduced the rhizome rot infection and improved the vigour and yield of ginger plants (Sarma, 2000).

19.3.2.2. Yellows, *Fusarium oxysporum* f. sp. *zingiberi*

(i) Symptoms: Fusarium yellows is a very common and serious fungal disease which is specific to ginger. Infected plants are stunted and yellow (Fig. 19.9), lower leaves dry out and turn brown. Eventually all above-ground shoots dry out completely. Plant collapse is very slow (up to several weeks) compared with the rapid collapse associated with bacterial wilt infection. Diseased rhizomes show a brown internal discolouration, are normally shriveled in appearance and eventually decay leaving the outer shell intact with fibrous internal tissue remaining. Increased nematode infestations are usually associated with *Fusarium* rhizome rot, accentuating yield losses. *Fusarium* is also responsible for serious loss of planting pieces and poor germination. The disease spreads rapidly in the field during wet weather.

Fig. 19.9. Ginger yellows symptoms

(ii) Integrated Management

(a) Bioagents and Botanicals: Application of *T. harzianum, T. viride, T. hamatum* and *T. virens* as seed treatment and soil application along with neem cake at 100 kg/ha gave protection and increased yield. Application of *T. viride* along with saw dust/neem cake was found to be highly effective in suppression of Fusarium yellows.

Soil application of *Trichoderma* bioformulation mixed with FYM gave highest yield of 12.274 MT/ha which was 56% more than the control plots (7.833 MT/ha) (Selvakumar *et al.*, 2009).

(b) Physical and Bioagents: Hot water treatment of ginger rhizomes along with soil application of *Trichoderma* spp. gave effective control of the disease and increased the yield (Selvakumar *et al.*, 2009). There was synergistic effect in protection where soil solarization followed by biocontrol was adopted.

19.3.2.3. Storage Rot, *Sclerotium rolfsii*

(i) Symptoms: This disease has caused serious losses on some farms. Infection first occurs beneath the scales on the rhizomes and may progress to produce a deep brown rot over the entire surface. Affected rhizomes become enveloped in a white fungal mycelial growth which may extend up to the basal stem. Mild infections usually produce no above ground symptoms. Severe infections cause the stems to turn yellow and slowly dry out.

(ii) Integrated Management

(a) Physical and Bioagents: The disinfected rhizomes (through solarization) when treated with *T. harzianum* and rhizobacterial strain consortia as seed treatment and soil application resulted in higher yields and growth promotion and storage rot suppression and minimal bacterial wilt. In addition, in combination with *Glomus* spp., the disease was absent probably through growth mediation.

(b) Two Bioagents: *T. harzianum* in combination with *Pseudomonas fluorescens* showed a synergistic effect in reducing the storage rot infection.

19.3.2.4. Bacterial Wilt, *Ralstonia solanacearum*

A bacterial wilt of ginger was recorded from Australia, Hawaii, India, Mauritius and Malaysia. Bacterial wilt is a very serious disease that can spread very rapidly through an area, causing complete destruction.

(i) Symptoms: The first symptoms of the disease are yellowing and wilting of the lower leaves, which quickly spread upwards, affecting the whole plant (Fig. 19.10). In advanced stages the base of the pseudo stem becomes water-soaked, readily breaking away from the rhizome at ground level and the plants eventually collapse. The vascular tissues become dark brown or black. When an infected stem or rhizome is cut transversely, and a little pressure applied, a milky white exudate flows freely from the cut surface. Diseased rhizomes are much darker than healthy ones. Biotype 4 was found to produce the most severe symptoms.

Fig. 19.10. Bacterial wilt symptoms in zinger

(ii) Integrated Management

(a) Botanicals and Bioagents: Lowest disease incidence (15.63) was recorded in the treatment of *P. fluorescens* mass cultured in the mixture of vermicompost (VC) and mustard oil cake (MOC) applied as seed treatment and soil application. It was followed by *T. harziamum* mass cultured in MOC (21.88%), which was statistically at par with the application of copper oxychloride (22.50%) (Bora and Bora, 2009).

(b) Physical, Bioagents and AMF: The disinfected rhizomes (through solarization) when treated with *T. harzianum* and rhizobacterial strain consortia as seed treatment and soil application resulted in higher yields and growth promotion and minimal bacterial wilt. In addition, in combination with *Glomus* spp., the disease was absent probably through growth mediation.

19.3.2.5. Soft Rot, *Erwinia* sp.

(i) Symptoms: Normally bacterial soft rot is only a storage rot. The bacteria are present in most soils but field infection usually occurs only in waterlogged areas. *Erwinia* sp. has caused severe losses of stored rhizomes on some farms, but is not considered a serious storage problem where precautions are taken. Softening of the tissue is accompanied by production of a strong odour and the rhizome eventually collapses completely. Bacterial soft rot differs from other rhizome rots in that putrid odour is produced in this soft rot.

(ii) Integrated Management

(a) Physical, Bioagents and AMF: The disinfected rhizomes (through solarization) when treated with *T. harzianum* and rhizobacterial strain consortia as seed treatment and soil application resulted in higher yields and growth promotion and soft rot suppression. In addition, in combination with *Glomus* spp., the disease was absent probably through growth mediation.

(b) Two Bioagents: *T. harzianum* in combination with *Pseudomonas fluorescens* showed a synergistic effect in reducing the soft rot infection.

(c) Physical, Botanicals and Bioagents: Solarization of beds before planting and addition of *T. harzianum* formulation (with 10 cfu/g) at 50 g/3 m^2 bed along with neem cake/FYM reduced the disease incidence.

19.3.3. Nematodes

19.3.3.1. Root-knot Nematodes, *Meloidogyne* spp.

M. incognita, M. javanica, M. arenaria and *M. hapla* have been reported to be associated with ginger from various countries. Under pot conditions, an initial inoculum level of 10,000 nematodes per plant over a period of six months caused 74% reduction in rhizome weight. A population level of one juvenile/30 g of soil was found to cause significant reduction in yield (Sukumaran and Sundararaju, 1986a). Kaur (1987) estimated 41-59% yield loss in ginger when the crop was raised using apparently healthy rhizomes in nematode infested fields.

(i) Symptoms: Heavily infested plants exhibit stunting and chlorotic leaves with marginal necrosis. The root-knot nematodes cause galling and rotting of roots and underground rhizomes. Infested rhizomes show brown, water-soaked lesions in the outer tissues, particularly in the angles between shoots. The J2 of *M. incognita* invade the rhizome through the axils of leaf sheaths in the shoot apex. In fibrous roots, penetration occurs in the area of differentiation and in fleshy roots, the entire length of root is invaded. In the rhizomes and fleshy roots extensive internal lesions develop.

(ii) Integrated Management

(a) Two Bioagents: *P. lilacinus* along with *T. harzianum/P. chlamydosporia* was effective in suppression of root-knot nematodes under field conditions.

(b) Botanicals and Chemicals: Application of neem cake at 1 t/ha at planting, followed by carbofuran at 1 kg a.i./ha at 45 DAP gave effective control of nematodes associated with ginger.

19.3.3.2. Lesion Nematode, *Pratylenchus coffeae* and *Fusarium Wilt, Fusarium oxysporum* f. sp. *zingiberi* Disease Complex

(i) Integrated Management

(a) Physical and Bioagents: The efficacy of biocontrol agents like *Aspergillus niger, Trichoderma harzianum, T. hamatum* and *T. virens* either alone or in combination with soil solarization has been reported for the management of yellows disease (*Fusarium + Pratylenchus coffeae)* in Himachal Pradesh (Dohroo, 1995). Similar results have been reported by Eapen and Ramana (1996).

19.4. TURMERIC, *CURCUMA LONGA*

19.4.1. Diseases

19.4.1.1. Rhizome Rot, *Pythium aphanidermatum*

(i) Symptoms: Rhizome rot shows progressive drying-up of the leaves of infected plants. The base of the aerial shoots shows water-soaked soft lesions. As the disease progresses infection gradually spreads to the rhizomes, which begin to rot and become soft. The bright orange colour of the rhizomes changes into brown. The disease may be confined to a few isolated plants or may occur in patches (Fig. 19.11). In severe attacks the yield is considerably reduced.

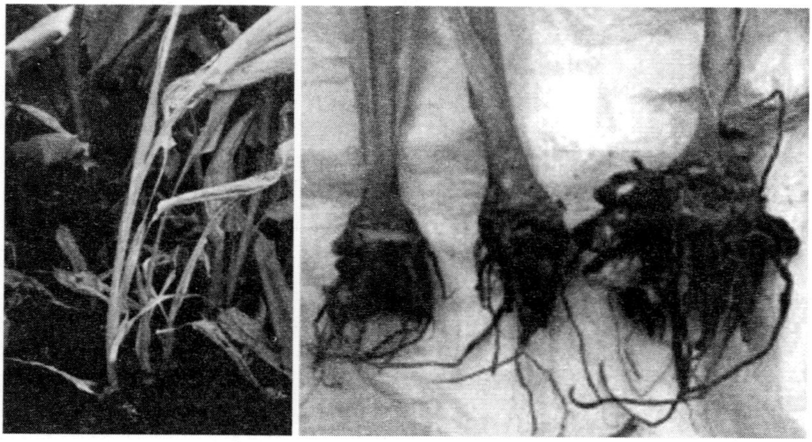

Fig. 19.11. Rhizome rot of turmeric.

(ii) Integrated Management

(a) Physical, Bioagents and Botanicals: Treatment of rhizomes with hot water at 51°C for 10 min and soil application of *Trichoderma harzianum* mixed with neem cake resulted in minimum incidence of rhizome rot and maximum yield.

(b) Botanicals and Bioagents: Application of *T. harzianum, T. viride, T. hamatum* and *T. virens* as seed treatment and soil application along with neem cake at 100 kg/ha gave protection and increased yield. Application of *T. viride* along with saw dust/neem cake was found to be highly effective in suppression of rhizome rot.

Seed treatment with *T. viride* + *P. fluorescens* at 4g/kg seed and soil application of *T. viride* (12.5 kg/ha) and *P. fluorescens* (25 kg/ha) along with FYM at 10 MT/ha resulted in minimizing rhizome rot to 11.7% compared to 37% in control, and higher yield of 28.6 MT/ha in treated compared to 10.3 MT/ha in control (Anon, 2005).

(c) Two Bioagents: *T. harzianum* in combination with *Pseudomonas fluorescens* showed a synergistic effect in reducing the rhizome rot infection.

(d) Physical and Bioagents: There was synergistic effect in protection where soil solarization followed by biocontrol (*T. harzianum*) was adopted.

19.4.1.2. Soft Rot, *Erwinia* sp.

(i) Integrated Management

(a) Physical, Bioagents and AMF: The disinfected rhizomes (through solarization) when treated with *T. harzianum* and rhizobacterial strain consortia as seed treatment and soil application resulted in higher yields and growth promotion and soft rot suppression and minimal bacterial wilt. In addition, in combination with *Glomus* spp., the disease was absent probably through growth mediation.

(b) Two Bioagents: *T. harzianum* in combination with *Pseudomonas fluorescens* showed a synergistic effect in reducing the soft rot infection.

19.4.2. Nematodes

19.4.2.1. Root-knot Nematodes, *Meloidogyne* spp.

M. incognita and *M. javanica* have been reported on turmeric of which *M. incognita* is more important. In pot experiments, 100,000 nematodes/plant resulted in 76.6% reduction in the rhizome weight after 6 months (Sukumaran and Sundararaju, 1986b).

(i) Symptoms: Turmeric plants infested with *M. incognita* exhibit stunting, yellowing, reduced tillering and marginal and tip drying of leaves. Galling and rotting of roots can also be noticed. Infested rhizomes have brown, water-soaked areas in the outer tissues and lose their bright yellow colour (Mani *et al.*, 1987). High populations of *M. incognita* in field cause stunting, yellowing and withering of plants in large patches. Premature death of plants takes place leaving a poor crop stand.

(ii) Integrated Management

(a) Botanicals and Bioagents: Soil application of *T. harzianum* + neem cake at 1 t/ha is recommended.

19.5. VANILLA, VANILLA PLANIFOLIA, V. ANDAMANICA

19.5.1. Diseases

19.5.1.1. Wilt, *Fusarium oxysporum* f. sp. *vanillae*

(i) Symptoms: *Fusarium* wilt is the most serious disease of vanilla. The disease is more prevalent in younger plantation especially during the monsoon season. The infection starts at leaf axil and spreads to the internodal region resulting in rotting and drying of the stem above the point of infection. The fungus also causes leaf rot on the plant.

(ii) Integrated Management

(a) Bioagents and Botanicals: Soil application of compost mounds enriched with *T. harzianum* twice resulted in substantial reduction in wilt.

(b) Two Bioagents: Soil application of rhizobacterial strain consortia (IISR-147 and IISR-148) with *T. harzianum* was effective in disease suppression of both *Phytophthora meadii* and *F. oxysporum* f. sp. *vanillae*.

Combined applications of *T. harzianum* and *Pseudomonas fluorescens* gave maximum reduction in percentage leaf infection (Table 19.6) (Athul Sandheep *et al.*, 2012).

Table 19.6. Evaluation of microbial antagonists against *Fusarium oxysporum* of vanilla plants

Pre inoculation with bio-control agents	Percentage of leaves infection*
T. virens	12.37
T . harzianum	8.49
P. flourescens	7.18
P. putida	20.74
P. flourescens + *T. harzianum*	7.03
T . harzianum (std)	8.50
P. fluorescens (std)	8.37
P. flourescens + *T. harzianum* (std)	8.42
Control (no biocontrol agent)	90.27
CD (P = 0.05%)	**1.95**

*Values are mean of three replicates

19.5.1.2. Phytophthora Rot, *Phytophthora meadii*

(i) Symptoms: The pathogen causes rotting of beans, leaves and stems (Fig. 19.12). In severe cases, all the beans in a bunch are completely rotten. The disease is more severe during the monsoon especially in shaded plantations and poorly drained soils.

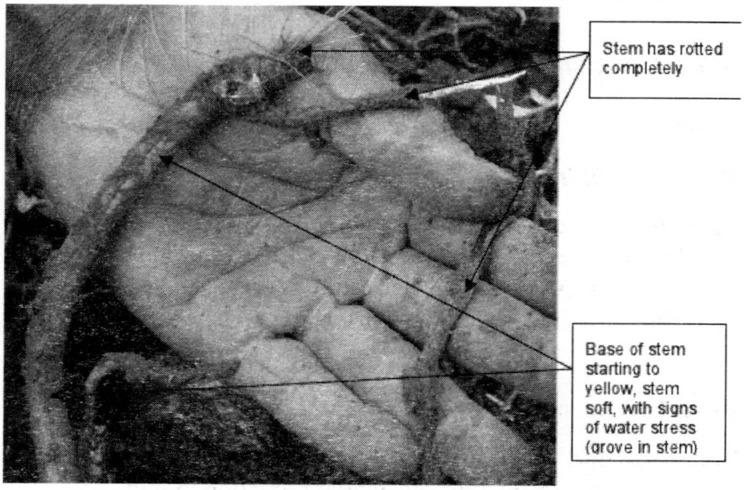

Fig. 19.12. *Phytophthora* rot symptoms on vanilla

(ii) Integrated Management

(a) Bioagents and Botanicals: Soil application of compost mounds enriched with *T. harzianum* twice resulted in substantial reduction in *Phytophthora* rot.

(b) Two Bioagents: Soil application of rhizobacterial strain consortia (IISR-147 and IISR-148) with *T. harzianum* was effective in disease suppression of both *P. meadii* and *F. oxysporum* f. sp. *vanillae*.

19.6. FENUGREEK, *TRIGONELLA FOENUM-GRAECUM*

19.6.1. Diseases

19.6.1.1. Root Rot, *Rhizoctonia solani*

(i) Integrated Management

(a) Botanicals and Bioagents: The efficacy of seed pelleting with *Trichoderma* spp. and soil application of *Trichoderma* spp. along with neem cake at 150 kg/ha showed greater protection against root rot in fenugreek and increased the yield (Table 19.7).

Table 19.7. Biocontrol of root rot disease of fenugreek

Treatment	Root rot incidence (%)		Yield (MT/ha)	
	Kharif 1992	Rabi 1992	Kharif 1992	Rabi 1992
T 1. Seed treatment + soil drenching with carbendazim	4.8	10.8	0.422	0.315
T 2. Seed treatment with *Trichoderma viride*	4.8	4.4	0.384	0.365
T 3. *T. viride* 20 days before sowing	26.3	20.4	0.288	0.285
T 4. Neem cake at 150 kg/ha	3.9	3.2	0.427	0.385
T 5. T4 + T2	3.2	3.4	0.424	0.360
T 6. T4 + T3	5.4	12.4	0.288	0.325
T 7. T4 + T1	12.8	14.3	0.345	0.340
T 8. Seed treatment with carbendazim	27.9	24.5	0.294	0.265
T 9. Control	36.2	32.8	0.163	0.184
CD (P = 0.05)	**5.7**	**4.3**	**52.4**	**27.0**

19.6.2. Nematodes

19.6.2.1. Root-knot Nematode, *Meloidogyne incognita*

(i) Integrated Management

(a) Botanicals and Bioagents: Seed treatment with *T. viride* (4 g/kg) followed by soil application (5 kg/ha) along with 150 kg/ha of neem cake consistently suppressed root -knot.

The better reduction in the root-knot nematode population was observed on the substrates (goat dung, sesame oil cake) + *P. lilacinus* treatment as compared to substrate alone. The root-knot index was lower in the sesame oil cake + *P. lilacinus* combination than in goat dung + *P. lilacinus*. The fungus (*P. lilacinus*) penetrated the eggs and fed upon their contents leaving empty cells. Invaded eggs were swollen in comparison with uncolonized ones (Sharma and Trivedi, 1989).

19.7. CUMIN, *CUMINUM CYMINUM*

19.7.1. Diseases

19.7.1.1. Wilt, *Fusarium oxysporum* f. sp. *cumini*

Wilt of cumin is an endemic problem in most of the cumin growing areas of Rajasthan and usually causes substantial yield losses.

(i) Symptoms: Infected plants show peculiar symptoms of drooping of tips and leaves, leading to mortality of the entire plant. Attack of wilt is severe in younger plants.

(ii) Integrated Management

(a) Botanicals and Bioagents: *T. harzianum* with neem cake as soil application reduced the incidence of wilt. Seed and soil treatment with *T. harzianum* was also found significantly effective for wilt disease.

Maximum reduction in disease incidence was recorded when *T. harzianum* was used as seed treatment at 4 g/kg seed + soil application at 5 g/kg soil along with soil amendment of farm yard manure at 10 g/kg soil (Table 19.8) (Gangopadhyay and Ram Gopal, 2010).

Table 19.8. Effect of farm yard manure on disease control efficacy of *Trichoderma* spp. against cumin *Fusarium* wilt

Treatment	Disease Incidence (%)	Disease control (%)	Dry weight of plant at 90 DAS (mg)	Increase in dry wt. over control (%)
T. harzianum ST 4 g/kg seed	36.7(37.3)*	45.0	284.3	78.5
T. harzianum ST 4 g/kg seed + SA 5 g/kg soil	28.5(32.3)	57.3	305.0	91.4
T. harzianum ST 4 g/kg seed + FYM 5 g/kg soil	29.4(32.8)	55.9	341.0	114.0
T. harzianum ST 4 g/kg seed + SA 5 g kg soil + FYM 5 g/kg soil	23.3(28.7)	65.0	369.7	132.0
T. harzianum ST 4 g/kg seed + FYM 10 g/kg soil	20.7(27.1)	69.0	392.3	146.2
T. harzianum ST 4 g/kg seed + SA 5 g kgsoil + FYM 10 g/kg soil	15.3(23.0)	77.1	417.0	162.0

Treatment	Disease Incidence (%)	Disease control (%)	Dry weight of plant at 90 DAS (mg)	Increase in dry wt. over control (%)
T. viride ST 4 g/kg seed	44.5(41.9)	33.3	229.0	43.7
T. viride ST 4 g/kg seed + SA 5 g/kg soil	34.1(35.7)	48.9	263.7	65.5
T. viride ST 4 g/kg seed + FYM 5 g/kg soil	34.2(35.8)	48.7	323.3	102.9
T. viride ST 4 g/kg seed + SA 5 g kg soil + FYM 5 g/kg soil	28.3(32.4)	57.6	345.0	116.5
T. viride ST 4 g/kg seed + FYM 10 g/kg soil	29.1(32.7)	56.3	367.7	130.7
T. viride ST 4 g/kg seed + SA 5 g kg soil + FYM 10 g/kg soil	20.3(26.7)	69.5	389.3	144.3
Control (Pathogen inoculated)	66.7(54.8)	-	158.3	-
CD (P=0.05)	**3.7**	-	**34.3**	-

*Figures in parentheses are angular transformed values

DAS = Days after sowing; FYM = Farm yard manure; ST = Seed treatment;
SA= Soil application

Section - VI

TRANSFER OF CROP PROTECTION TECHNOLOGY AND CONCLUSIONS

Chapter 20

TRANSFER OF CROP PROTECTION TECHNOLOGY AND CONCLUSIONS

20.1. TRANSFER OF CROP PROTECTION TECHNOLOGY

The transfer of new crop protection technologies from research to practice is running unsatisfactorily in India. Many promising new technologies are not picked up by the growers and consequently the national targets for pesticide reduction are not met. There is scope for reduction in pesticide usage.

The constraint for wider adoption of eco-friendly crop protection technologies is transfer of technology. Crop protection technologies developed have not reached the small and marginal farmers. Unless these technologies are assessed in farmers' fields and refined to suit local conditions, the fruits of research will not benefit the farmers. Researchers and extension personnel should work hand in hand for successful transfer of crop protection technologies. Communication media such as radio, TV, audio and video cassettes, agri-portals, farmer's field schools and KVK's should be used for effective transfer of technologies.

Agromedical training, pest management workshops, adequate libraries, onsite demonstration projects, and crop protection research and extension and others will be required to develop the necessary knowledge bank and to implement vigorous, effective BIPM programmes. It is estimated that 50 per cent of the extremely high pest-caused losses in the developing world may be prevented through application of appropriate BIPM systems.

20.1.1. Crop Protection Technology Transfer Methods

The aim of transfer of technology programmes is to promote client oriented on-farm research and technology assessment, refinement and transfer through participatory approaches and by promoting the Institute-Village Linkage Programmes (IVLP). Transfer of technology involves organizing the field demonstrations, farmer's-scientist meetings, krishi melas, and other activities.

20.1.1.1. Institutional and Off Campus Training Programmes: Institutional and off campus training programmes on various aspects of crop protection should be organized at the HQ, regional stations and research centres of the SAU's/ ICAR Institutes to benefit farmers and extension personnel. Besides, training programmes are also conducted on selected topics on request from individuals and organizations for which training fee may be charged (Fig. 20.1).

Fig. 20.1. Training programme on crop protection

20.1.1.2. Front Line Demonstration Programmes: Front line demonstration programmes on selected technologies relating to BIPM to be organized in farmers' field for convincing the farmers about the viability of the technologies and to obtain proper feedback from farmers on the constraints in adoption of the recommended technologies (Fig. 20.2).

Fig. 20.2. Front line demonstration on crop protection

20.1.1.3. Utilization of Media and Other Extension Methods for Transfer of Technology: The following activities are to be regularly taken up to effectively utilize various media and methods of extension for effective TOT:

- **Publication of Extension Pamphlets, CD-ROM etc..** Extension literature in the form of books, leaflets, pamphlets, bulletins and articles etc. is prepared and printed. Extension literature are distributed to farmers to provide horticulture information related to crop protection (Fig. 20.3).

Fig. 20.3. Publication of books, bulletins on crop protection

Organization of Film Shows: Film and CD show of a recommended package of practices on crop protection plays an important role in transfer of technology among a group of farmers. After watching a particular technique/method and guidance of scientist, farmers can adopt that beneficial practice (Fig. 20.4).

Fig. 20.4. Organization of film shows

- **Organizing Kisan Melas:** The Kisan Melas should be organized to provide latest crop protection information, guide/motivate farmers to adopt new scientific and profitable practices. Farmers get benefited to know about latest eco-friendly technologies in crop protection.

Through the Kisan Melas, the farmers get interaction with scientists which helps for transfer of new crop protection technologies in different areas. Farmers get acquainted with latest and recent developments in crop protection sector (Fig. 20.5).

Fig. 20.5. Organization of Kisan Melas

- **Organizing Field Days:** Farmers-Scientist-Extension workers discussion is an important feature of Field Day. Field Days are arranged to demonstrate new crop protection technologies in front of a large manageable group of interested farmers. Through this activity farm experts, extension workers and farmers are involved and learn from each other (Fig. 20.6).

Fig. 20.6. Organization of field days

- **Organizing Campus Exhibitions:** Campus exhibitions should be organized on different themes for dissemination of new crop protection technologies, to increase their knowledge and interest about agriculture information/technique (Fig. 20.7).

Fig. 20.7. Organization of campus exhibitions on crop protection

- **Technology Dissemination through News Papers, Farm Journals and Radio/TV Talk Programmes:** The popular articles of scientists on crop protection in local languages are published in agriculture news papers and magazines.
- Radio and TV talk plays a major role in dissemination of crop protection and allied technologies among the farming community through mass communication. All the scientists of the SAU's/ICAR Institutes should prepare programmes on different crop protection topics and give answers to farmers questions in Q&A session. The programmes are telecasted on Doordarshan and TV channels and broadcasted on All India Radio.
- **Organizing Exposure Visit-Cum-Training Programmes for Farmers:** Institutes should arranges farmers meet from time to time for discussion to a group of farmers. Scientist replies are given to the queries of farmers about current issues, crop problems, or other agricultural activity.
- **Institute Visit by Farmers and Other Clients.** Farmers visit to Institute and its instructional farm, demonstration plots/units and information centre for acquiring knowledge about latest agricultural practices and to solve their queries. KVK Scientists also visit farmers field to solve their problems.
- **Production of video programmes on selected technologies.**
- **Farm advisory services.**
- **Participation in agricultural seminars.**

20.1.1.4. Cyber Extension Activities: The Information and Communication Technology (ICT) initiatives on the lines of web-based systems, interactive software and cyber extension activities should be implemented (Fig. 20.8). To place SAU's/ICAR Institutes in the arena of internet to communicate research result to the end users, websites should be developed and hosted under the URL. Farmers and extension personnel seek clarifications on queries through online query facility provided at the website.

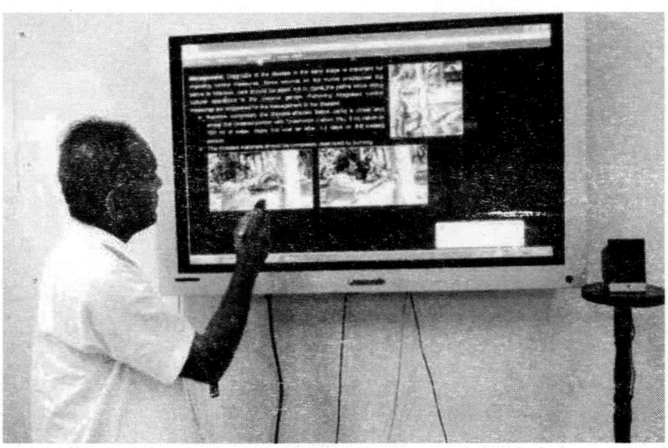

Fig. 20.8. Cyber extension activities on crop protection

ATIC also should utilize the advances in ICT in transfer of technology to farmers in single window concept. Kiosks, web based consultancies and e-mail and mobile based queries for technological clarification of farmers are some of the initiatives to be implemented through ATIC.

20.1.1.5. Software Development

- Interactive and auto executable soft wares should be developed on pest and disease management technologies for the benefit of farmers and extension workers.
- IPM soft wares should be developed to include description of pests, life cycle, symptoms (with photographs) and management methods (chemical, physical, cultural, chemical, biological, host resistance and IPM). A few video clippings could also be included.

20.1.1.6. Research-Extension-Farmer Interface Facilitated through Video Conferencing: Cyber extension activities should be launched utilizing the videoconferencing facility covering several states as part of strengthening the technology transfer programmes of the SAU's/ICAR Institutes in mandate crops. A group video conferencing system through ISDN can also be installed at the ATIC to facilitate interaction between various stakeholders for enhancing technology utilization. The videoconferencing facility should be effectively utilized for scheduling and implementing interface programmes at regular intervals involving various stake holders including researchers, extension personnel, farmers and entrepreneurs. It enables interaction of farmers at remote villages with the subject matter specialists of the SAU's/ICAR Institutes thereby reducing the time, effort and cost in transfer of know-how from lab to field (Fig. 20.9).

Fig. 20.9. Interface of research-extension-farmer facilitated through video conferencing

20.1.2. Extension Research

Besides organizing extension activities, research studies also should be undertaken on various aspects of technology generation, transfer and utilization as follows:

- Studies on knowledge and adoption of recommended technologies and feedback to research system.
- Action research projects on the performance of technologies in farmers' fields.
- Collection and documentation of indigenous technical knowledge.
- Studies on women empowerment through micro enterprises.
- Action research projects on group approach for enhancing income through effective technology integration.

Trade publications were identified by both researchers and producers as a good way to introduce a new technology to the producers, but limited in detailed information.

The internet was used the most by producers in remote locations. Internet usage, and its importance as a source of information, may increase as advances are made and more producers use computers.

In general, field days and workshops were considered by principal investigators to be one of the best methods for transferring information to the producers. Yet producers tended to think that they (especially workshops) were redundant and held at inopportune times. Many producers said that they carefully chose which field days or workshops to attend. One of the primary benefits of attending the workshops was the interaction with the other producers. Farmers receive the greatest satisfaction when they are able to share experiences with peers in group interaction.

Demonstrations located on a producers' farm were most favoured by producers, followed by field days in which producers conducted some of the presentations. Although there was great respect for university studies, concerns were often expressed that the small plots and conditions associated with research farms were not indicative of what they might encounter on their own farm. When the practice was tried on a producers' farm, they tended to believe what they saw.

Principal investigators found that on-farm demonstration projects were difficult to implement and monitor for research purposes. Often the researcher has limited control: the producer may decide to plant a different crop than originally planned or alter the proposed management procedure. Successful on-farm trial demonstrations are the result of a coordinated effort between researchers and producers. Research studies conducted on a strategically located producers' farm seemed to have the attention of all the neighbours and were highly effective in transferring information.

Before a new technology was adopted, almost all of the producers contacted someone else (even if located across the country) who was using the technology before they made the decision to adopt the new practice. This was especially true for those technologies that were costly or required a major shift in the farming operation. This extreme need to see the new practice in operation, or talk to someone who is using it, indicates that one of the most effective ways to speed up the technology transfer process is to use producers in demonstrations and field days. This is especially true for those technologies that are costly, complex, or require a major shift in the operation.

The qualitative study described here examined the technology transfer preferences of early adopters of sustainable agricultural practices, which may or may not represent the preferences of all producers (Table 20.1).

Table 20.1. Technology transfer methods used by principal investigators and producers in descending order

Rank	Principal Investigators	Hawaii Producers	California Producers	Wyoming Producers
1	Workshops/Field days	University & agricultural professionals	Peers	Trade publications
2	Periodicals/ Handouts	Peers	University professionals	University & agricultural professionals
3	University professionals	Trade publications	Trade publications	Internet
4	Internet	Workshops/Field days	Workshops/ Field days	Peers
5	Books	Internet	Internet	Workshops/ Conferences

20.2. CONCLUSIONS

Globalization driven by WTO is opening up fantastic opportunities for export of horticultural products and processed food from India. It is a revolution, which is taking place, and our farmers' will miss this golden opportunity if they are not equipped with the right crop protection technologies to produce horticultural products of international standards without pesticide residues. The challenge before the crop protection scientists is to prevent crop losses due to pests before and after harvest without harming the environment. There is a need to develop low input and eco-friendly crop protection technologies so as to be very competitive in the international market.

There is a need to revamp the extension system in SAU's/ICAR Institutes to bridge the gap between technology generation and technology dissemination, since 60 percent of farmers have no access to technology as revealed by the latest National Sample Survey Organization Report (Suryamurthy, 2005).

Padma Vibhushan Prof. M.S. Swaminathan, an eminent agricultural scientist of international repute, stated that "The ever-green revolution will be triggered by farming systems that can help produce more from the available land, water and labour resources without either ecological or social harm" (Swaminathan, 2000). Let us rededicate ourselves to achieve Prof. Swaminathan's dream of "ever-green revolution".

REFERENCES

Abu-Elamayem, M.M., Shehata, M.R.A., Tantaway, G.A., Ibrahim, I.K. and Schuman, M.A. 1978. Effect of CGA 12223 and benomyl against *Meloidogyne javanica* and *Rhizoctonia solani. Phytopathol. Zeist* **92**: 289-293.

Adams, S. 1997. Seein' red: colored mulch starves nematodes. *Agri. Res*. October. p. 18.

Ali, S.S. 1982. Occurrence of root-knot nematodes in cardamom plantations of Tamil Nadu. *Proc. of Fifth Symp. on Plantation Crops*, Kasaragod, Kerala, India, pp. 615-620.

Ali, S.S. 1984. Preliminary observations on the effect of some systemic nematicides and neem oil cakes in a cardamom field infected with root-knot nematodes (Abstr.). *Proc. of the Sixth Symposium on Plantation Crops*, Kottayam, Kerala, India, pp. 215-223.

Ali, S.S. 1986. Root-knot nematode problem in cardamom and its management. *Proc. of the Second Group Discussion on the Nematological Problems of Plantation Crops,* Central Coffee Res. Inst., Balehonnur, pp. 10-12.

Ali, S.S. and Koshy, P.K. 1982. Occurrence and distribution of root-knot nematodes in cardamom plantations of Kerala. *Nematol. Medit.* **10**: 107-110.

Ali, S.S. and Venugopal, M.N. 1992. *Nematol. Medit.* **20**: 65-66.

Ali, S.S. and Venugopal, M.N. 1993. *Curr. Nematol.* **4**: 19-24.

Amenduni, M., D'Amico, M., Colella, C. and Cirulli, M. 2004. Effect of Brassicaceae green manures and soil-solarization on corky root of tomato. *First Int. Biofumigation Symp.*, Florence, Italy.

Anandraj, M. and Eapen, S.J. 2003. Achievements in biological control of diseases of spice crops with antagonistic organisms at Indian Institute of Spices Research, Calicut. In *Current Status of Biological Control of Plant Diseases Using Antagonistic Organisms in India* (B. Ramanujam and R.J. Rabindra, eds.), pp. 189-215. Project Directorate of Biological Control, Bangalore.

Anitha, B. and Subramanian, S. 1998. Management of the reniform nematode *Rotylenchulus reniformis* in tomato. In *Nematology - Challenges & Opportunities in 21st Century* (Usha K. Mehta, ed.), pp. 249-250. Sugarcane Breeding Inst., Coimbatore.

Anitha, B. and Vadivelu, S. 1997. Management of root-knot nematode, *Meloidogyne hapla* on scented geranium, *Pelargonium graveolans*. *Indian J. Nematol.* **27**:123-125

Anon. 1993. Biennial Report (1991-93) of AICRP on Plant Parasitic Nematodes with Integrated Approach for their Control, Dept. of Nematology, Haryana Agri. Univ., Hisar, 80 pp.

Anon. 2005. Annual Report for 2003-04 of All India Coordinated Research Project on Spices, Indian Inst. of Spices, Calicut, Kerala, India.

Anon. 2012. Annual Report (2011-12), Indian Inst. of Hort. Res., Bangalore, pp. 23-24.

Anusuya, K. and Vadivelu, S. 2002. Mutualistic symbiosis of VAM fungi on plant growth promoting rhizobacteria (PGPR) on micro propagated carnation. *J. Microbiol. World* **4**: 105-107.

Athul Sandheep, R., Aju, K. and Asok Ans Jisha, M.S. 2012. Biocontrol of *Fusarium* wilt of vanilla (*Vanilla planifolia*) using combined inoculation of *Trichoderma* sp. and *Pseudomonas* sp. *Int. J. Pharm. Bio. Sci.* **3**(B): 706-716.

Baby, U.I. and Chandramouli. 1996. Biological antagonism of *Trichoderma* and *Gliocladium* spp. against certain primary rot pathogens of tea. *J. Plantn. Crops* **24**: 249-255.

Baghel, P.P.S. 1995. Nematode problems in citrus. In *Nematode Pest Management-An Appraisal of Eco-friendly Approach* (G. Swarup, D.R. Dasgupta and J.S. Gill, eds.), pp. 203-210. Nematol. Soc. of India, New Delhi, India.

Baghel, P.P.S. and Bhatti, D.S. 1983a. Evaluation of pesticides for the control of phytonematodes on citrus. *Third Nematol. Symp.*, Y.S. Parmar Univ. of Hort. & Forestry, Solan, India, pp. 38-39.

Baghel, P.P.S. and Bhatti, D.S. 1983b. Relative efficacy of nematicides for control of phytonematodes on grapevine varieties. *Third Nematol. Symp.,* Y.S. Parmar Univ. of Hort. & Forestry, Solan, p. 39.

Baghel, P.P.S., Bhatti, D.S. and Chauhan, K.S. 1980. *Haryana J. Hort. Sci.* **9**: 136-137.

Balakrishnan, P., Usman, N.M. and Sarma, Y.R. 1996. Management of rhizome rot disease of ginger by soil solarization. *J. Plantation Crops* **24** (Suppl.): 192-199.

Balasuriya, A. and Kalaichelvan, J. 2000. Is there potential in natural tea phylloplane microorganisms in the control of blister blight leaf disease of tea (*Camellia sinensis*)? *The Planter, Kaula Lumpur* **76**: 409-417.

Barber, C.A. 1901. *Department Land Records and Agriculture*. Madras Agri. Branch 2, Bull. No. 45, pp. 227-234.

Baskaran, P. and Kumar, A. 1980. Further studies on Dipel-insecticide combinations against the insect pests of brinjal. *Pesticides* **14**: 9-11.

Bélair G. and Coulombe, J. 2008. Green manure and biofumigation for root lesion nematode and Verticillium wilt management in strawberry production. *Third Int. Biofumigation Symp*., Canberra, Australia.

Bello, A. 1998. Biofumigation and integrated crop management. In *Alternatives to Methyl Bromide for the Southern European Countries* (A. Bello, J.A. Gonzalez, M. Arias and R. Rodriguez-Kabana, eds.), pp. 99-126. Phytoma-Espania, DG XI EU, CSIC, Valencia, Spain.

Bello, A., Escuer, M., Sanz, R., López-Pérez, J.A. and Guirao, P. 1997. In: *Posibilidad de Alternativas Viables al Bromuro de Metilo en Pimiento de Invernadero* (A. López and J.A. Mora, eds.), pp. 67-108. Consejería de Medio Ambiente, Agricultura y Agua de Murcia, Spain.

Bello, A., López-Pérez, J.A., Garcia-Alvarez, A., Sanz, R. And Lacasa, A. 2004. Biofumigation and nematode control in the Mediterranean region. In *Proc. of the*

Fourth Int. Cong. of Nematol. (R.C. Cook and D.J. Hunt, eds.), pp. 133-149. Nematology Monographs and Perspectives, Vol. 2, Brill, Leiden and Boston.

Benbrook, C.M. 1996. *Pest Management at the Crossroads*. Consumers Union, Yonkers, NY. 272 pp.

Besri, M. and Drame, A. 1982. Control of *Fusarium oxysporum* f. sp. *lycopersici* and *Verticillium dahliae* by soil solarization. Proc. of the First Meeting of the Arab Soc. for Plant Prot., Amman, p. 113.

Bhagawati, B., Choudhury, B.N and Sinha, A.K. 2009. Management of *Meloidogyne incognita-Rhizoctonia solani* complex on okra through bioagents. *Indian J. Nematol.* **39**: 156-161.

Bhagawati, B., Das, B.C. and Sinha, A.K. 2007. Interaction of *Meloidogyne incognita* and *Rhizoctonia solani* on okra. *Ann. Plant Prot. Sci.* **15**: 533-535.

Bhagawati, B., Goswami, B.K. and Singh, C.S. 2000. Management of disease complex of tomato caused by *M. incognita* and *Fusarium oxysporum* f. sp. *lycopersici* through bioagents. *Indian J. Nematol.* **30**: 16-22.

Bharadwaj, L.N. and Sharma, S.K. 2000. Y.S. Parmar University of Horticulture and Forestry, Solan, Himachal Pradesh. In Progress Report for the Period Sept. 1997 - Aug. 2000 of National Network Project on Phytophthora Diseases of Horticultural Crops (S.R. Sarma, ed.), pp. 26-40. Indian Inst. of Spices Research, Calicut, Kerala.

Bhatti, D.S. and Jain, R.K. 1977. Estimation of loss in okra, tomato and brinjal yield due to *Meloidogyne incognita. Indian J. Nematol.* **7**: 37-41.

Bird, G.W. 1981. Integrated nematode management for plant protection. In *Plant Parasitic Nematodes, Vol. 3* (B.M. Zuckerman and R.A. Rohde, eds.), pp. 355-375. Academic Press, New York.

Blake, C.D. 1966. *Nematologica* **12**: 129-132.

Boby, B.U. and Bagyaraj, D.J. 2003. Biological control of root rot of *Coleus forskohlii* Briq. using microbial inoculants. *World J. Microbiol. Biotechnol.* **19**: 175-180.

Bora, P. and Bora, L.C. 2009. Bioformulation of antagonists for management of bacterial wilt of ginger. *Int. Conf. on Hort.*, Bangalore.

Borah, A. and Phukan, P.N. 2000. Effect of VAM fungus, *Glomus fasciculatum* and root-knot nematode, *Meloidogyne incognita* on brinjal. *J. Agri. Sci. Soc. North-Eastern India* **13**: 212-214.

Borkakaty, D. 1993. Management of Root-knot Nematode, Meloidogyne incognita (Kofoid and White, 1919) Chitwood, 1949 on Brinjal by Integrated Approach. M. Sc. (Agri.) Thesis, Assam Agri. Univ., Jorhat.

Burchill, R.T. and Cook, R.T.A. 1971. The interaction of urea and microorganisms in suppressing development of perithecia of *Venturia inequalis* (Cke.) Wint. In *Ecology of Leaf Surface Microorganisms* (T.F. Preace and C.H. Dickensen, eds.), pp. 471-483. Academic Press, New York.

Butler, E.J. 1906. The wilt disease of pigeon-pea and pepper. *Agric. J. India* **1**: 25-36.

Calvo, J., Calvente, V., De Orellano, M.E., Benuzzi, D. And De Tosetti, M.I.S. 2003. Improvement in the biocontrol of post-harvest diseases of apples with the use of yeast mixtures. *BioControl* **48**: 579-593.

Cartia, G. and Greco, N. 1987. Effetti della solarizzazione del suolo su una coltura di peperone in serra. *Colture Protette* **16**(5): 61-65.

Cartia, G., Greco, N. and Cipriano, T. 1989. Effect of soil solarization and fumigants on soil-borne pathogens of pepper in greenhouse. *Acta Hort.* **255**: 111-116.

Cartia, G., Greco, N. and Cirvilleri, G. 1988. Solarizzazione e bromuro di mettle nella difesa dai parassiti del pomodoro in ambiente protetto. *Proc. Giornate Fitopatologiche*, Lecce, Italy. **1**: 437-488.

Chaaya, M.K., Rao, M.S. and Ramachandran, N. 2010. Biomanagement of disease complex in okra (*Abelmoschus esculentus* L. Moench) using microbial antagonists. *National Conf. on Innovations in Nematol. Res. for Agri. Sustainability – Challenges and A Roadmap Ahead*, Tamil Nadu Agri. Univ., Coimbatore, pp. 96-97.

Chaitali, L., Singh, S. and Goswami, B.K. 2003. Effect of cakes with *Trichoderma viride* for the management of disease complex caused by *Rhizoctonia bataticola* and *Meloidogyne incognita* on okra. *Ann. Plant Prot. Sci.* **11**: 178-180.

Chakraborti, S. 2000. Management approaches for root knot nematode in pointed gourd. *Indian J. Nematol.* **30**:136-140.

Chakraborty, B.N., Chakraborty, U., Das, G. and Das, S.K. 1994. Phyllosphere micro-flora of tea and their interaction with *Glomerella cingulata* the causal agent of brown blight disease. *Tea* **15**: 27-34.

Chakravarti, B.P., Porwal, S. and Rangarajan, M. 1966. Studies on citrus canker in Rajasthan. I. Disease incidence and survival of the pathogen. *Labdev. J. Sci. Tech.* **4**: 262-265.

Chandawani, G.H. and Reddy, T.S.N. 1967. The host range of root-knot nematode, *Meloidogyne javanica* in tobacco nurseries at Rajahmundry, Andhra Pradesh. *Indian Phytopath.* **20**: 383-384.

Chandel, S.S., Khanna, A.S. and Khan, M.L. 1997. Faunistic studies on nematodes associated with gladiolus. *J. Agric. Res.* **18**: 44-47.

Chand-Goyal, T. and Spotts, R.A. 1996a. Post-harvest biological control of blue mold of apple and brown rot of sweet cherry by natural saprophytic yeasts alone or in combination with low doses of fungicides. *Biol. Control* **6**: 252-259.

Chand-Goyal, T. and Spotts, R.A. 1996b. Biological control of postharvest diseases of apple and pear under semi-commercial and commercial conditions using three saprophytic yeasts. *Biol. Control* **10**: 199-206

Chandrashekara, K.N. and Prasannakumar, M.K. 2010. New host plants for *Ralstonia solanacearum* from India. *New Dis. Reports* **22**: 6.

Channabasappa, B.S., Krishnappa, K. and Reddy, B.M.R. 1995. Utilization of ecofriendly biological agents and biocomponents in the integrated management of *Radopholus similis* on banana. *Natl. Symp. on Nematode problems of India – An Appraisal of the Nematode Management with Ecofriendly Approaches and Biocomponents,* Indian Agri. Res. Inst., New Delhi.

Charles, J.S.K., Venkitesan, T.S., Thomas, Y. and Varkey, P.A. 1985. Correlation of plant growth components to bunch weight in banana infested with burrowing nematode, *Radopholus similis* (Cobb). *Indian J. Nematol.* **15** : 186-190.

Chattopadhyaya, C. and Sen, B. 1996. Integrated management of *Fusarium* wilt of muskmelon caused by *Fusarium oxysporum*. *Indian J. Mycol. Pl. Path.* **26**: 162-170.

Chaudhari, S. and Ramakrishnan, N. 1980. Field efficacy of baculovirus and its combination with sub-lethal dose of DDT and endosulfan on cauliflower against tobacco caterpillar *Spodoptera litura* (Fabricus). *Indian J. Ent.* **42**: 592-596.

Chhabra, H.K. and Sharma, J.K. 1981. Combined effect of *Meloidogyne incognita* and *Rhizoctonia bataticola* on pre-emergence damping-off of okra and brinjal. *Sci. Cult.* **47**: 256-257.

Chitwood, B.G. and Toung, M.C. 1960. Host parasite interactions of the Asiatic pyroid citrus nematode. *Plant Dis. Reptr.* **48**: 848-854.

Chona, B.L., Sethi, C.L., Swarup, G. and Gill, J.S. 1965. Citrus nematode (*Tylenchulus semipenetrans* Cobb, 1913) associated with die-back disease of citrus. *Indian J. Hort.* **22**: 370-373.

Cooksey, D.A. 1988. Reduction in infection by *Pseudomonas syringae* pv. *tomato* on tomato using a nonpathogenic, copper-resistant strain combined with a copper bactericide. *Phytopathology* **78**: 601-603.

Couch, G.J. 1994. The use of growing degree days and plant phenology in scheduling pest management activities. *Yankee Nursery Quarterly*. Fall. p. 12-17.

Curto, G., Lazzeri, L., Dallavalle, E. and Santi, R. 2008. Management of *Meloidogyne incognita* (Kofoid *et* White) Chitw. in organic horticulture. *Third Int. Biofumigation Symp.*, Canberra, Australia.

Dabbas, M.R., Singh, D.P., Yadav, J.R. and Katiyar, R.P. 2009. Management of rhizoctonia root rot of cauliflower through IDM practice. *Int. Conf. on Hort.*, Bangalore.

Darekar, K.S. and Mahse, N.L. 1988. Assessment of yield losses due to root-knot nematode, *Meloidogyne incognita* Race 3 in tomato, brinjal and bitter gourd. *Int. Nematol. Network Newsl.* **5**(4): 7-9.

Darekar, K.S. and Patil, N.G. 1985. *Journal of Maharashtra Agricultural Universities* **10**: 104.

Das, N. and Sinha, A.K. 2005. Integrated management of root-knot nematode (*Meloidogyne incognita*) on okra (*Abelmoschus esculentus* (L.) Moench). *Indian J. Nematol.* **35**: 175-182.

Devapriyaga, R., Jonathan, E.I., Sankarimeena, K. and Kavitha, P.G. 2012. Bioefficacy of *Pseudomonas* and *Bacillus* isolates against root-knot nematode, *Meloidogyne incognita* in black pepper cv. Paniyur 1. *Indian J. Nematol.* **42**: 57-65.

Devi, Geetanjali. 1993. Pathogenicity, Crop Loss Assessment and Management of *Meloidogyne incognita* (Kofoid and White, 1919) Chitwood, 1949 on Carrot (*Dacus carota* L.). M. Sc. (Agri.) thesis, Assam Agri. Univ., Jorhat.

Devi, T.P. and Goswami, B.K. 1992. Effect of VA mycorrhiza on the disease incidence due to *Macrophomina phaseolina* and *Meloidogyne incognita* on cowpea. *Ann. Agric. Res.* **13**: 253-265.

Dey, R.K. and Muhkopadhyay, A.N. 1994. Biological control of tomato damping-off by *Gliocladium virens*. *J. Biol. Control* **8**: 34-40.

Dohroo, N.P. 1995. Integrated management of yellows of ginger. *Indian Phytopath.* **48**: 90-92.

Droby, S., Wisniewski, M., El-Ghaouth, A. and Wilson, C. 2003. Influence of food additives on the control of post-harvest rots of apple and peach and efficacy of the yeast-based biocontrol product Aspire. *Postharvest Biol. Technol.* **27**: 127-135.

Dutt, R. and Bhatti, D.S. 1986. Determination of effective doses and time of application of nematicides and castor leaves for controlling *Meloidogyne javanica* in tomato. *Indian J. Nematol.* **16**: 8-11.

Eapen, S.J. 1987. Plant parasitic nematode problem in small cardamom. *Proc. of the Third Group Discussion on the Nematological Problems of Plantation Crops*, Sugarcane Breeding Institute, Coimbatore, pp. 43-44.

Eapen, S.J. and Ramana, K.V. 1996. Biological control of plant nematodes of spices. In *Biological Control in Spices* (M. Anandraj and K.V. Peter, eds.), pp. 20-22. Indian Inst. of Spices Res., Calicut.

Eapen, S.J. and Venugopal, M.N. 1995. Field evaluation of *Paecilomyces lilacinus* and *Trichoderma* spp. in cardamom nurseries for the control of root-knot nematodes and rhizome rot disease. *National Symp. on Nematode Problems of India – An Appraisal of the Nematode Mangmt. with Eco-friendly Approaches and Biocomponents,* Indian Agri. Res. Inst., New Delhi.

Eckert, J.W. and Eaks, I.L. 1989. Post-harvest disorders and diseases of citrus fruits. In *The Citrus Industry* Vol. 5 (W. Reuther, E.C. Calavan and G.E. Carman, eds.), pp. 179-250. Univ. of Calif. Press, Oakland, USA.

Elad, Y., Katan, J. and Chet, I. 1980. Physical, biological, and chemical control integrated for soil borne diseases in potatoes. *Phytopathology* **70**: 418-422.

Elad, Y., Zimand, G., Zaqs, Y., Zuriel, S. and Chet, I. 1993. Use of *Trichoderma harzianum* in combination or alternation with fungicides to control cucumber grey mould (*Botrytis cinerea*) under commercial greenhouse conditions. *Plan Pathol.* **42**: 324-332.

El-Ghaouth, A., Smilanick, J.L., Brown, G.E., Ippolito, A., Wisniewski, M. and Wilson, C.L. 2000. Application of *Candida saitoana* and glycolchitosan for the control of postharvest diseases of apple and citrus fruits under semi-commercial conditions. *Plant Dis.* **84**: 243-248.

Ganga Visalakshy, P.N. and Krishnamoorthy, A. 2009. Development of bio-intensive integrated management for brinjal shoot and fruit borer, *Leucinodes orbonalis* Guene. *Int. Conf. on Hort.*, Bangalore.

Gangopadhyay, S. and Ram Gopal. 2010. Evaluation of *Trichoderma* spp. along with farm yard manure for the management of *Fusarium* wilt of cumin (*Cuminum cyminum* L.). *J. Spices and Aromatic Crops* **19**: 57–60.

Ganguli, R.N., Singh, V.V., Dixit, S.A. and Kowshik, U.K. 1997. Efficacy of NPV (nuclear polyhedrosis virus) and endosulfan against tomato fruit borer, *Heliothis armigera. Curr. Res.* **26**: 210212.

Gardner, P.D., Eckert, J.W., Baritelle, J.L. and Bancroft, M.N. 1986. Management strategies for control of *Pencillium* decay in lemon packing houses: Economic benefits. *Crop Prot.* **5**: 26-32.

Girija Ganeshan. 1999. Biological control of collar rot disease of French bean using *Glomus mosseae* and *Trichoderma viride*. *Int. Symp. on Biol. Control Agents in Crop and Animal Protection,* Univ. of Wales, Swansea, U.K.

Gopalakrishnan, C. and Ajith Kumar, K. 2006. Increasing the field efficacy of *Pseudomonas fluorescens* in the management of bacterial wilt in tomato, *Lycopersicon esculentum* Mill. *National Symp. on Improving Input Use Efficiency in Hort.*, Indian Inst. of Hort. Res., Bangalore, pp. 197-198.

Gopinatha, K.V., Nanje Gowda, D.N. and Nagesh, M. 2002. Management of root-knot nematode *Meloidogyne incognita* on tomato using bioagent *Verticillium chlamydosporium*, neem cake, marigold and carbofuran. *Indian J. Nematol.* **32**: 179-181.

Goswami, B.K., Pandey, R.K., Goswami, J. and Tiwari, D.D. 2007. Management of disease complex caused by root knot nematode and root wilt fungus on pigeon pea through soil organically enriched with Vesicular Arbuscular Mycorrhiza, karanj (*Pongamia pinnata*) oilseed cake and farmyard manure. *J. Environ. Sci. Health* **42**: 899-904.

Goswami, B.K. and Singh, S. 2002. Effect of *Aspergillus niger* and *Cladosporium oxysporum* on plant vigour and root-knot nematode, *Meloidogyne incognita* multiplication on egg plant. *Indian J. Nematol.* **32**: 94-96.

Goswami, B.K., Uma Rao and Singh, S. 1998. Potentiality of some fungal bioagents against root-knot nematode, *Meloidogyne incognita* infecting tomato. *Proc. of First National Symp. on Pest Management in Horticultural Crops*, Bangalore, pp. 304-307.

Govindaiah, Dandin, S.B., Philip, T. and Datia, R.K. 1990. Effect of marigold (*Tagetes patula*) intercropping against *Meloidogyne incognita* infecting mulberry. *Indian J. Nematol.* **20**: 96-99.

Govindaiah, Dandin, S.B. and Sharma, D.D. 1991. Pathogenicity and avoidable leaf yield loss due to *Meloidogyne incognita* in mulberry (*Morus alba* L.). *Indian J. Nematol.* **21**: 52-57.

Gupta, G.K. 1985. Apple Scab (*Venturia inaequalis*). R. Merck (India) Ltd., Bombay, 36 pp.

Gupta, P.R. and Raja Ram Mohan Babu, B. 1998. Management of *Helicoverpa armigera* on tomato with *Trichogramma pretiosum* and *Bacillus thuringiensis* var. *kurstaki*. In *Advances in Pest Management of Horticultural Crops* (P. Parvatha Reddy, N.K. Krishna Kumar and A. Verghese, eds.), pp. 75-80. Association for Advancement of Pest Management in Horticultural Ecosystems, Bangalore.

Gupta, V.K. and Sharma, K. 2004. Integration of chemicals and biocontrol agents for managing white root rot of apple. *Acta Hort.* **635**:141-149

Gupta, V.P. 2000. Biological control of mulberry diseases – The current status. In *Diseases of Plantation Crops, Spices, Betelvine and Mulberry* (M.K. Dasgupta, ed.), pp. 145-148. Visva-Bharati Publications, Srineketan, India.

Gupta, V.P. 2001. Diseases of mulberry and their management. In *Plant Pathology* (P.C Trivedi, ed.), pp. 130-164. Pointer Publishers, Jaipur, India.

Gupta, V.P., Mishra, R.K. and Sarkar, A. 1998. Management of nursery diseases of mulberry through integrated approach. *7th Int. Cong. of Pl. Path,* Edinburgh, Scotland, U.K., Vol. 3, Abstr. 3.7.80.

Gupta, V.P., Sharma, D.D., Rekha, M. and Chandrasekar, D.S. 1999. Integration of *Trichoderma pseudokoningii* with agrochemicals for disease management and plant development in mulberry. *Archiv fur Phytopasthologie und Pflanzenschutz.* **32**: 521-529.

Hadar, Y., Chet, I. and Henis, Y. 1979. Biological control of *Rhizoctonia solani* damping -off with wheat bran culture of *Trichoderma harzianum*. *Phytopathology* **69**: 64-68.

Haider, M.G., Nath, R.P., Thakur, S.C. and Ojha, K.L. 1987. Interaction of *Meloidogyne incognita* and *Pseudomonas solanacearum* on tomato plants. *Indian J. Nematol.* **17**: 174-176.

Harman, G.E., Taylor, A.G. and Stasz, T.E. 1989. Combining effective strains of *Trichoderma harzianum* and solid matrix priming to improve biological seed treatments. *Plant Dis.* **73**: 631-637.

Hasan, A. 1985a. Synergistic interaction between *Pythium aphanidermatum* and *Rhizoctonia solani* with *Meloidogyne incognita* on chilli. *Nematologica* **31**: 210-217.

Hasan, A. 1985b. Breaking resistance in chilli to root-knot nematode by fungal pathogens. *Nematologica* **31**: 210-217.

Hasan, A. and Nehal Khan, M. 2004. Interaction of root-knot nematode, VAM fungus and neem cake in amended soils on gladiolus. *Nematol. Medit.* **32**: 95-99.

Hasan, N. and Jain, R.K. 1992. Field application of *Paecilomyces lilacinus* in combination with certain organic matters for controlling *Meloidogyne incognita* infecting cowpea followed by lucerne. *First Afro-Asian Nematol. Symp.*, Aligarh Muslim Univ., Aligarh, pp. 10-11.

Haseeb, A. 2003. Management of root-knot nematode and wilt complex and their interaction in vegetable crops using organic amendments and biocontrol agents. Report of the UPCAR Sponsored Project, Lucknow, 52 pp.

Haseeb, A. and Pandey, R. 1989. Observation on *Meloidogyne* spp. infesting Japanese mint - New disease records. *Nematropica* **19**: 93-97.

Haseeb, A. and Pandey, R. 1990. Root-knot nematodes - a constraint to cultivation of Davana, *Artemisia pallens*. *Trop. Pest Mangmt.* **36**: 317-319.

Haseeb, A. and Shukla, P.K. 2005. Wilt disease complex of pigeonpea and its management. In *Plant Diseases: Biocontrol Management* (S. Nehra, ed.), pp. 84-96. Avishkar Publishers & Distributors, Jaipur.

Haseeb, A., Shukla, P.K., Kumar, V. and Khan, R.U. 2004. Comparative efficacy of biocontrol agents, neem seed powder and pesticides against *Meloidogyne incognita* and *Fusarium oxysporum* on brinjal. *National Symp. on Paradigms in Nematol. Res. for Biodynamic Farming*, Univ. of Agri. Sci., Bangalore, p. 80.

Hashim, I. 1990. Possible integration of *Trichoderma* with fungicides for the control of white root disease of rubber. Root diseases of *H. brasiliensis*. *Proc. of IRRDB Symp.*, Kunning, Chiang, pp. 1-8.

Hassan, M.G. and Sobita Devi 2004. Efficacy of bioagents against root-knot nematode, (*Meloidogyne incognita*) of tomato. *National Symp. on Paradigms in Nematol. Res. for Biodynamic Farming*, Univ. of Agri. Sci., Bangalore, p. 88.

Heald, C.M. and Robinson, A.F. 1987. Effect of soil solarization on *Rotylenchulus reniformis* in the lower Rio Grande Valley of Texas. *J. Nematol.* **19**: 93-103.

Henis, Y., Ghaffar, A. and Barker, R. 1978. Integrated control of *Rhizoctonia solani* of radish: Effect of successive plantings, PCNB, and *Trichoderma harzianum* on pathogen and disease. *Phytopathology* **68**: 900-907.

Hussain, Z. and Bora, B.C. 2008. Integrated management of *Meloidogyne incognita* and *Ralstonia solanacearum* complex in brinjal. *Indian J. Nematol.* **38**: 159-164.

Ichniohe, M. 1975. *Japanese J. Nematol.* **5**: 36-40.

Jadon, K.S. 2009. Eco-friendly management of brinjal collar rot caused by *Sclerotium rolfsii* Sacc. *Indian Phytopath.* **62**: 345-347.

Jahagirdar, S. and Siddaramaiah, A.L. 2000. Association of pathogens in disease complex of black pepper (*Piper nigrum* L.) in Karnataka. *Karnataka J. Agri. Sci.* **31**: 757-760.

Jahagirdar, S. and Siddaramaiah, A.L. 2003. Etiology and management of foot rot of black pepper (*Piper nigrum*). *National Symp. on Plant Pathogens Diversity in Relation to Plant Health*, Hyderabad.

Jain, R.K. and Bhatti, D.S. 1985. Effect of summer ploughing alone and in combination with some other effective practices on the incidence of root-knot nematode (*Meloidogyne javanica*) on tomato (Cultivar HS 101). *Indian J. Nematol.* **15**: 262.

Jain, R.K. and Gupta, D.C. 1991. Integrated management of root-knot nematode, *Meloidogyne javanica* infecting *Solanum melongena*. *Afro-Asian J. Nematol.* **1**: 156-160.

Jain, R.K. and Hasan, N. 1995. Nematode problems in forage crops. In *Nematode Pest Management-An Eco-friendly Approach* (Gopal Swarup *et al.*, eds.), pp. 217-227. Nematol. Soc. of India, Indian Agri. Res. Inst., New Delhi.

Jain, R.K., Paruthi, I.J., Gupta, D.C. and Dhankar, B.S. 1986. Appraisal of losses due to root-knot nematode, *Meloidogyne javanica* on okra under field conditions. *Trop. Pest Mangmt.* **32**: 341-342.

Jakobsen, I. 1999. Effects of *Pseudomonas fluorescens* DF 57 on growth and P uptake of two arbuscular mycorrhizal fungi in symbiosis with cucumber. *Mycorrhiza* **8**: 319-334.

Janisiewicz, W.J. 1988. Biocontrol of postharvest diseases of apples with antagonist mixtures. *Phytopathology* **78**: 194-198.

Janisiewicz, W.J. and Bors, B. 1995. Development of a microbial community of bacterial and yeast antagonists to control wound-invading postharvest pathogens of fruits. *Appl. Environ. Microbiol.* **61**: 3261-3267.

Jatala, P. 1985. Biological control of nematodes. In *Advanced Treatise on* Meloidogyne. *Biology and Control, Vol. 1* (J.N. Sasser and C.C. Carter, eds.), pp. 303-308. North Carolina State Univ. Graphics, Raleigh.

Jayanthi, P.D.K. and Verghese, A. 1998. Hourly trap catch of the mamngo fruit fly (*Bactrocera dorsalis* Hendel) using methyl eugenol bottle trap. *Insect Envt.* **4**(2): 60.

Jayaraman, V., Rajendran, G. and Muthukrishnan, T.S. 1975. Occurrence of root-knot nematodes in *Polianthes tuberosa* L. in Tamil Nadu. *Indian J. Nematol.* **5**: 101-102.

Jenkins, S.F. and Averre, C.W. 1986. Problems and progress in integrated control of southern blight of vegetables. *Plant Dis*. **70**: 614-619.

Jenkins, W.R. and Coursen, B.W. 1957. The effect of root-knot nematodes, *Meloidogyne incognita acrita* and *M. hapla*, on *Fusarium* wilt of tomato. *Pl. Dis. Reptr.* **41**: 182-186.

Jonathan, E.I. and Cannayane, I. 2002. Field application of biocontrol agents for management of spiral nematode, *Helicotylenchus multicinctus* in banana (Abstr.). *Global Conf. Banana and Plantain,* Bangalore, p. 193.

Jonathan, E.I., Nagalakshmi, S. and Padmanabhan, D. 1990. Estimation of yield losses in betel vine due to *Meloidogyne incognita. Int. Nematol. Network Newsl.* **7** (4): 26.

Jonathan, E.I. and Rajendran, G. 2000. Assessment of avoidable yield loss in banana due to root-knot nematode, *Meloidogyne incognita*. *Indian J. Nematol.* **30**: 162-164.

Jonathan, E.I., Umamaheswari, R. and Bommaraju, P. 2006. Bioefficacy of native plant growth promoting rhizobacteria against *Meloidogyne incognita* and *Phytophthora capsici. Indian J. Nematol.* **36**: 230-233.

Jones, F.C.E. 1969. Integrated control of the potato cyst nematode. *Proc. Fifth British Insecticide Fungicide Conf.* **3**: 646-656.

Joseph Thomas, Suseela Bhai, R. Vijayan, A.K and Naidu, R. 1993. Evaluation of antagonists and their efficacy in managing rot disease of small cardamom. *J. Biol. Control* **7**: 29-36.

Kabayashi, N. 1989a. Charcoal compost fixed antagonistic microorganisms and its effect to soil-borne diseases (Abstr.). *Ann. Phytopath. Soc. Japan* **55**.

Kabayashi, N. 1989b. Suppression of *Rhizoctonia* and *Pythium* damping-off of cucumber by microorganisms in charcoal and VAM Fungi. In *Research Advances in Microbial Ecology* (H. Hattori *et al*., eds.), pp. 242-246. Japan Scientific Press, Japan.

Kalita, S. and Bora, B.C. 2006. Effect of nematicides and organic amendments in the management of *Meloidogyne incognita* in tea nursery. *Indian J. Nematol.* **36**: 148-149.

Kapil, R. and Kapoor, A.S. 2002. Management of white rot (*Sclerotinia sclerotiorum*) using *Trichoderma* spp. and biopesticides. *Annual Meeting of the Indian Phytopath. Soc. (North Zone)*, Punjab Agri. Univ., Ludhiana.

Kapoor, A.S. and Sharma, C.L. 2000. Integrated management of white rot of pea. *IPS Int. Golden Jubilee Conf. on Integrated Plant Dis. Mangmt. for Sustainable Agri.,* Indian Agri. Res. Inst., New Delhi, Vol. II, p. 712.

Kapoor, V.C. (ed.). 1993. *Indian Fruit Flies (Insecta: Diptera: Tephritidae)*. Oxford & IBH Publishing Co. Pvt. Ltd., New Delhi, 228 pp.

Karthikeyan, G., Duraisamy, S., Sivakumar, C.V. and Duraisamy, S. 1999. Biological control of *Pythium aphanidermatum*, *Meloidogyne incognita* disease complex in chilli with organic amendments. *Madras Agric. J.* **86**: 320-323.

Kaur, D.J. 1987. Studies on Nematodes Associated with Ginger (*Zingiber officinale* Rosc.) in Himachal Pradesh. Ph. D. thesis, Y.S. Parmar Univ. of Hort. and Forestry, Solan, India, 198 pp.

Kaur, P. 1999. *Integrated Management Studies on Pea Diseases*. M. Sc. thesis, Y.S. Parmar Uni. of Hort. and Forestry, Solan, 75 pp.

Khaderkhan, H., Nataraju, M.S. and Nagaraja, G.N. 1998. Economics of IPM in tomato. In: *Advances in IPM for Horticultural Crops* (P. Parvatha Reddy, N.K. Krishna Kumar and A Verghese, eds), pp 151-152. Association for Advancement of Pest Management in Horticultural Ecosystems, Bangalore.

Khan, M.R. 2001. White-tip nematode, Aphelenchoides besseyi in rice-tuberose cropping system. In: Proc. National Seminar on Frontiers of Crop Management (S.K. Mukhopadhyay, D.C. Ghosh and G.C. De, eds.), pp. 142-144. Visva-Bharati, Sriniketan, India.

Khan, M.R. 2006. Managing foliar nematode-a threat for tuberose cultivation. Indian Hort. 51(1):17.

Khan, M.R., Bhattacharya, I., Chattopadhyay, S.B. and Ghosh, S. 2009. Integrated management of root-knot (*Meloidogyne incognita*) and other nematodes in pointed gourd (*Trichosanthes dioica*). *Indian J. Nematol.* **39**: 25-28.

Khan, M.R. and Ghadipur, M.H. 2004. Management of root-knot disease of some Solanaceous vegetables by soil application of fly ash. *National Symp. on Paradigms in Nematol. Res. for Biodynamic Farming,* Univ. of Agri. Sci., Bangalore, p. 92.

Khan, M.R. and Pal, A.K. 2001. Plant parasitic nematodes associated with tuberose (*Polianthes tuberosa* L.) in West Bengal. *Ann. Pl. Prot. Sci.* 9: 357-359.

Khan, M.R., Shit, S., Pal, A.K. and Biswas, A.B. 2004. Studies on tuberose foliar nematode, *Aphelenchoides besseyi* in West Bengal. *National Symp. on Paradigms in Nematol. Res. for Biodynamic Farming*, Univ. of Agri. Sci., Bangalore, p. 135-136.

Khan, R.M. and Parvatha Reddy, P. 1992. Nematode pests of ornamental crops. In *Nematode Pests of Crops* (D.S. Bhatti and R.K. Walia, eds.), pp. 250-257. CBS Publishers & Distributors, Delhi, 381 pp.

Khan, R.M. and Parvatha Reddy, P. 1994. Nematode problems of ornamental crops and their management. In *Floriculture-Technology, Trades and Trends* (J. Prakash and K.R. Bhandary, eds.), pp. 468-472. Oxford & IBH Publishing Co. Pvt. Ltd., New Delhi, 667 pp.

Khan, T.A. 1991. Studies on a Disease Complex of Papaya caused by *Meloidogyne incognita* and *Fusarium solani*. Final Tech. Rept., DST Project, Aligarh Muslim Univ., Aligarh, 47 pp.

Khan, T.A. and Hussain, S.I. 1990. Effect of interaction of various inoculum levels of *Meloidogyne incognita* and *Fusarium solani* on seedling emergence and post-emergence damping-off of papaya seedlings. *New Agric.* **1**: 17-20.

Khan, T.A., Khan, S.T., Fazal, M. and Siddiqui, Z.A. 1997. Biological control of *Meloidogyne incognita* and *Fusarium solani* disease complex in papaya using *Paecilomyces lilacinus* and *Trichoderma harzianum*. *Indian J. Nematol.* **7**: 127-132.

Kishore, R., Kamalwanshi, R.S. and Pandey, M. 2005. Effect of root-knot nematode with *Fusarium solani* on damping-off of papaya. *Indian J. Nematol.* **35**: 96-97.

Koshy, P.K. 1986. The burrowing nematode, *Radopholus similis* (Cobb, 1893) Thorne, 1949. In *Plant Parasitic Nematodes of India-Problems and Progress.* (G. Swarup and D.R. Dasgupta, eds.), pp. 223-248. Indian Agri. Res. Inst., New Delhi, India.

Koshy, P.K. and Geetha, S.M. 1992. Nematode pests of palms and cocoa. In *Nematode Pests of Crops* (D.S. Bhatti and R.K. Walia, eds.), pp. 214-227. CBS Publishers & Distributors, Delhi.

Koshy, P.K., Nair, R.C.P., Sosamma, V.K. and Sundararaju, P. 1976. On the incidence of root-knot nematode in cardamom nurseries. *Indian J. Nematol.* **6**: 174-175.

Koshy, P.K. and Sosamma, V.K. 1978. Studies on the population fluctuations of *Radopholus similis* in coconut and areca nut roots. *Indian Phytopath.* **31**: 180-185.

Koshy, P.K. and Sosamma, V.K. 1987. Pathogenicity of *Radopholus similis* on coconut (*Cocos nucifera* L.) seedlings under greenhouse and field conditions. *Indian J. Nematol.* **17**: 108-118.

Koshy, P.K., Sundararaju, P. and Sosamma, V.K. 1978. Occurrence and distribution of *Radopholus similis* (Cobb, 1893) Thorne, 1949 in South India. *Indian J. Nematol.* **8**: 49-58.

Krishna Moorthy, P.N. and Krishna Kumar, P.N. 2002. Advances in the use of botanicals for the IPM of major vegetable pests. *Proc. of the Int. Conf. on Vegetables,* Bangalore. Dr. Prem Nath Agri. Sci. Foundn., Bangalore, pp. 262-272.

Krishna Moorthy, P.N., Krishna Kumar, N.K., Girija, G., Varalakshmi, B., Prabhakar, M. 2003. *Integrated Pest Management in Cabbage Cultivation*. Extn Bull No 1, Indian Inst of Hort Res, Bangalore, 10 pp.

Krishnamoorthy, A., Gopalakrishnan, C. and Mani, M. 1999. Integration of egg parasitoid and *Ha*NPV for the control of *Helicoverpa armigera* Hubn. on tomato. *Int. Seminar on IPM,* Indian Inst. of Chem. Technol., Hyderabad.

Krishna Rao, A.B.V., Padhi, N.N. and Acharya, A. 1987. Effect of different nematicides, oilcakes and urea in the control of *Rotylenchulus* in okra. *Indian J. Nematol.* **17**: 171-173.

Kulkarni, M.S., Ramprasad, S., Hedge, Y., Laxminarayan, H. and Hedge, N.K. 2007. Management of collar rot complex disease of *Coleus forskohlii* (Wild) Briq. using bioagents, organic amendments and chemicals. *Biomed.* **2**: 37-40.

Kumar, A.C., Viswanathan, P.R.K. and D'Souza, G.I. 1971. A study of plant parasitic nematodes of certain commercial crops in coffee tracts of South India. *Indian Coffee* **36**: 1-3.

Kumar, B. 2008. Studies on Root-knot and Wilt Complex in *Coleus forskohlii* (Wild.) Briq. Caused by *Meloidogyne incognita* (Kofoid and White) Chitwood and *Fusarium chlamydosporum* (Frag. and Cif.) Booth. M.Sc. (Agri) thesis in Plant Pathology, University of Agricultural Sciences, Dharwad.

Kumar, V. and Haseeb, A. 2009. Interactive effect of *Meloidogyne incognita* and *Rhizoctonia solani* on the growth and yield of tomato. *Indian J. Nematol.* **39**: 178-181.

Kumar, V., Haseeb, A. and Sharma, A. 2009a. Integrated management of *Meloidogyne incognita-Fusarium solani* disease complex of brinjal cv. Pusa Kranti. *Ann. Plant Prot. Sci.* **17**:

Kumar, V., Haseeb, A. and Sharma, A. 2009b. Integrated management of *Meloidogyne incognita* and *Fusarium solani* disease complex of chilli. *Indian Phytopath.* **62**: 324.327.

Kumutha, K. 2001. Symbiotic Influence of AM Fungi and Rhizobacteria on Biochemical and Nutritional Changes in Mulberry (*Morus alba*). Ph. D. thesis, Tamil Nadu Agri. Univ., Coimbatore.

Lakshmanan, P.L. and Sivakumar, C.V. 1981. Control of *Rotylenchulus reniformis* in Bhendi by integrated method. *Second Nematol. Symp.*, Tamil Nadu Agri. Univ., Coimbatore, p.54.

Laqman Khan, M. 2001. Dr. Y.S. Paramar University of Horticulture and Forestry, Solan, Himachal Pradesh. In *Indian Nematology-Progress and Prospectives*. (S.C. Dhawan *et al.*, eds.), pp. 153-158. Division of Nematology, Indian Agri. Res. Inst., New Delhi.

Leibinger, W., Breuker, B., Hahn, M. and Mendgen, K. 1997. Control of post-harvest pathogens and colonization of the apple surface by antagonistic microorganisms in the field. *Phytopathology* **87**: 1103-1110.

Lewis, J.A. and Papavizas, G.C. 1980. Integrated control of *Rhizoctonia* root rot of cucumber. *Phytopathology* **70**: 85-89.

Lifshitz, R., Lifshitz, F. and Baker, R. 1985. Decrease in incidence of *Rhizoctonia* pre-emergence damping-off by use of integrated chemical and biological control. *Plant Dis.* **60**: 431-434.

Lindow, S.E., McGourty, G. and Elkins, R. 1996. Interactions of antibiotics with *Pseudomonas fluorescens* strain A506 in the control of fire blight and frost injury to pear. *Phytopathology* **86**: 841-848.

Lodha, B.C. and Mathur, K. 1997. Management of rhizome rot of ginger - Progress and Perspective. In *Management of Threatening Plant Diseases of National Importance* (V.P. Agnihotri, A.K. Sarabhoy and D.V. Singh, eds.), pp. 249-262. Malhotra Publishing House, New Delhi.

Lodha, B.C., Mathur, K. and Shekawat, P.S.1994. Management of rhizome rot of ginger. In *Vistas in Seed Biology, Vol. 1* (U. Kant, T. Singh and P.C. Trivedi, eds.), pp. 334-335. Printwell Publications, Jaipur.

Loganathan, M., Swarnakumari, N., Sivakumar, M., Prakasam, V., Mohan, L., Ramraj, B. and Samiyappan, R. 2001. Biological suppression of fungal nematode complex diseases of major cruciferous vegetables. *S. Indian Hort.* **49**: Special issue.

Loos, C.A. 1959. Symptom expression of *Fusarium* wilt disease of the Gros Michel banana in the presence of *Radopholus similis* (Cobb, 1893) Thorne, 1949 and *Meloidogyne incognita acrita* Chitwood, 1949. *Proc. Helminthol. Soc. Wash.* **26**: 103 –111.

Loos, C.A. 1961. Eradication of the burrowing nematode, *Radopholus similis*, from bananas. *Plant Dis. Reptr.* **45**: 457-461.

Mahalingam, C.A. and Saminathan, V.R. 2003. Efficacy of *Bt*, *Ha*NPV and endosulfan individually and in combination for the management of the tomato fruit borer, *Helicoverpa armigera* (Hubner) (Noctuidae: Lepidoptera). In *Biological Control of Insect Pests* (S. Ignacimuthu and S. Jayaraj, eds.), pp. 88-91. Phoenix Publishing House Pvt. Ltd., New Delhi.

Maheswari, T.U. and Mani, A. 1988. Combined efficacy of *Pasteuria penetrans* and *Paecilomyces lilacinus* on the biocontrol of *Meloidogyne javanica* on tomato. *Int. Nematol. Network Newsl.* **5**(3): 10-11.

Maheswari, T.U., Mani, A. and Rao, P.K. 1987. Combined efficacy of the bacterial spore parasite *Pasteuria penetrans* (Thorne, 1940) and nematicides in the control of *Meloidogyne javanica* on tomato. *J. Biol. Control* **1**: 53-57.

Maicykutty, P.M. and Gopalakrishnan, T.R. 2003. Eco-friendly management of major pests of bitter gourd. In *Biological Control of Insect Pests* (S. Ignacimuthu and S. Jayaraj, eds.), pp.192-196. Phoenix Publishing House Pvt. Ltd., New Delhi.

Maiti, D. and Sen, C. 1998. Integrated biological control of *Sclerotium rolfsii* with nitrogenous fertilizers and *Trichoderma harzianum*. *Indian J. Agric. Sci.* **55**: 464-468.

Mani, A. 1986. Occurrence of *Meloidogyne javanica* on citrus in Andhra Pradesh (India). *Int. Nematol. Network Newsl.* **3**(2): 9-10.

Mani, A., Naidu, P.H. and Madhavachari, S. 1987. Occurrence and control of *Meloidogyne incognita* on turmeric in Andhra Pradesh, India. *Int. Nematol. Network Newsl.* **4**: 13-18.

Mani, M. 2001. Biological control of fruit crop pests. In *Integrated Pest Management in Horticultural Ecosystems* (P. Parvatha Reddy, A. Verghese and N.K. Krishna Kumar, eds.), pp. 93-107. Capital Publishing Co., New Delhi.

Manoj Kumar, R., Rao, M.S. and Ramachandran, N. 2010. Biomanagement of disease complex caused by *Meloidogyne incognita* and *Phytophthora parasitica* Dastur in protected cultivation of gerbera using formulation of *Trichoderma harzianum* and *Pseudomonas fluorescens*. *National Conf. on Innovations in Nematol. Res. for Agri. Sustainability – Challenges and A Roadmap Ahead*, Tamil Nadu Agri. Univ., Coimbatore, pp. 97-98.

Mao, W., Lewis, J.A., Lumsden, R.D. and Hebber, K.P. 1998. Biocontrol of selected soilborne disease of tomato and pepper plants. *Crop. Prot.* **17**: 535-542.

Martinez, G. and Pinzon, L. 1999. New strategies in the integrated disease management of the vascular wilt of carnation (*Dianthus caryophyllus* L.) caused by *Fusarium oxysporum* f. sp. *dianthi* (Pril *et* Del.) Snyd. *et* Hans. *Acta Hort.* **482**: 139-144.

Mathur, N.M., Sharma, G.K. and Qureshi, Q.G. 1998. Fruit borer (*Earias* spp.) management on okra in semi-arid region of Rajasthan. In *Advances in IPM for Horticultural Crops* (P. Parvatha Reddy, N.K. Krishna Kumar and Abraham Verghese, eds.), pp. 121-123. Association for Advancement of Pest Management in Horticultural Ecosystems, Bangalore.

Matsubara, Y., Hasegawa, N. and Fukui, H. 2002. Incidence of Fusarium root rot in asparagus seedlings infected with arbuscular mycorrhizal fungus as affected by several soil amendments. *J. Japanese Soc. Hort. Sci.* **71**: 370-374.

McLaughlin, R.J., Wisniewski, J.I., Wilson, C.L. and Chalutz, E. 1990. Effect of inoculum concentration and salt solutions on the biological control of post-harvest diseases of apples with *Candida* sp. *Phytopathology* **80**: 456-461.

McLean, K.L., Hunt, J. and Stewart, A. 2001. Compatibility of the biocontrol agent *Trichoderma harzianum* (C52) with fungicides. *New Zealand Plant Prot.* **54**: 84-88.

Mesta, R.K., Mohankumar, H.D., Shivaprasad, M., Tatagar, M.H. and Rao, M.S.L. 2009. Management of fruit rot and powdery mildew of chilli through *Pseudomonas fluorescens*. *Int. Conf. on Hort.*, Bangalore.

Mishra, P.K., Mukhopadhyay, A.N. and Fox, R.T.V. 2002. Integrated and biological control of *Gladiolus* corm rot and wilt caused by *Fusarium oxysporum* f. sp. *gladioli*. *Ann. Appl. Biol.* **137**: 361-364.

Mohanasundaram, A. and Dhandapani, N. 2009. Gibberellic acid (GA_3) and *Pseudomonas fluorescens* as components of biointensive pest management strategy against *Plutella xylostella* in cauliflower. *Pest Mangmt. Hortil. Ecosystems* **15**: 1-8.

Mohandas, C. and Ramana, K.V. 1991. Pathogenicity of *Meloidogyne incognita* and *Radopholus similis* on black pepper (*Piper nigrum* L.). *J. Plantation Crops* **19**: 41-43.

Mukherjee, P.K., Upadhyay, J.P. and Mukhopadhyay, A.N. 1989. Biological control of *Pythium* damping-off of cauliflower by *Trichoderma harzianum*. *J. Biol. Control* **3**: 119-124.

Mukhopadhyay, A.N. 1994. Biological control of soil-borne fungal plant pathogens – Current status, future prospects and potential limitations. *Indian Phytopath.* **47**: 119-126.

Mukhopadhyaya, M.C. and Dalal, M.R. 1971. Effect of two nematicides on *Tylenchulus semipenetrans* and on sweet lime yield. *Indian J. Nematol.* **1**: 95-97.

Mukhopadhyaya, M.C. and Suryanarayana, D. 1969. Citrus decline in Haryana : Role of *Tylenchulus semipenetrans* and its control. *Indian Phytopath.* **22**: 495-497.

Muthulakshmi, M., Devarajan, K. and Jonathan, E.I. 2010. Biocontrol of root-knot nematode, *Meloidogyne incognita* (Kofoid and White) Chitwood in mulberry (*Morus alba* L.). *J. Pesticides* **3**: 479-482.

Nagesh, M., Hussaini, S.S., Ramanujam, B. and Chidanandaswamy, B.S. 2006. Management of *Meloidogyne incognita* and *Fusarium oxysporum* f. sp. *lycopersici* wilt complex using antagonistic fungi in tomato. *Nematol. Medit.* **34**: 63-68.

Nagesh, M. and Janakiram, T. 2004. Root-knot nematode problem in polyhouse roses and its management using dazomet, neem cake and *Pochonia chlamydosporia* (*Verticillium chlamydosporium*). *J. Orn. Hort.* **7**: 147-152.

Nagesh, M. and Parvatha Reddy, P. 1995. Comparative efficacy of *Paecilomyces lilacinus* (Thom.) Samson and *Verticillium lecanii* (A. Zimmerman) Viegas in combination with botanicals against *Meloidogyne incognita* (Kofoid and White, 1919) Chitwood, 1949 infecting *Crossandra undulaefolia* L. *J. Biol. Control* **9**: 109-112.

Nagesh, M. and Parvatha Reddy, P. 1996a. Management of root-knot nematodes on carnation using antagonistic fungi and neem cake. *National Seminar on Integrated Crop Disease Management for Sustainable Agriculture*, Aurangabad, P. G105.

Nagesh, M. and Parvatha Reddy, P. 1996b. Eco-friendly management of *Meloidogyne incognita* on *Crossandra undulaefolia* using vesicular arbuscular mycorrhiza, *Glomus mosseae* and oil cakes. *Mycorrhiza News* **9**(1): 12-14.

Nagesh, M. and Parvatha Reddy, P. 2000. Crop loss estimation in carnation and gerebera due to the root-knot nematode, *Meloidogyne incognita* (Kofoid & White) Chitwood. *Pest Mangmt. in Hortil. Ecosystems* **6**: 158-159.

Nagesh, M. and Parvatha Reddy, P. 2005. Management of carnation and gerbera to control the root-knot nematode, *Meloidogyne incognita*, in commercial polyhouses. *Nematol. Medit.* **33**: 157-162.

Nagesh, M., Parvatha Reddy, P. and Ramachandran, N. 1998. Integrated management of *Meloidogyne incognita* and *Fusarium oxysporum* f sp. *gladioli* in gladiolus using antagonistic fungi and neem cake. In *Nematology - Challenges & Opportunities in 21st Century* (Usha K. Mehta, ed.), pp. 263-266. Sugarcane Breeding Inst., Coimbatore.

Nagesh, M., Parvatha Reddy, P. and Rao, M.S. 1995. Management of *Meloidogyne incognita* on *Crossandra undulaefolia* using *Paecilomyces lilacinus* and *Pasteuria penetrans* formulations. National Symp. on Perspectives in Eco-friendly Approaches to Plant Prot., Indian Agri. Res. Inst., New Delhi, pp. 47-48.

Nagesh, M., Parvatha Reddy, P. and Rao, M.S. 1997. Integrated management of *Meloidogyne incognita* on tuberose using *Paecilomyces lilacinus* in combination with plant extracts. *Nematol. Medit.* **25**: 3-7.

Nagesh, M., Parvatha Reddy, P. and Rao, M.S. 1998. Comparative efficacy of VAM fungi in combination with neem cake against *Meloidogyne incognita* on *Crossandra undulaefolia. Mycorrhiza News* **11**: 11-13.

Nagesh, M., Parvatha Reddy, P., Vijaya Kumar, M.V. and Nagaraju, B.M. 1998. Management of *Meloidogyne incognita* on *Polianthes tuberosa* using parasitic fungus, *Paecilomyces lilacinus* in combination with oil cakes. In "*Advances in IPM for Horticultural Crops*" (P. Parvatha Reddy, N.K. Krishna Kumar and A. Verghese, eds.), pp. 339-344. Association for Advancement of Pest Management in Hortil. Ecosystems, Bangalore.

Nageswari, S. and Mishra, S.D. 2005. Integrated nematode management schedule incorporating neem products, VAM and soil solarization against *Heterodera cajani* infecting pigeonpea. *Indian J. Nematol.* **35**: 68-71.

Naik, D. 2004. Biotechnological Approaches for the Management of Wilt Disease Complex in Capsicum (*Capsicum annum* L.) and Egg plant (*Solanum melongena*) with Special Emphasis on Biological Control. Ph. D. thesis, Kuvempu Univ., Shimoga.

Naik, D., Rao, M.S., Shylaja, M., Reddy, M.S. and Rahiman, B.A. 2003. Management of wilt disease complex in capsicum using *Bacillus pumilus* and *Paecilomyces lilacinus. Proc. of Int. Plant Growth Promoting Rhizobacteria Workshop*, Calicut, India, pp. 276-281.

Nair, K.K.R. 1979. Studies on the chemical control of banana nematodes. *Agri. Res. J. Kerala* **17**: 232-235.

Nair, M.R.G.K., Das, N.M. and Menon, M.R. 1966. On the occurrence of the burrowing nematode *Radopholus similis* (Cobb, 1893) Thorne, 1949 on banana in Kerala. *Indian J. Ent.* **28**: 553-554.

Napiere, C.M. 1980. Varying inoculum levels of bacteria, nematodes and the severity of tomato bacterial wilt. *Ann. Trop. Res.* **2**: 129-134.

Napiere, C.M. and Quinio, A.J. 1980. Influence of root-knot nematode on bacterial wilt severity in tomato. *Ann. Trop. Res.* **2**: 29-39.

Naqvi, S.A.M.H. 2001. Diagnosis and management of fungal diseases of citrus. In *Citrus* (Shyam Singh and S.A.M.H. Naqvi, eds.), pp. 375-391. Intn. Book Distributing Co., Lucknow, India.

Narayanan, K. 1987. Field efficacy of nuclear polyhedrosis virus of *Adisura atkinsoni* Moore on field beans. *J. Biol. Control* **1**: 44-47.

Narayanan, K. and Gopalakrishnan, G. 1988. *Microbial Control of* Heliothis armigera. Tech. Bull. No. 6, Indian Inst. of Hort. Res., Bangalore, pp. 13-14.

Narayanan, K. and Gopalakrishnan, G.1990. Integrated control *Heliothis armigera* with the use of pathogens and pesticides in certain horticultural crops. In *Heliothis Management-Proc. Natl. Workshop* (S. Jayaraj, S. Uttamasamy, M. Gopalan and R.J. Rabindra, eds.), pp. 165-176. Centre for Plant Protection Studies, Tamil Nadu Agri. Univ., Coimbatore, 328 pp.

Newhall, A.G. 1958. The incidence of Panama disease of banana in the presence of the root-knot and the burrowing nematodes (*Meloidogyne incognita* and *Radopholus similis). Pl. Dis. Reptr.* **42**: 853-856.

Nirmal Johnson, S.B. 2000. Studies on Nematode Pests of Cut Flowers with Special Reference to Carnation, Cystera, Gladiolus and Asiatic Lily. Ph. D. thesis, Tamil Nadu Agri. Univ., Coimbatore, Tamil Nadu, 160 pp.

Nirmala Kannan, Devasikhamani, S., Sudhakar Bhat, Naidu, R. And Sreenivasan, C.S. 1997. Biological control of root disease in coffee (Abstr.). *Int. Conf. on Integrated Plant Dis. Mangmt. for Sustainable Agri.,* Indian Phytopath. Soc., Indian Agri. Res. Inst., New Delhi, p. 210.

Nirvan, R.S. 1961. Citrus canker and its control. *Hort. Adv.* **5**: 124-125.

Nisha, M.S and Sheela, M.S. 2006. Bio-management of *Meloidogyne incognita* on Coleus, *Solenostemon rotundifolius* by integrating solarization, *Paecilomyces lilacinus, Bacillus macerans* and neem cake. *Indian J. Nematol.* **36**: 136-138.

Nyczepir, A.P. and Rodriguez-Kabana, R. 2004. Effectiveness of biofumigation for ring nematode control in a young peach orchard. *First Int. Biofumigation Symp.*, Florence, Italy.

Oostenbrink, M. 1972. Evaluation and integration of nematode control methods. In *Economic Nematology* (J.M. Webster, ed.), pp. 497-514. Academic Press, New York.

Ordentlich, A., Nachmias, A. and Chet, I. 1990. Integrated control of *Verticillium dahliae* in potato by *Trichoderma harzianum* and captan. *Crop. Prot.* **9**: 363-366.

Palaniswami, M.S. 2002. Advances in integrated approaches for sweet potato weevil management. *Proc. of the Int. Conf. on Vegetables,* Bangalore, India, pp. 304-308. Dr. Prem Nath Agricultural Science Foundation, Bangalore.

Pandey, R. 1990. Pathogenicity and reproduction of *Meloidogyne incognita* on *Hyoscyamus albus* and rhizospheric association of other phytoparasitic nematodes. *Int. Nematol. Network Newsl.* **7** (4): 34-37.

Pandey, R. 2001. Influence of *Meloidogyne incognita, Pratylenchus thornei* and *Tylenchorhynchus vulgaris* on growth and oil yield of menthol mint. *Indian J. Nematol.* **31**: 111-114.

Pandey, R. 2005. Field application of bio-organics in the management of *Meloidogyne incognita* in *Mentha arvensis*. *Nematol. Medit.* **33**: 51-54.

Pandey, R., Gupta, M.L., Singh, H.B., Singh, H.N. and Kumar, S. 2000. Disease management potentiality of VAM fungi to *Meloidogyne incognita* on *Hyoscyamus muticus*. In *Proc. Int. Conf. on Integrated Plant Dis. Mangmt. for Sustainable Agri.*", Indian Agri. Res. Inst., New Delhi, pp. 1239-1240.

Pandey, R. and Kalra, A. 2003. Root-knot disease of Ashwagandha, *Withania somnifera* and its ecofriendly and cost effective management. *J. Mycol. Pl. Pathol.* **33**: 240-245.

Pandey, R., Mishra, A.K., Tiwari, S., Singh, H.N. and Kalra, A. 2011. Enhanced tolerance of *Mentha arvensis* against *Meloidogyne incognita* (Kofoid and White) Chitwood through mutualistic endophytes and PGPRs. *J. Plant Interactions* (In Press).

Pandey, R., Singh, H.B. and Gupta, M.L. 1997. Antagonistic impact of vesicular arbuscular mycorrhizae (VAM) on *Meloidogyne incognita* population development in Japanese mint. *Int. J. Trop. Pl. Dis.* **15**:237-245.

Pandey, R., Singh, H.B. and Kumar, S. 1999. Pathogenicity of *Meloidogyne incognita* (Kofoid & White) Chitwood on *Matricaria chamomilla* L. *J. Spices and Aromatic Crops* **8**: 201-203.

Pani, A.K. and Das, S.N. 1972. Studies on etiological complexes in plant disease. I. Association of root-knot nematodes in bacterial wilt of tomato. *J. Res. Orissa Univ. Agri. Technol.* **2**: 54-59.

Paramasivan, M., Mohan, S. and Muthukrishnan, N. 2007. Management of *Coleus* dry root rot pathogen *Macrphhomina phaseolina* by fungal and bacterial antagonist. *Indian J. Plant Prot.* **35**: 133-135.

Paruthi, L.J., Jain, R.K. and Gupta, D.C. 1987. Effect of different periods of degradation of subabool leaves alone and in combination with nematicides in okra. *Indian J. Nematol.* **17**: 30-32.

Parvatha Reddy, P. and Khan, R.M. 1988. Evaluation of *Paecilomyces lilacinus* for the biological control of *Rotylenchulus reniformis* infecting tomato as compared with carbofuran. *Nematol. Medit.* **16**: 113-116.

Parvatha Reddy, P. and Khan, R.M. 1989. Evaluation of biocontrol agent *Paecilomyces lilacinus* and carbofuran for the management of *Rotylenchulus renfiromis* infecting brinjal. *Pakistan J. Nematol.* **7**: 55-59.

Parvatha Reddy, P. and Khan, R.M. 1991. Integrated management of root-knot nematodes infecting okra. *Curr. Nematol.* **2**:115-116.

Parvatha Reddy, P., Khan, R.M. and Rao, M.S. 1991. Integrated management of the citrus nematode, *Tylenchulus semipenetrans* using oil cakes and *Paecilomyces lilacinus*. *Afro-Asian J. Nematol.* **1**: 221-222.

Parvatha Reddy, P., Khan, R.M. and Rao, M.S. 1993. Integrated management of the citrus nematode, *Tylenchulus semipenetrans* using neem cake and the parasitic fungus, *Paecilomyces lilacinus*. *Proc. of World Neem Conf.*, Bangalore, pp. 647-650.

Parvatha Reddy, P. and Nagesh, M. 2000. Integrated management of the citrus nematode using bacterial (*Pasteuria penetrans)* and fungal (*Paecilomyces lilacinus)* biocontrol agents. *Proc. of Int. Symp. on Citriculture*, Nagpur, pp. 825-829.

Parvatha Reddy, P. and Nagesh, M. 2002. *Avermectins: Isolation, Fermentation, Preliminary Characterization and Screening for Nematicidal Activity*. Tech. Bull. – 17, Indian Inst. of Hort. Res., Bangalore, 28 pp.

Parvatha Reddy, P., Nagesh, M. and Devappa, V. 1997. Effect of integration of *Pasteuria penetrans, Paecilomyces lilacinus* and neem cake for the management of root-knot nematode infecting tomato. *Pest Managmt. Hortil. Ecosystems* **3**: 100-104.

Parvatha Reddy, P., Nagesh, M. and Devappa, V. 1998. Management of *Meloidogyne incognita* on tomato by integration of *Trichoderma harzianum, Glomus fasciculatum* and neem cake. In "*Advances in IPM for Horticultural Crops*" (P. Parvatha Reddy, N. K. Krishna Kumar & A. Verghese, eds.), pp. 349-352. Association for Advancement of Pest Management in Hortil. Ecosystems, Bangalore.

Parvatha Reddy, P., Nagesh, M., Devappa, V. and Vijaya Kumar, M.V. 1998. Management of *Meloidogyne incognita* on tomato by integrating endomycorrhiza, *Glomus mosseae* with oil cakes under nursery and field conditions. *J. Plant Dis. Prot.* **105**: 53-57.

Parvatha Reddy, P., Nagesh, M., Rao, M.S. and Rama, N. 2000. Management of *Tylenchulus semipenetrans* by integration of *Pseudomonas fluorescens* with oil cakes. *Proc. Int. Symp. on Citriculture*, Nagpur, pp. 830-833.

Parvatha Reddy, P., Rao, M.S., Mohandas, S. and Nagesh, M. 1995. Integrated management of the citrus nematode, *Tylenchulus semipenetrans* Cobb using VA mycorrhiza, *Glomus fasciculatum* (Thaxt.) Gerd & Trappe and oil cakes. *Pest Mangmt. Hort. Ecosystems* **1**: 37-41.

Parvatha Reddy, P., Rao, M.S. and Nagesh, M. 1996a. Management of the citrus nematode, *Tylenchulus semipenetrans* by integration of *Trichoderma harzianum* with oil cakes. *Nematol. Medit.* **24**: 265-267.

Parvatha Reddy, P., Rao, M.S. and Nagesh, M. 1996b. Management of the citrus nematode on acid lime by integration of parasitic fungi and oil cakes. *Pest Mangmt. Hort. Ecosystems* **2**: 15-18.

Parvatha Reddy, P., Rao, M.S. and Nagesh, M. 1996c. Integrated management of the citrus nematode, *Tylenchulus semipenetrans* using pesticides and parasitic fungus, *Paecilomyces lilacinus*. *Pest Mangmt. Hort. Ecosystems* **2**: 61-63.

Parvatha Reddy, P., Rao, M.S. and Nagesh, M. 1996d. Crop loss estimation in banana due to the burrowing nematode, *Radopholus similis*. *Pest Mangmt. Hort. Ecosystems* **2**: 85-86.

Parvatha Reddy, P., Rao, M.S. and Nagesh, M. 1998. Effect of bare-root dip treatment of tomato seedlings in plant leaf extracts mixed with *Paecilomyces lilacinus* spores for the management of *Meloidogyne incognita*. In "*Advances in IPM for Horticultural Crops*" (P. Parvatha Reddy, N.K. Krishna Kumar and A. Verghese, eds.), pp. 334-338. Association for Advancement of Pest Management in Hortil. Ecosystems, Bangalore.

Parvatha Reddy, P., Rao, M.S. and Nagesh, M. 1999. Eco-friendly management of *Meloidogyne incognita* on tomato by integration of *Verticillium chlamydosporium* with neem and calotropis leaves. *J. Plant Dis. Prot.* **106**: 530-533.

Parvatha Reddy, P., Rao, M.S. and Nagesh, M. 2002. Integrated management of burrowing nematode (*Radopholus similis*) using endomycorrhiza (*Glomus mosseae*) and oil cakes. In *Banana* (H.P. Singh and K.L. Chadha, eds.), pp. 344-348. AIPUB, Trichy.

Parvatha Reddy, P. and Singh, D.B. 1981. Assessment of avoidable yield loss in okra, brinjal, French bean and cowpea due to root-knot nematodes. *Third Int. Symp. Pl. Path.*, New Delhi, pp.93-94.

Patel, S.K. Patel, H.V. and Patel, A.D. 2006. Integrated management of root-knot nematode in tomato nursery through botanical plant materials. *Indian J. Nematol.* **36**: 307-308.

Pathak, B. and Khan, M.R. 2008. Yield loss potential of foliar nematode, Aphelenchoides besseyi in tuberose. *Indian J. Nematol.* 38(1): In press

Patil, N.M., Dharne, P.K. and Deshmukh, S.S. 1999. Sucker dip treatment in monocrotophos with intercrops for the management of plant parasitic nematodes in banana. *Seminar on Technological Advancement in Banana Production, Handling and Processing Management*, Jalgaon, Maharashtra, pp. 126-128.

Pedroche, N.B., Villanueva, L.M. and De Waele, D. 2009. Management of root-knot nematode, *Meloidogyne incognita* in carrot. *Commun. Agric. Appl. Biol. Sci.* **74**: 605-615.

Pereira, C.M.F., Chaves, G.M., Zambolim, L., Matsuoka, K., Acuna, R.S. and DoVale, F.X.R. 1996. Control of *Sclerotium cepivorum* by the use of vermicompost, solarization, *Trichoderma harzianum* and *Bacillus subtilis. Summa Phytopathologica* **22**: 228-234.

Perveen, S., Ehteshamul-Haque, S. and Ghaffar, A. 1998. Efficacy of *Pseudomonas aeruginosa* and *Paecilomyces lilacinus* in the control of root rot-root knot disease complex on some vegetables. *Nematol. Medit.* **26**: 209-212.

Philip, T., Sharma, D.D. and Govindaiah 1996. Biological control of mulberry root-rot disease. *Indian Silk* **34**: 6-8.

Ponmurugan, P. and Baby, U.I. 2005. Management of *Phomopsis* canker of tea with fungicides and biocontrol agents. *J. Plantation Crops* **33**: 175-178.

Ponte, J.J.D.A. 1980. *Meloidogyne* - importanica and control no nordeste. *Sociedade Brasileira Nematologia* **4**: 1-14.

Porras, M., Barrau, C., Arroyo, F.T., Santos, B., Blanco, C. and Romero, F. 2007. Reduction of *Phytophthora cactorum* in strawberry fields by *Trichoderma* spp. and soil solarization. *Plant Dis.* 91: 142-146.

Porras, M., Barrau, C., Romero, E., Zurera, C. and Romero, F. 2009. Effect of biofumigation with *brassica carinata* and soil solarization on *phytophthora* spp. and strawberry yield. *Acta Hort.* **842**: 969-972.

Prasad, K.S.K. 1989. Nematological problems and progress of research on potato in India. *Fourth Group Meeting on Nematol. Problems of Plantation Crops*, Univ. Agri. Sci., Bangalore.

Prasad, P.R.K 1978. Studies on the Root-knot Nematodes Infecting Patchouli (*Pogostemon cablin* Bench). M. Sc. (Agri.) thesis, Univ. of Agri. Sci., Bangalore.

Prasad, P.R.K. and Reddy, D.D.R. 1979. Study on the root-knot nematode infecting patchouli (*Pogostemon cablin* Benth.). *Mysore J. Agric. Sci.* **13**: 367-368.

Prasad, P.R.K. and Reddy, D.D.R. 1984. Pathogenicity and analysis of crop losses on patchouli due to *Meloidogyne incognita. Indian J. Nematol.* **14**: 36-38.

Prasad, R.D., Rangeswaran, R., Sunanda, C.R. and Vinita, J. 2002. Biological control of black spot of rose caused by *Diplocarpon roseae. Ann. Pl. Prot. Sci.* **10**: 256-259.

Prasad, S.K. and Chawla, M.L. 1965. Observations on the population fluctuations of citrus nematode, *Tylenchulus semipenetrans* Cobb, 1913. *Indian J. Ent.* **27**: 450-455.

Praveen, P.M. and Dhandapani, N. 2001. Development of biocontrol based pest management in okra (*Abelmoschus esculentus* Moench). In *IPM in Horticultural Crops: Emerging Trends in the New Millennium* (A. Verghese and P. Parvatha Reddy, eds.), pp. 192-193. Association for Advancement of Pest Management in Hortil. Ecosystems, Bangalore.

Pusey, P.L., Wilson, C.L., Hotchkiss, M.W. and Franklin, J.D. 1986. Compatibility of *Bacillus subtilis* for postharvest control of peach brown rot with commercial fruit waxes, dicloran, and cold-storage conditions. *Plant Dis.* **70**: 587-590.

Putter, J.G., Mac Connell, F.A., Preiser, F.A., Haidri, A.A., Rishich, S.S. and Dybas, R.A. 1981 Avermectins: Novel class of insecticides, acaricides and nematicides from a soil microorganism. *Experientia* **37**: 963-964.

Quereshi, Q.G., Mathur, N.M. and Srivastava, R.C. 1998. Management of brinjal shoot and fruit borer *Leucinodes orbonalis* Guenee using *Bacillus thuringiensis* var *kurstaki*. In *Advances in Pest Management of Horticultural Crops* (P. Parvatha Reddy, N.K. Krishna Kumar and A. Verghese, eds.), pp. 81-83. Association for Advancement of Pest Management in Hortil. Ecosystems, Bangalore.

Raghumoorthi, K.N. and Armugam, R. 1992. Control of moringa fruitfly *Gitona* sp., and leaf caterpillar *Noorda blitealis* with insecticides and botanicals. *Indian J. Plant Prot.* **20**: 61-65.

Rahman, M.A., Vijaya, M., Chiranjeevi, C.H., Sai Reddy, C. and Reddy, I.P. 2002. Disease management of Solanaceous vegetable crops at nursery stage with bio-control agents. *Proc. of the Int. Conf. on Vegetables,* Bangalore, India, pp. 352-355. Dr. Prem Nath Agri. Sci. Foundn., Bangalore.

Raja, J., Rajendran, B. and Papaiah, C.M. 1998a. Management of egg plant shoot and fruit borer, *Leucinodes orbonalis* Guen. In *Advances in Pest Management of Horticultural Crops* (P. Parvatha Reddy, N.K. Krishna Kumar and A. Verghese, eds.), pp. 84-86. Association for Advancement of Pest Management in Hortil. Ecosystems, Bangalore.

Raja, J., Rajendran, B. and Papaiah, C.M. 1998b. Management of bhendi fruit borer, *Earias vitella* F. In *Advances in Pest Management of Horticultural Crops* (P. Parvatha Reddy, N.K. Krishna Kumar and A. Verghese, eds.), pp. 118-120. Association for Advancement of Pest Management in Hortil. Ecosystems, Bangalore.

Rajagopalan, P. and Naganathan, T.G. 1977a. Studies on nematode parasites of grape-vine. *Tamil Nadu Agri. Univ. Ann. Rept.* **6**: 129.

Rajagopalan, P. and Naganathan, T.G. 1977b. Studies on nematode parasites of banana. *Tamil Nadu Agri. Univ. Ann. Rept.* **6**: 131.

Rajan, P.P. and Sarma, Y.R. 1997. Compatibility of potassium phosphonate (Akomin-40) with different species of *Trichoderma harzianum* and *Gliocladium virens*. In *Biotechnology of Spices, Medicinal and Aromatic Plants* (S. Edison, K.V. Ramana, B. Saikumar, K. Nirmal Babu and S.J. Eapen, eds.), pp. 150-155. Indian Soc. for Spices, Indian Inst. of Spices Res., Calicut.

Rajendran, B. 1998. Effect of neem oil and insecticide on the population and parasitization of eggplant spotted beetle, *Henosepilachna vignitioctopunctata* F. (Cocinellidae : Coleoptera). In *Advaces in IPM for Horticultural Crops* (P. Parvatha Reddy, N.K. Krishna Kumar and A. Verghese, eds.), pp. 102-105. Association for Advancement of Pest Management in Horticultural Ecosystems, Bangalore.

Rajendran, B. and Rajendran, G. 1979. Report of *Meloidogyne incognita* in *Jasminum flexile*. *S. Indian Hort.* **27**: 70.

Rajendran, G., Vadivelu, S. and Muthukrishnan, T.S. 1976. Pathogenicity of *Meloidogyne incognita* on crossandra. *Indian J. Nematol.* **6**: 115-116.

Ram, P., Mathur, K. and Lodha, B.C. 1999. Integrated management of rhizome rot of ginger involving biocontrol agents and fungicides. *J. Mycol. Plant Pathol.* **29**: 416-417.

Ramachandran, N., Sarma, Y.R., Anandraj, M. and Jose, A. 1988. Effect of climatic factors on *Phytophthora* leaf infection in black pepper mixed cropping system. *J. Plantn. Crops* **16**: 140-148.

Ramakrishnan, T.S. 1954. *Common Diseases of Citrus in Madras*. Govt. of Madras Publn.

Ramakrishnan, T.S. and Damodaran, S. 1956. *Proc. Indian Acad. Sci.* **43**: 213-222.

Ramana, K.V., Mohandas, C. and Balakrishnan, R. 1987. Role of plant parasitic nematodes in the slow wilt disease complex of black pepper (*Piper nigrum* L.) in Kerala. *Indian J. Nematol.* **17**: 225-230.

Ramanujam, B., Nambiar, K.K.N. and Iyer, R. 1997. Management of stem bleeding disease of coconut with fungicides and a biocontrol agent. *J. Plantn. Crops* **25**: 175-179.

Ramanujam, B., Prasad, R.D. and Rangeswaran, R. 2003. Achievements in biological control of diseases with antagonistic organisms at Project Directorate of Biological Control, Bangalore. In *Current Status of Biological Control of Plant Diseases using Antagonistic Organisms in India* (B. Ramanujam and R.J. Rabindra, eds.), pp. 258-271. Project Directorate of Biological Control, Bangalore.

Ramaprasad Shresti, A.Y. 2005. Studies on Collar Rot Complex of *Coleus forskohlii* (Wild.) Briq. M. Sc. (Agri.) thesis, Uni. of Agri. Sci., Dharwad, Karnataka, India.

Rangaswamy, S.D., Parvatha Reddy, P., Nagesh, M. and Nanje Gowda, D.N. 1999. Integrated management of *Meloidogyne incognita* using *Pasteuria penetrans, Trichoderma viride* and oil cakes in tomato. *Pest Managmt. Hort. Ecosystems* **5**: 122-126.

Rao, M.S. 2005. Management of *Meloidogyne javanica* on acid lime nursery seedlings by using formulations of *Pochonia chlamydosporia* and *Paecilomyces lilacinus*. *Nematol. Medit.* **33**: 145-148.

Rao, M.S. 2007a. Papaya seedlings colonized by the bio-agents *Trichoderma harzianum* and *Pseudomonas fluorescens* to control root-knot nematodes. *Nematol. Medit.* **35**: 199-203.

Rao, M.S. 2007b. Management of root-knot nematode, *Meloidogyne incognita* (Kafoid & White) Chitwood, on crossandra (*Crossandra undulaefolia* Salisb.) using *Pochonia chlamydosporia* and *Pseudomonas fluorescens*. *J. Orn. Hort.* **10**: 110-114.

Rao, M.S., Kerry, B.R., Gowen, S.R., Bourne, J.M. and Parvatha Reddy, P. 1997. Management of *Meloidogyne incognita* in tomato nurseries by integration of *Glomus deserticola* with *Verticillium chlamydosporium*. *J. Plant Dis. Prot.* **104**: 419-422.

Rao, M.S., Mohandas, S., Parvatha Reddy, P. and Khan, R.M. 1993a. Synergistic effect of endomycorrhizal fungus (*Glomus fasciculatum*) and biocontrol fungus (*Paecilomyces lilacinus*) in the management of root-knot nematode infecting tomato. In *"Soil Organisms and Sustainability"* (D. Rajagopal, R. D. Kale and K. Bano, eds.), pp. 119-123. Indian Soc. of Soil Biol. and Ecol., Univ. of Agri. Sci., Bangalore.

Rao, M.S., Mohan Das, S., Parvatha Reddy, P. and Khan, R.M. 1993b. Management of reniform nematode on egg plant by integrating a botanical (castor cake), mycorrhiza (*Glomus fasciculatum*) and bio-agent (*Paecilomyces lilacinus*). *Golden Jubilee Symp. on Hortil. Res. - Changing Scenario*, Bangalore, pp. 259-260.

Rao, M.S. and Naik, D. 2003. Effect of *Trichoderma harzianum* and *Paecilomyces lilacinus* on *Meloidogyne incognita* on papaya (*Carica papaya* L.) nursery seedlings. Pest Mangmt. Hort. Ecosystems **9**: 155-160.

Rao, M.S., Naik, D. and Shylaja, M. 2004. Bio-intensive management of root-knot nematodes on bell pepper using *Pochonia chlamydosporia* and *Pseudomonas fluorescens*. *Nematol. Medit.* **32**: 159-163.

Rao, M.S., Naik, D., Shylaja, M. and Parvatha Reddy, P. 2002a. Prospects for the management of nematode disease complex in capsicum using biological control agents. *Proc. of Int. Conf. on Vegetables,* Bangalore, pp. 347-351.

Rao, M.S., Naik, D., Shylaja, M. and Parvatha Reddy, P. 2002b. Interaction of *Meloidogyne incognita* and *Fusarium oxysporum* f. sp. *dianthi* in the manifestation of the disease complex on tuberose. *Zonal Meeting of the Southern Chapter of Indian Phytopath. Soc.*, Bangalore.

Rao, M.S. and Parvatha Reddy, P. 1992. Innovative approaches of utilization of biocontrol agents-*Paecilomyces lilacinus* and *Verticillium chlamydosporium* against root-knot nematode on tomato (Abstr.). *First Afro-Asian Nematology Symposium,* Aligarh Muslim University, Aligarh, pp. 25-26.

Rao, M.S. and Parvatha Reddy, P. 1993a. Effective use of 'Welgro' and 'RD-9 Repelin' for the management of root-knot nematode on tomato. In *"Neem and Environment"* (R. P. Singh, M. S. Chari, A. K. Raheja and W. Kraus, eds.), **2**: 651-656. Oxford & IBH Publishing Co. Pvt. Ltd., New Delhi.

Rao, M.S. and Parvatha Reddy, P. 1993b. Interactive effect of *Verticillium chlamydosporium* and castor cake on the control of root-knot nematode on eggplant. In *"Soil Organisms and Sustainability"* (D. Rajagopal, R.D. Kale and K. Bano, eds.), pp. 114-118. Indian Soc. of Soil Biol. and Ecol., Univ. of Agri. Sci., Bangalore.

Rao, M.S. and Parvatha Reddy, P. 1994. A method for conveying *Paecilomyces lilacinus* to soil for the management of root-knot nematodes on eggplant. *Nematol. Medit.* **22**: 265-267.

Rao, M.S. and Parvatha Reddy, P. 2001. Control of *Meloidogyne incognita* on eggplant using *Glomus mosseae* integrated with *Paecilomyces lilacinus* and neem cake. *Nematol. Medit.* **29**: 153-157.

Rao, M.S., Parvatha Reddy, P. and Mohandas, S. 1995. Studies on development of a biorational management strategy against root-knot nematode attacking tomato. *National Symp. on Nematode Problems of India – An Appraisal of the Nematode Management with Eco-friendly Approaches* (G. Swarup, D.R. Dasgupta and J.S. Gill, eds.), p. 14. Nematol. Soc. of India, New Delhi.

Rao, M.S., Parvatha Reddy, P. and Mohandas, S. 1996. Effect of integration of *Calotropis procera* leaf and *Glomus fasciculatum* on the management of *Meloidogyne incognita* infesting tomato. *Nematol. Medit.* **24**: 59-61.

Rao, M.S., Parvatha Reddy, P. and Nagesh, M. 1997a. Integrated management of *Meloidogyne incognita* on okra by castor cake suspension and *Paecilomyces lilacinus*. *Nematol. Medit.* **25**: 17-19.

Rao, M.S., Parvatha Reddy, P. and Nagesh, M. 1997b. Integration of *Paecilomyces lilacinus* with neem leaf suspension for the management of root-knot nematodes on eggplant. *Nematol. Medit.* **25**: 249-252.

Rao, M.S., Parvatha Reddy, P. and Nagesh, M. 1997c. Management of root-knot nematode, *Meloidogyne incognita* on tomato by integration of *Trichoderma harzianum* with neem cake. *J. Plant Dis. Prot.* **104**: 423-425.

Rao, M.S., Parvatha Reddy, P. and Nagesh, M. 1998a. Integrated management of *Meloidogyne incognita* on tomato using *Verticillium chlamydosporium* and *Pasteuria penetrans*. *Pest Mangmt. Hort. Ecosystems* **4**: 32-35.

Rao, M.S., Parvatha Reddy, P. and Nagesh, M. 1998b. Integrated management of *Meloidogyne incognita* on tomato using oil cakes and a bioagent, *Verticillium chlamydosporium*. In "*Advances in IPM for Horticultural Crops*" (P. Parvatha Reddy, N.K. Krishna Kumar and A. Verghese, eds.), pp. 345-348. Association for Advancement of Pest Management in Hortil. Ecosystems, Bangalore.

Rao, M.S., Parvatha Reddy, P. and Nagesh, M. 1998c. Evaluation of plant based formulations of *Trichoderma harzianum* for the management of *Meloidogyne incognita* on eggplant. *Nematol. Medit.* **26**: 59-62.

Rao, M.S., Parvatha Reddy, P. and Nagesh, M. 1998d. Evaluation of *Paecilomyces lilacinus* cultured on neem cake extract for the management of root-knot nematodes on eggplant. *Pest Mangmt. Hort. Ecosystems* **4**:116-119.

Rao, M.S., Parvatha Reddy, P. and Nagesh, M. 1999. Bare root-dip treatment of tomato seedlings in calotropis or castor leaf extracts mixed with *Paecilomyces lilacinus* spores for the management of *Meloidogyne incognita*. *Nematol. Medit.* **27**: 323-326.

Rao, M.S., Parvatha Reddy, P., Somasekhar, N. and Nagesh, M. 1997. Management of root-knot nematodes, *Meloidogyne incognita* in tomato nursery by integration of endomycorrhiza, *Glomus fasciculatum* with castor cake. *Pest Mangmt. Hort. Ecosystems* **3**: 31-35.

Rao, M.S., Parvatha Reddy, P. and Sukhada, M. 1998. Biointensive management of *Meloidogyne incognita* on eggplant by integrating *Paeciloymces lilacinus* and *Glomus mosseae*. *Nematol. Medit.* **26**: 213-216.

Rao, M.S., Parvatha Reddy, P., Sukhada, M., Nagesh, M. and Pankaj 1998. Management of root-knot nematodes on eggplant by integrating endomycorrhiza (*Glomus fasciculatum*) and castor (*Ricinus communis*) cake. *Nematol. Medit.* **26**: 217-219.

Rao, M.S., Ramachandran, N., Sowmya, D.S., Ratnamma, K., Chaya, M.K. and Manoj Kumar, R. 2009. Biological control of nematode induced disease complex in certain vegetable crops. *Int. Conf. on Hort.*, Bangalore.

Rao, M.S., Shylaja, M. and Naik, D. 2003. Management of nematode induced wilt disease complex in tuberose (*Polianthes tuberosa* L.) cultivar Prajwal using *Pochonia chlamydosporia* (*Verticillium chlamydosporium*) and *Trichoderma harzianum*. *J. Orn. Hort.* **6**: 341-346.

Rao, Y.P. and Hingorani, M.K. 1963. Survival of *Xanthomonas citri* (Hasse) Dowson in leaves and soil. *Indian Phytopath.* **16**: 362-364.

Ravi, K., Nanje Gowda, D.N. and Parvatha Reddy, P. 2001. Integrated management of disease complex in banana involving borrowing nematode and Panama wilt – an ecofriendly approach. In *IPM in Horticultural Crops: Emerging Trends in the New Millenium* (A. Verghese and P. Parvatha Reddy, eds.), p. 154. Association for Advancement of Pest Management in Horticultural Ecosystems, Bangalore.

Ravi Kumar, M.R., Somasekhara, Y.M., Jahagirdar, S. and Ryagi, Y.H. 2001. Field evaluation of antibiotics against citrus canker caused by *Xanthomonas axonopodis* pv. *citri*. *Agri. Sci. Digest* **21**: 253-255.

Reddy, D.D.R. 1985. Analysis of crop losses in tomato due to *Meloidogyne incognita. Indian J. Nematol.* **15**: 55-59.

Reichert, S.E. and Leslie, B. 1989. Prey control by an assemblage of generalist predators: Spiders in garden test systems. *Ecology*. Fall. p. 1441-1450.

Riga, E., Pierce, F. and Collins, H. 2006. The use of arugula on its own and in combination with systemic nematicides against plant parasitic nematodes of potatoes (Abstr.). *Second Int. Biofumigation Symp.,* Idaho, USA, p. 42.

Rishbeth, J. 1960. *Emp. J. Exp. Agric.* **28**: 109-113.

Ristaino, J.B., Perry, K.B. and Lumsden, R.D. 1991. Effect of solarization and *Gliocladium virens* on sclerotia of *Sclerotium rolfsii*, soil microflora, and the incidence of southern blight of tomato. *Phytopathology* **91**: 1117-1124.

Ristaino, J.B., Perry, K.B. and Lumsden, R.D. 1996. Soil solarization and *Gliocladium virens* reduce the incidence of southern blight (*Sclerotium rolfsii*) in bell pepper in the field. *Biocontrol Sci. Technol.* **6**: 583-593.

Routaray, B.N. and Sahoo, H. 1985. Integrated control of root-knot nematode, *Meloidogyne incognita* with neem cake and granular nematicides on tomato. *Indian J. Nematol.* **15**: 261.

Saikia, B. 1992. Pathogenicity, Crop Loss Assessment and Biological Control of *Meloidogyne incognita* (Kofoid and White, 1919) Chitwood, 1949 on Betel Vine (*Piper betle* L.). M. Sc. (Agri.) thesis, Assam Agri. Univ., Jorhat.

Saikia, M.K. and Borah, A. 2008. Comparative efficacy of fungal and bacterial bioagents with nematicides against *Meloidogyne incognita* on brinjal. *Indian J. Nematol.* **38**: 159-164.

Saju, K.A. 2004. Factors Affecting the Biological Control of *Phytophthora capsici* Infections in Black Pepper (*Piper nigrum* L.). Ph. D. thesis, Univ. of Calicut, 232 pp.

Saju, K.A., Anandraj, M. and Sarma, Y.R. 2003. Evaluation of *Trichoderma* spp. and *Pseudomonas fluorescens* for suppression of *Phytophthora capsici* infecting black pepper (Abstr.). *VIth Int. PGPR Workshop* (M.S. Reddy, M. Anandraj, S.J. Eapen, Y.R. Sarma and J.W. Kloepper, eds.), pp. 52-58. Indian Inst. of Spices Res., Calicut, Kerala, India.

Samiyappan, R. 2003. Biological control of fungal and nematode complex diseases by plant growth promoting rhizobacteria (PGPR). *Winter School on Biological Control of Plant Parasitic Nematodes*, Tamil Nadu Agri. Univ., Coimbatore, pp. 69-77.

Sankarimeena, K., Jonathan, E.I., Devarajan, K. and Raghuchander, T. 2012. *Pseudomonas fluorescens* induced systemic resistance in tomato against *Meloidogyne incognita. Indian J. Nematol.* **42**: 5-10.

Sardana, H.R., Trivedi, T.P., Bambawale, O.M., Naved Sabir, Singh, R.V., Arora, S., Krishna Moorthy, P.N., Pandey, P.K., Pandey, K.K. and Sengupta, A. 2004. Validated IPM Technologies: Vegetable crops. In *Validated IPM Technologies for Selected Crops* (Amerika Singh, H.R. Sardana and Naved Sabir, eds.), pp. 119-148. National Centre for Integrated Pest Management (ICAR), New Delhi.

Sarma, Y.R. 1997. Antagonistic fungi and their role in plant disease management (Abstr.). In *TCDC Int. Workshop on Application of Biotechnology in Biofertilizers and Biopesticides,* IIT, New Delhi, pp. 60-64.

Sarma, Y.R. 2000. Integrated management of rhizome rot of ginger. Final Report of ICAR AP Cess Fund Scheme, Indian Inst. of Spices Res., Calicut, Kerala, India.

Sarma, Y.R. 2003. Recent trends in the use of antagonistic organisms for the disease management in spice crops. In *Current Status of Biological Control of Plant Diseases Using Antagonistic Organisms in India* (B. Ramanujam and R.J. Rabindra, eds.), pp. 49-73. Project Directorate of Biological Control, Bangalore.

Sarma, Y.R., Anandraj, M. and Venugopal, M.N. 1996. Biological control of diseases of spices. In *Biological Control in Spices* (M. Anandraj and K.V. Peter, eds.), pp. 1-19. Indian Inst. of Spices Res., Calicut.

Sarma, Y.R. and Saju, K.A. 2004. Biological control for the management of foot rot and slow decline diseases of black pepper. Focus on Pepper (*Piper nigrum* L.) **1**: 25-51. Int. Pepper Community, Jakarta, Indonesia.

Sastry, K.S.M. and Singh, S.J. 1974. Effect of yellow vein mosaic virus infection on growth and yield of okra crop. *Indian Phytopath.* **27**: 316-317.

Satish. 1996. Management of Panama Disease of Banana Caused by *Fusarium oxysporum* f. sp. *cubense* by Using Microorganisms. M.Sc. thesis, Univ. of Agri. Sci., Bangalore, 98 pp.

Sawant, I.S. and Sawant, S.D. 1989. Coffee fruit skin and cherry husk as substrates for mass multiplication of *Trichoderma harzianum* an antagonist of citrus *Phytophthora. Indian Phytopath.* **42**: 336.

Seenivasan, N. 2010. Biointensive management of *Meloidogyne incognita* and *Macrophomina phaseolina* disease complex in medicinal coleus. *Indian J. Pl. Prot.* **38**: 186-192.

Seenivasan, N. and Deevrajan, K. 2008. Integrated approach for the management of root-knot nematode, *Meloidogyne incognita* in medicinal coleus. *Indian J. Nematol.* **38**: 154-158.

Selvakumar, R., Singh, A.K., Sureja, A.K. and Raja, P. 2009. Evaluation of *Trichoderma* isolates against rhizome rot of ginger incited by *Fusarium* spp. (Abstr.). *Indian Phytopath.* **62**: 408.

Sen, K. and Dasgupta, M.K. 1977. Additional hosts of the root-knot nematode, *Meloidogyne* spp. from India. *Indian J. Nematol.* **7**: 74.

Senthamari, K., Poornima, K., Subramanian, S. and Sudheer, J. 2008. Nematode-fungal disease complex involving *Meloidogyne incognita* and *Macrophomina phaseolina* on medicinal coleus, *Coleus forskohlii* Briq. *Ind. J. Nematol.* **38**: 30-33.

Senthil Kumar, T. and Rajendran, G. 2004. Biocontrol agents for the management of disease complex involving root-knot nematode, *Meloidogyne incognita* and *Fusarium moniliforme* on grapevine (*Vitis vinefera*). *Indian J. Nematol.* **34**: 49-51.

Shanthi, A., Rajeswari, S. and Sivakumar, C.V. 1998. Soil application of *Pseudomonas fluorescens* (Migula) for the control of root-knot nematode (*Meloidogyne incognita*) on grapevine (*Vitis vinefera* Linn.). In *Nematology – Challenges and Opportunities in 21st Century,* (Usha K. Mehta, ed.), pp. 203-206. Sugarcane Breeding Institute, Coimbatore.

Sharma, A. and Trivedi, P.C. 1989. Influence of inoculum levels of fungus *Paecilomyces lilacinus* (Thom.) Samson on the biocontrol of root-knot nematode, *Meloidogyne incognita* (Chitwood). *Int. Nematol. Network Newsl.* **6**(2): 27-29.

Sharma, H.K., Pankaj and Mishra, S.D. 2005. Polyethylene mulching in the management of plant parasitic nematodes. *Indian J. Nematol.* **35**: 82-85.

Sharma, O.P., Sehgal, M., Jeswani, M.D., Trividi, T.P., Garg, D.K., Srivastava, C.P., Benagi, V.I., Nigam, G.L. and Lavekar, R.C. 2004. Validated IPM Technologies: Pigeonpea. In *Validated IPM Technologies for Selected Crops* (Amerika Singh, H.R. Sardana and Naved Sabir, eds.), pp. 77-89. National Centre for Integrated Pest Management (ICAR), New Delhi.

Shashidhara, S. 2007. *Studies on Foot Rot of Black Pepper Caused by* Phytophthora capsici *Leonian, emend, Alidezeh and Tsao.* M. Sc. (Agri.) thesis in Plant Pathology, Uni. of Agri. Sci., Dharwad.

Sheela, M.S., Jiji, T. and Nisha, M.S. 2002. Evaluation of different control strategies for the management of nematodes associated with vegetables (brinjal) (Abstr.). *Int. conf. Vegetables,* Bangalore, p. 268.

Shyla, M. 1998. *Etiology and Management of a Root-rot of* Coleus forskohlii. M.Sc. (Agri.) thesis, Univ. of Agri. Sci., Bangalore.

Shylaja, M. 2004. Studies on the Development of Bio-Intensive Management Strategies of Nematode Induced Wilt Disease Complex in Tuberose (*Polianthes tuberosa*) and Carnations (*Dianthus caryophyllus*) using Biological Control Agents. Ph. D. thesis, Kuvempu Univ., Shimoga.

Shylaja, M., Rao, M.S. and Rahiman, A.B. 2004. Studies on the development of bio-intensive disease management strategies in tuberose (*Polyanthes tuberosa* L.) and carnations (*Dianthus caryophyllus* L.) using biological control agents. *National Symp. on Paradigms in Nematol. Res. for Biodynamic Farming,* Univ. of Agri. Sci., Bangalore, p. 136-137.

Siddiqi, M.R. 1961. Occurrence of the citrus nematode, *Tylenchulus semipenetrans* Cobb. 1913 and the reniform nematode, *Rotylenchulus reniformis* in India. *Proc. of 48th Indian Sci. Cong.* **3**: 504.

Siddiqui, I.A. and Ehteshamul-Haque, S. 2001. Suppression of the root rot–root knot disease complex by Pseudomonas aeruginosa in tomato: The influence of inoculum density, nematode populations, moisture and other plant-associated bacteria. *Plant and Soil* **237**: 81-89.

Siddiqui, I.A., Ehteshamul-Haque, S. and Ghaffar, A. 1999. Root-dip treatment with *Pseudomonas aeruginosa* and *Trichoderma* spp. in the control of root rot-root knot disease complex in chilli (*Capsicum annuum* L.). *Pakistan J. Nematol.* **17**: 67-75

Siddiqui, I.A., Qureshi, S.A., Sultana, V., Ehteshamul-Haque, S. and Ghaffar, A. 2000. Biological control of root rot-root knot disease complex of tomato. *Plant and Soil* **227**: 163-169

Siddiqui, Z.A. and Mahmood, I. 1995. Biological control of *Heterodera cajani* and *Fusarium udum* by *Bacillus subtilis, Bradyrhizobium japonicum* and *Glomus fasciculatum* on pigeonpea. Fund. Appl. Nematol. **18**: 559-566.

Singh, A., Mehta, S., Singh, M.H.B. and Nautiyal, C.S. 2003. Biocontrol of collar rot disease of betelvine (*Piper betle* L.) caused by *Sclerotium rolfsii* by using rhizosphere-competent *Pseudomonas fluorescens* NBRI-N6 and P. fluorescens NBRI-N. *Curr. Microbiol.* **47:** 153-158.

Singh, B., Ali, S.S., Naimuddin and Askary, T.H. 2004. Combined effect of *Fusarium udum* and *Meloidogyne javanica* on wilt resistant accessions of pigeon pea. *Ann. Pl. Prot. Sci.* **12**: 130133.

Singh, D. 1991. *Studies on the Effect of Tagetes Species on Meloidogyne javanica* Infectivity in Brinjal (*Solanum melongena*). Ph. D. thesis, Haryana Agri. Univ., Hisar, 83 pp.

Singh, H. and Singh, G. 1975. Biological studies on *Heliothis armigera* (Hubner) in Punjab. *Indian J. Ent.* **37**: 154-164.

Singh, R.S., Singh, P.P. and Bedi, J.S. 2002. Final Report of DBT Scheme on Biocontrol of Seed and Soil-borne Diseases of Vegetables, Punjab Agri. Univ., Ludhiana, 90 pp.

Singh, R.V. and Gill, J.S. 1998. Important nematode problems on vegetables and their management. In *Pathological Problems of Economic Crop Plants and Their Management* (S.M. Paul Khurana, ed.), pp. 557-570. Scientific Publishers (India), Jodhpur, India.

Singh, S., Dhawan, S.C. and Goswami, B.K. 2007. Histopathology of cowpea roots co-infected with root-knot nematode, *Meloidogyne incognita* and wilt fungus *Fusarium oxysporum. Indian J. Nematol.* **37**: 156-160.

Singh, S., Goswami, B.K. and Sharma, H.K. 2007. Role of reniform nematode, *Rotylenchulus reniformis* in reducing disease-complex on tomato caused by root-knot nematode, *Meloidogyne incognita* and wilt fungus *Fusarium oxysporum* f. sp. *lycopersici*. *Indian J. Nematol.* **37**: 172-175.

Singh, S.J. 1990. Etiology and epidemiology of whitefly-transmitted virus diseases of okra in India. *Plant Dis. Res.* **5**: 64-70.

Singh, S.P. 1995. *Technology for Production of Natural Enemies*. Tech. Bull. No. 4, Project Directotate of Biological Control, Bangalore, India, 221 pp.

Singh, S.P. 2000. Application of biotechnology in bio-intensive integrated pest management. In *Biotechnology in Horticultural and Plantation Crops* (K.L. Chadha, P.N. Raveendran and Leela Sahijram, eds.). Malhotra Publishing House, New Delhi, 836 pp.

Singh, V., Pathak, S.P. and Singh, S.P. 2002. Integrated disease management (IDM) for vine and fruit of parwal (*Trichosanthes dioica* Roxb.) (Abstr.). *J. Mycol. Plant Pathol.* **32**: 140.

Sitaramaiah, K. and Naidu, H. 2003. Plant parasitic nematodes-management opportunities and challenges. In *Nematode Management in Plants* (P.C. Trivedi, ed.), pp. 57-80. Scientific Publishers (India), Jodhpur.

Sivakumar, M. and Vadivelu, S. 1999. Management of *Meloidogyne incognita* on grapevine using biocontrol agents, botanicals and biofertilizers. *Pest Mangmt. Hort. Ecosystems* **5**: 127-131.

Smits, B.G. and Noguera, R. 1982. Effect of *Meloidogyne incognita* on the pathogenicity of different isolates of *Fusarium oxysporum* on brinjal (*Solanum melongena* L.). *Agronomia Tropical* **32**: 284-290.

Sosamma, V.K. and Koshy, P.K. 1995. Effect of *Pasteuria penetrans* and *Paecilomyces lilacinus* on population build up of root-knot nematode, *Meloidogyne incognita* on black pepper. National Symp. on Nematode Problems of India – An Appraisal of the Nematode Mangmt. with Eco-friendly Approaches and Biocomponents, Indian Agri. Res. Inst., New Delhi, p. 47.

Sowmya, D.S., Rao, M.S., Gopalakrishnan, C. and Ramachandran, N. 2010. Biomanagement of *Meloidogyne incognita* and *Erwinia carotovora* s. sp. carotovora infecting carrot (*Dacus carota* L.). National Conf. on Innovations in Nematol. Res. for Agri. Sustainability – Challenges and A Roadmap Ahead, Tamil Nadu Agri. Univ., Coimbatore, p. 97.

Sreenivasa, M.N. 1997. Influence of VAM and *Trichoderma harzianum* on *Sclerotium* in three varieties of chilli. *Environ. Ecol.* **15**: 343-345.

Srinivasan, K. and Krishna Moorthy, P.N. 1991. Indian mustard as a trap crop for management of major lepidopterous pests on cabbage. *Trop. Pest Mangmt.* **37**: 26-32.

Srinivasan, K., Krishna Moorthy, P.N. and Raviprasad, T.N. 1994. African marigold as a trap crop for the management of the fruit borer *Helicoverpa armigera* on tomato. *Int. J. Pest Mangmt.* **40**: 56-63.

Srinivasan, P.M. 1958. Control of fruit borer, *Heliothis armigera* (Hub) on tomato. *Indian J. Hort.* **16**: 187-188.

Srivastava, S.S. and Swarup, G. 1986. Biomanagement of root-knot nematode (*Meloidogyne incognita*) through the predatory fungus, *Arthrobotrys conoides*.

National Conf. on Plant Parasitic Nematodes of India-Problems and Progress, Indian Agri. Res. Inst., New Delhi, p. 85.

Subramanyam, S., Rajendran, G. and Vadivelu, S. 1990. Estimation of loss in tomato due to *Meloidogyne incognita* and *Rotylenchulus reniformis*. *Indian J. Nematol.* **20**: 239-240.

Sudha, S. and Sundararaju, P. 1998. Effect of neem oil cake and nematicide for the control of burrowing nematode, *Radopholus similis* in the areca nut based cropping system. In *Nematology - Challenges & Opportunities in 21st Century* (Usha K. Mehta, ed.), pp. 251-257. Sugarcane Breeding Inst., Coimbatore.

Sugar, D. and Spotts, R.A. 1999. Control of postharvest decay in pear by four laboratory-grown yeasts and two registered biocontrol products. *Plant Dis.* **83**: 155-158.

Sukumaran, S. and Sundararaju, P. 1986a. Pathogenicity of *Meloidogyne incognita* on ginger (*Zingiber officinale* Rosc.). *Indian J. Nematol.* **16**: 258.

Sukumaran, S. and Sundararaju, P. 1986b. Effect of root-knot nematode, *Meloidogyne incognita* on the growth of turmeric (Abstr.). *Seventh Symp. of Plantation Crops*, Coonoor, Nilgiris, Tamil Nadu, India, p. 33.

Sundarababu, R. 1992. Nematodes in jasmine. *Tamil Nadu Agri. Univ. Newsl.* **21**(11): 2.

Sundarababu, R. and Vadivelu, S. 1988. Pathogenicity of *Meloidogyne* species to tuberose (*Polianthes tuberosa* L.). *Indian J. Nematol.* **18**: 146-148.

Suseela Bhai, R. 1998. *Studies on Capsule Rot (Azhakul) Disease of Cardamom*. Ph. D. thesis, Univ. of Calicut, Calicut, Kerala, India.

Suseela Bhai, R., Joseph Thomas and Naidu, R. 1993. Biological control of 'Azhakul' disease of small cardamom caused by *Phytophthora meadii* Mc Rae. *J. Plantation Crops* **21**(Suppl.): 134-139.

Suseela Bhai, R., Joseph Thomas and Naidu, R. 1994. Evaluation of carrier media for the field application of *Trichoderma* sp. in cardamom growing soils. *J. Plantation Crops* **22**: 50-52.

Swaminathan, M.S. 2000. For an evergreen revolution. *The Hindu Survey of Indian Agriculture 2000*, pp. 9-15.

Swarup, G. and Seshadri, A. R. 1974. Nematology in India - Problems and Progress. In *Current Trends in Plant Pathology* (S.P. Raychaudhuri and J. P. Verma, eds.), pp. 303-311. Dept. of Botany, Univ. of Lucknow, Lucknow.

Talam, D.K. and Otieno, W. 2002. Temperature relations in *Armillaria* spp. associated with root rot of tea in Kenya: Implications in disease management by soil solarization. *Tea* **23**(1): 22-28.

Taylor, A.L., Sasser, J.N. and Nelson, L.A. 1982. *Relationship of Climate and Soil Characteristics to Geographical Distribution of Meloidogyne Species in Agricultural Soil*. Co-op. Pubn., Dept. of Plant Path., North Carolina State Univ. & U. S. Agency for Intn. Devp., Raleigh, N.C., 65 pp.

Tewari, G.C. and Krishna Murthy, P.N. 1984. Yield loss in tomato caused by fruit borer. *Indian J. Agric. Sci.* **54**: 341-343.

Thankamani, C.K., Anandraj, M., Diby Paul and Kandiannan, K. 2003. Compatibility of *Pseudomonas fluorescens*, VAM (*Glomus fasciculatum*), *Trichoderma harzianum*, pesticides on establishment and growth of black pepper under nursery conditions

(Abstr.). *VIth Int. Workshop on PGPR* (M.S. Reddy, M. Anandraj, S.J. Eapen, Y.R. Sarma and J.W. Kloepper, eds.), pp. 82-88. Indian Inst. of Spices Res., Calicut, Kerala, India.

Thankamani, C.K., Sreekala, K. and Anandraj, M. 2005. Effect of *Pseudomonas fluorescens* (IISR 6) and *Trichoderma harzianum* (P-26) on the growth of black pepper (*Piper nigrum* L.) in the nursery. *J. Spices Aromatic Crops* **14**: 112-116.

Thirumala Rao, V. 1956. Stray notes on some pest outbreaks of South India. *Indian J. Ent.* **18**: 123-126.

Tripathi, P.K. and Singh, C.S. 2006. Effect of some compatible biocontrol agents along with mustard cake and furadan on *Meloidogyne incognita* infecting tomato plant. *Indian J. Nematol.* **36**: 309-311.

Trivedi, T.P., Sardana, H.R., Shukla, R.P. and Misra, A.K. 2004. Validated IPM Technologies: Mango. In *Validated IPM Technologies for Selected Crops* (Amerika Singh, H.R. Sardana and Naved Sabir, eds.), pp. 149-162. National Centre for Integrated Pest Management (ICAR), New Delhi.

Trivedi, T.P., Sardana, H.R. and Thakur, V.S. 2004. Validated IPM Technologies: Apple. In *Validated IPM Technologies for Selected Crops* (Amerika Singh, H.R. Sardana and Naved Sabir, eds.), pp. 163-177. National Centre for Integrated Pest Management (ICAR), New Delhi.

Troncoso-Rojas, R., Espinoza, C., Sánchez-Estrada, A., Sergio-García, H., Pryor, B., Baez-Flores, M.E. and Tiznado-Hernández, M.E. 2008. Evaluation of isothiocyanates to control postharvest fruit and vegetable losses by fungal infections in Mexico. *Third Int. Biofumigation Symp.*, Canberra, Australia.

Trutmann, P., Keane, P.J. and Merriman, P.R. 1982. Biological control of *Sclerotinia sclerotiorum* on aerial parts of plants by the hyperparasite *Coniothyrium minitans*. *Trans. British Mycol. Soc.* **78**: 521-529.

Tsror, L., Lebiush, S., Meshulam, M., Matan, E. and Lazzeri, L. 2006. Biofumigation for controlling soil-borne diseases of tomato, potato and olive. *Second Int. Biofumigation Symp.*, Idaho, USA, p. 46.

Tzortzakakis, E.A. and Gowen, S.R. 1994. Evaluation of *Pasteuria penetrans* alone and in combination with oxamyl, plant resistance and solarization for control of *Meloidogyne* spp. on vegetables grown in greenhouses at Crete. *Crop Prot.* **13**: 455-462.

Upadhyay, J.P and Roy, S. 1997. Third Annual Report of ICAR adhoc scheme on *Biological Control of Pigeonpea Wilt* submitted to ICAR, New Delhi. 35 pp.

Usman, N.M., Balakrishnan, P. and Sarma, Y.R. 1996. Biocontrol of rhizome rot of ginger. *J. Plantation Crops* **24** (Suppl.): 184-191.

Utkhede, R.S. and Gupta, V.K. 1996. *Management of Soil Borne Diseases.* Kalyani Publishers, New Delhi, pp. 100-122.

Vadivelu, R.S., Rajendran, G., Naganathan, T.G. and Jayaraj, S. 1987. Studies on the nematode pests of banana and their management - A review. *Third Group Discussion on the Nematol. Problems of Plantn. Crops*, Sugarcane Breeding Institute, Coimbatore, pp.32-33.

Venkitesan, T.S. and Jacob, A. 1985. Integrated control of root-knot nematode, infecting pepper vines in Kerala (Abstr.). *Indian J. Nematol.* **15**: 261-262.

Verghese, A., Jayanthi, P.D.K. and Sudha Devi, K. 2000. Developing an IPM based export protocol and forewarning model for mango fruit fly, *Bactrocera dorsalis* Hendel. *National Seminar on Hitech Hort.,* Bangalore.

Verma, A.C., Singh, H.K. and Khan, N. 2005. Management of root-knot nematode, *Meloidogyne incognita* through antagonistic approaches in pointed gourd. *Indian J. Nematol.* **35**: 78-79.

Vidya, K. and Reddy, B.M.R. 1998. Integrated management of *Radopholus similis* infecting banana using plant product, nematicide and biocontrol agents. *Mysore J. Agri. Sci.* **32**: 186-190.

Vishwadhar and Chaudhary, R.G. 2000. Disease resistance in pulse crops – Current status and future approaches. In *Proc. of Nat. Symp. on Role of Resistance in Intensive Agri.*, Directorate of Wheat Res., Karnal, pp. 144-157.

Vishwakarma, R., Prasad, P.H. and Ghatak, S.S. 2009. Bio-efficacy of botanicals and bio-pesticide against diamondback moth (*Plutella xylostella* Linn.) infesting red cabbage (*Brassica oleracea* L. var. *capitata* f. *rubra* L.). *Int. Conf. on Hort.*, Bangalore.

Vyas, R.V., Patel, B.A., Patel, B.N. and Patel, J.G. 2009. Integrated management of root-knot nematode in brinjal under field conditions. *Indian J. Nematol.* **39**: 35-37.

Walode, N.B., Sinha, A.K. and Neog, P.P. 2008. Biological control of citrus nematode, *Tylenchulus semipenetrans* on *Citrus jambhiri. Indian J. Nematol.* **38**: 244-245.

Weischer, B. 1967. *Report to the Government of India on Plant Parasitic Nematodes.* PL 24/52 Report No. TA 2332, FAO, Rome, 15 pp.

Whitehead, A.G. 1968. Taxonomy of *Meloidogyne* (Nematodea: Heteroderidae) with descriptions of four new species. *Trans. of Zool. Soc. of London* **31**: 263-401.

Winslow, R.D. and Willis, R.J. 1972. Nematode diseases of potato. In *Economic Nematology* (J.M. Webster, ed.), pp. 17-48. Academic Press, London.

Wisniewski, M., Droby, S., Chalutz, E. and Eilam, Y. 1995. Effect of Ca^{2+} and Mg^{2+} on *Botrytis cineria* and *Pencillium expansum in vitro* and on the biocontrol activity of *Candida oleophila. Plant Pathol.* **44**: 1016-1024.

Zaki, F.A. 1998. Biological control of *Meloidogyne javanica* in tomato by *Paecilomyces lilacinus* and castor leaves. *Indian J. Nematol.* **28**: 132-139.

Zaki, F.A. and Bhatti, D.S. 1989. Effect of castor (*Ricinus communis*) leaves in combination with different fertilizer doses on *Meloidogyne javanica* infecting tomato. *Indian J. Nematol.* **19**: 171-176.

Zaki, F.A. and Bhatti, D.S. 1991. Effect of castor *Ricinus communis* and the biocontrol fungus, *Paecilomyces lilacinus* on *Meloidogyne javanica. Nematologica* **36**: 114-122.

Zhou, T., Northover, J. and Schneider, K. 1999. Biological control of postharvest diseases of peach with phyllosphere isolates of *Pseudomonas syringae. Can. J. Plant Pathol.* **21**: 375-381.

Zhou, T., Northover, J., Schneider, K. and Lu, X. 2002. Interaction between *Pseudomonas syringae* MA-4 and cyprodinil in the control of blue mold and gray mold of apple. *Can. J. Plant Pathol.* **24**: 154-161.

Zhou, T. and Schneider, K. 1998. Control of peach brown rot by preharvest applications of an isolate of *Pseudomonas syringae* (Abstr.). 7th *Int. Cong. of Plant Pathol.* **3**: 20.

SUBJECT INDEX

D